Green Energy and Technology

Climate change, environmental impact and the limited natural resources urge scientific research and novel technical solutions. The monograph series Green Energy and Technology serves as a publishing platform for scientific and technological approaches to "green"—i.e. environmentally friendly and sustainable—technologies. While a focus lies on energy and power supply, it also covers "green" solutions in industrial engineering and engineering design. Green Energy and Technology addresses researchers, advanced students, technical consultants as well as decision makers in industries and politics. Hence, the level of presentation spans from instructional to highly technical.

Indexed in Scopus.

Indexed in Ei Compendex.

Muhammad Adly Rahandi Lubis
Editor

Proceedings of the 1st International Conference of Lignocellulose

ICON-LIG 2021

Springer

Editor
Muhammad Adly Rahandi Lubis
Research Center for Biomass
and Bioproducts
National Research and Innovation Agency
Cibinong, Indonesia

ISSN 1865-3529 ISSN 1865-3537 (electronic)
Green Energy and Technology
ISBN 978-981-19-1595-6 ISBN 978-981-19-1596-3 (eBook)
https://doi.org/10.1007/978-981-19-1596-3

This Springer imprint is published by the registered company Springer Nature Singapore Pte Ltd.
The registered company address is: 152 Beach Road, #21-01/04 Gateway East, Singapore 189721, Singapore

Preface

The 1st International Conference of Lignocellulose 2021 (ICON-LIG 2021) was held virtually on September 13–14, 2021. The forum was organized by the JSPS Alumni Association of Indonesia (JAAI) and Research Center for Biomaterials of Indonesian Institute of Sciences Indonesia (LIPI), and supported by JSPS Bangkok Office, Research Institute for Sustainable Humanosphere (RISH), Kyoto University, IPB University, University of Padjajaran, University of Jambi, University of Islam Indonesia, University of Lambung Mangkurat, and the University of North Sumatera.

As a scientific community, we aim to design and assess effective solutions for a sustainable future as one of the actions we can actively contribute to the sustainable development goals and climate change mitigation. Thereforeò, our step-up action is to gather all the researchers, scientists, academia, and practitioners in the field of "*biomass conversion and bio-based smart materials*" in a symposium entitled "The 1st International Conference of Lignocellulose 2021."

The event showcased a "Frontier Innovation of Lignocellulose and Humanosphere Sciences Towards Sustainable Development Goals and Climate Change Mitigation" tagline. Together with climate change mitigation, science, technology, and innovation are two thematic-ongoing activities alongside other sectors such as urbanization, transportation, water, ocean, and energy. In a global context, the bio-economy strategy connects the sustainable use of renewable biological resources for food, bio-based products, and bioenergy while maintaining the safety and restoration of biodiversity, ecosystems, and natural habitat.

The conference proceedings contain original research articles and review articles in biomass conversion and bio-based smart materials. In total, 30 articles were accepted for publication after the peer-review process. Together, the papers presented here represent a set of high-quality contributions to the literature on frontier innovation of lignocellulose sciences towards sustainable development goals.

Finally, We want to thank everyone who contributed to this effort, including paper authors, session presenters, track chairs, Program Committee members, volunteers, and sponsors. Without their support, the event would not have been as successful.

Bogor, Indonesia — Dr. Muhammad Adly Rahandi Lubis
Kyoto, Japan — Prof. Dr. Kenji Umemura
Sofia, Bulgaria — Prof. Dr. Petar Antov
Zvolen, Slovakia — Dr. Ľuboš Krišťák
Hyderabad, India — Dr. Sumit Manohar Yadav
Medan, Indonesia — Prof. Dr. Apri Heri Iswanto
January 2022

Editorial Board

Contents

Bio-based Smart Materials (Biocomposite, Nanomaterials, Biopolymer, Bioplastic, etc.)

Coconut Milk-Cells Encapsulation: A Model for Protecting *Lactobacillus* spp.

Achmad Dinoto and Gatot Adi Nugroho

Abstract This study aims to develop a model of cells encapsulated with coconut-milk, a natural emulsion of protein-carbohydrate-oil, for protecting *Lactobacillus* spp. Cells of plant origin *Lactobacillus plantarum* AP1 and intestinal *Lactobacillus fermentum* AA0014 used in this study were encapsulated by spray drying with coconut milk. Morphological observation of encapsulated cells was conducted by scanning electron microscopy. The coconut milk-encapsulated cells was evaluated for the viability against environmental barriers including very low pH and high bile salts concentration under in vitro assays, and the oxygen exposures during four week storage at room temperature. The encapsulation process contributed on decreasing the cell viability, however, coconut milk spray dried cells of assayed *L. plantarum* AP1 and intestinal *L. fermentum* AA0014 survived in high number at about 10^8 CFU/g. The encapsulated cells were relatively more stable in viability than those of non-encapsulated cells under the exposures of low pH and bile salts. Survivability of those encapsulated bacterial strains for four-week storage at room temperature was only reduced less than 0.4log. To our knowledge, this study firstly reported the use of coconut milk as natural encapsulation matrix for sprayed dried *Lactobacillus* spp. cells against the environmental barriers. This study contributes on the practical aspects of the use of natural resource that is widely available in the tropical countries for development microbial encapsulation technology.

Keywords Coconut milk · Encapsulation · *Lactobacillus fermentum* · *Lactobacillus plantarum* · Spray drying

---

A. Dinoto (✉)
Research Center for Applied Microbiology, National Research and Innovation Agency (BRIN), Cibinong, Jawa Barat, Indonesia
e-mail: achmaddinoto@yahoo.com

G. A. Nugroho
Study Program of Pharmacy, Faculty of Mathematics and Sciences, University of Muhammadiyah Prof. Dr. Hamka (UHAMKA), Jakarta, Indonesia

M. A. Rahandi Lubis (ed.), *Proceedings of the 1st International Conference of Lignocellulose*, Green Energy and Technology,
https://doi.org/10.1007/978-981-19-1596-3_1

1 Introduction

Many bacterial strains were reported to be used in the probiotic foods such as *Bifidobacterium* spp. and *Lactobacillus* spp. Several *Lactobacillus* strains have been documented to deliver benefits for human including *L. casei, L. acidophilus, L. reuteri, L. curvatus, L. gasseri, L. fermentum*, and *L. plantarum* [1]. Among the group of lactobacilli, *L. plantarum* is a known species related to the plant materials and be considered as probiotic in the activity of lowering serum cholesterol [2, 3]. In addition, *L. fermentum* was also reported on reducing hypercholesterolemia by altering gut microbiota [4]. Unfortunately *Lactobacillus* was reported to have limitation on the cell viability under extreme environments such as low acid environment, bile acids, and enzymes representing the condition in the gastrointestinal tract [5].

In order to protect lactic acid bacteria (LAB) during storage and digestion, several encapsulation methods have been developed such as extrusion and internal emulsion [6]. Microencapsulation is useful for improving the survivability of probiotic bacteria under environmental stresses, i.e. heat during drying, oxygen during storage, and low pH and bile salts exposure during gut transit [7]. Microencapsulation was developed by common application of spray drying technique for LAB. Many substances have been investigated in the cell encapsulation of potential probiotic strains including starch [8], alginate [9, 10], chitosan [11], whey [12, 13], hydrated gellan and xanthan gums [14], and soy protein isolate-alginate [15]. In our previous study, gelatine-based encapsulation successfully delivered infant origin *Bifidobacterium breve* to passage through the upper part of gastrointestinal tract and live in caecum of rats [16], leading the promising study to explore other encapsulating materials.

Spray drying is one method for the encapsulation that is mainly used for flavor compounds [17], but currently also applied for microbial cells [13, 18–20]. Principally, the aqueous solution containing cells is fed as droplets to the heated chamber of the spray drier to form dehydrated capsules with diameters at the scale of micrometer [17] depending on the apparatus arrangements such as nozzle size. In some cases, an ideal encapsulant used in encapsulation should be stable on passing through the upper part of gastrointestinal tract and have a food grade quality. The innovative strategy was demonstrated by applied encapsulation of bacterial cells within film-forming protein-carbohydrate-oil emulsion. Cells were encapsulated with casseinate, fructooligosaccharide or glucose syrups or starch, and vegetable oil in the form of emulsion after Maillard reaction and spray drying processes [21]. In nature, dried coconut endosperm contains approximately 7% (w w^{-1}) protein, 20% (w w^{-1}) carbohydrates, and 64% fat (w w^{-1}) [22]. Extraction of coconut endosperm provided a coconut milk that is actually a mimic of protein-carbohydrate-oil emulsion, however, report of the using coconut milk as protectant of cells was not available.

In our preliminary study, two strains *L. plantarum* AP1 and *L. fermentum* AA0014 were evaluated for anticholesterol activity (unpublished). *L. plantarum* AP1 was also characterized for its capability on hydrolyzing isoflavone glycosides of soybean through enzymatic reaction [23]. However, these strains were relatively unstable in the viability due to pH changes and required protective material for the application as probiotic candidates. Exploring new natural protective materials for cells is necessary to obtain the effective application of probiotic. Considering the psychochemical properties of emulsion with protein-carbohydrate-oil, a coconut milk is promising to be used as natural ingredient for protection of cells during passing thorugh in the upper part of gastrointestinal tract. Although spray drying method have been applied for coconut milk, most studies were focused on the psychochemical properties of coconut milk as food ingredients [24–27]. Use of coconut milk as protective material for cells through spray drying remains unclear. This study aims to develop a model of cells encapsulated with coconut milk as natural emulsion of protein-carbohydrate-oil for protecting *Lactobacillus* spp. The effects of exposures of low acid and bile salts, and the subsequent storage at room temperature on the viability of spray dried coconut milk-based encapsulated lactobacilli were investigated. To our knowledge, this study firstly reported the use of coconut milk as natural film-forming protein-carbohydrate-oil emulsion for *Lactobacillus* spp. cell protection by spray drying method.

2 Materials and Method

2.1 Bacterial Strains Used in This Study

The strains of *L. plantarum* AP1 and *L. fermentum* AA0014 were originally isolated from *nira lontar* (palm sap) of *Arenga pinnata* (Wurmb) Merr and from human faeces, respectively. Those strains were the collection of Research Center for Biology, Indonesian Institute of Sciences (LIPI), Indonesia. The bacterial cultures used in the this study were maintained at 4 °C on Lactobacilli MRS agar medium (10 g peptone, 10 g meat extract, 5 g yeast extract, 20 g glucose, 2 g K_2HPO_4, 5 g $CH_3COONa.3H_2O$, 2 g triammoniumcitrate, 0.2 g $MgSO_4$. $7H_2O$, 0.05 g $MnSO_4.7H_2O$, 0.25 g L-cysteine, and 1 mL Tween 80 per litre) supplemented with agar [28].

2.2 Preparation of Coconut Milk

Coconut was obtained from Cibinong, West Java, Indonesia. Coconut milk was prepared by water extraction in which about 200 g cutted and milled coconut endosperms was added with 800 mL of hot water. The obtained coconut milk was then pasteurized. No microbial contamination in sterilized coconut milk was observed by platting on both nutrient agar (Merck) and MRS agar (Merck).

2.3 Pretreatment of Bacterial Cells

Prior to encapsulation, the cells of *L. plantarum* AP1 and *L. fermentum* AA0014 were propagated separately in a pre-reduced MRS broth with CO_2 gas injection in three set of 250 mL screwed Schott glass bottles (total 750 mL in each strain) at 37 °C for 12 h. Cell suspension from three set cultures was pooled, homogenized and determined for cell counts. Furthermore, a centrifugation (at 9500 × g, −4 °C for 6 min) was conducted the pellets were then washed twice with sterilized 0.85% NaCl to obtain the free cells [21].

2.4 Encapsulation of Bacterial Cells

Cell encapsulation was performed by following the method of [21] with modification. Modification included to the different type of encapsulant materials where the protein-carbohydrate-oil emulsion was coconut milk as natural emulsion substance. For encapsulation, the cells of *L. plantarum* AP1 and *L. fermentum* AA0014 were added separately into pre-cooled coconut milk at 10% (w v^{-1}) and homogenized prior to spray drying process. Spray drying was conducted using a laboratory scale spray dryer EYELA SD-1000 (Tokyo Rikakikai Co. Ltd., Japan). The parameters included the inlet temperature of 165 °C, outlet temperature of 35 °C, feed rate of 10 mL min^{-1}, nozzle size of 0.4 mm, atomization air pressure of 200 kPa, and drying air flow rate of 0.30 m^3 min^{-1}. The powder of spray dried bacterial cells were kept in the sealed aluminium package at 4 °C until use.

2.5 Measurement of Viable Counts

Viability (colony forming units) of free cells and spray dried *Lactobacillus* was determined by the spread plate method using MRS agar. Resuspended free cells or capsules of *Lactobacillus* suspension were spread onto MRS agar plates by serial dilution with sterilized 0.85% NaCl. Incubation was conducted anaerobically at 37 °C for 48 h in a 1.5 L Anaerobic Jar (Merck) supplemented with Anaerocult (Merck).

2.6 Observation of Scanning Electron Microscopy (SEM)

Samples of powder of encapsulated bacterial cells were mounted on sample holders by using carbon adhesive tabs (Electron Microscopy Sciences, Washington, D.C.). The sample position of holders was inside the preparation chamber. The samples were coated with gold and analysed with a field emission scanning electron microscope

JEOL JSM 5310 LV (JEOL, Tokyo, Japan) at 2.5 kV. Digital images were then processed and visualized with Adobe PhotoShop 7.0 (Adobe Inc.).

2.7 *Survival of Encapsulated Cells Under Exposure of Very Low pH*

The ability of the encapsulating material to protect *L. plantarum* AP1 and *L. fermentum* AA0014 during gastrointestinal transit was assessed using an in vitro model simulating conditions in the human gastrointestinal tracts. Aliquot of 0.15 mL free cells (non-encapsulated) at about 10^{10-11} CFU mL^{-1} in sterilized 0.85% (w v^{-1}) NaCl solution and 0.15 g spray dried bacterial cell within coconut milk (encapsulated) at about 10^8 CFU g^{-1} of each *L. plantarum* AP1 and *L. fermentum* AA0014 were added separately into 1.35 mL MRS medium with pH 6.5 and pH 1.2, and incubated anaerobically for 6 h at 37 °C. Following incubation, cells washed carefully centrifuged at 9500 × g, −4 °C for 6 min and suspended with equal volume of sterilized 0.85% (w v^{-1}) NaCl solution. Viable counts of the bacteria were compared between the culture on MRS broth pH 6.5 (control) and MRS broth pH 1.2 (acid condition) of the in vitro model. Survival rates were assessed in three repeat experiments.

2.8 *Survival of Encapsulated Cells Under Exposure of Bile Salts*

Aliquot of 0.15 mL free cells (non-encapsulated) at about 10^{10-11} CFU mL^{-1} in sterilized 0.85% (w v^{-1}) NaCl solution and 0.15 g spray dried bacterial cell within coconut milk (encapsulated) of each *L. plantarum* AP1 and *L. fermentum* AA0014 at about 10^8 CFU g^{-1} were added separately into 1.35 mL MRS medium with or without addition of 0.3% (w v^{-1}) Oxgall bile salts (Sigma) and incubated anaerobically for 6 h at 37 °C. Following incubation, cells washed carefully by centrifugation at 9500 × *g*, −4 °C for 6 min and suspended with equal volume of sterilized 0.85% (w v^{-1}) NaCl. Viable counts of the bacteria were compared between the cultures on no added-bile salts (control) and bile salts-added MRS broth of the in vitro model. Survival rates were assessed in three repeat experiments.

2.9 *Survival of Encapsulated Cells During Storage*

The viability of encapsulated bacteria was investigated during storage following the method of [29] with modification. The spray dried capsules of *L. plantarum* AP1 and

L. fermentum AA0014 were transferred into 50 mL conical centrifuge tubes covered with an aluminium foil. Storage was applied at room temperature (at about 22 °C). Total of 0.15 g encapsulated cells were collected from the tubes every week for four weeks of storage. At predetermined time points (0W, 1W, 2W, 3W, and 4W), the encapsulated cells were taken out and added to each tube containing 1.35 mL of sterilized 0.85% (w v^{-1}) NaCl solution. After homogenization using vortex, serial dilution was applied and then spread onto MRS agar for determining the viable cells after storage.

2.9.1 Statistical Analysis

All means ± SD of each treatment of viable cell (n = 3) were compared before and after environmental treatments of the in vitro model. The statistical significance of differences between treatments was assessed using student *T-test* using the SPSS software (version 10.0). The differences were considered significant at $p \leq 0.05$.

3 Results and Discussion

In the 12 h MRS broth-cell cultivation for at 37 °C under anaerobic condition, *L. fermentum* AA0014 grew faster than *L. plantarum* AP1. Total counts in 750 mL cell suspension of *L. plantarum* AP1 and *L. fermentum* AA0014 in after 12 h incubation were 3.8×10^{10} CFU mL^{-1} and 2.3×10^{11} CFU mL^{-1} as free cells, respectively. Encapsulation technique of spray drying contributed to the reduced cell viability in both strains which most probably due to an impact of relatively high temperature (inlet temperature of 165 °C). Total of 4.75 g spray dried capsule of *L. plantarum* AP1 was obtained with the final cell concentration of 2.0×10^8 CFU g^{-1}. Similarly, total of 4.95 g spray dried capsule of *L. fermentum* AA0014 was obtained, however, a slightly higher in the cell concentration (7.0×10^8 CFU g^{-1}) as compared to *L. plantarum* AP1. Briefly, the process encapsulation through spray drying reduced the cell concentration up to 2.28 log and 2.52 log for *L. plantarum* AP1 and *L. fermentum* AA0014, respectively. Phenomena of decrease in cell viability of lactobacilli due to drying process were also observed in several studies [30, 31]. Although reduction of cells was observed, final concentration cells of 10^8 CFU g^{-1} microcapsules was still acceptable. As illustration, to consume at least 10^6 CFU g^{-1} or mL^{-1} as generally accepted a minimum concentration of probiotic microorganisms [32], it requires about 0.01 g capsule of encapsulated *L. plantarum* AP1 and *L. fermentum* AA0014 in single serving.

Spray drying demonstrated great advantages in comparison to other drying techniques. Spray drying of cells within coconut milk conducted in this study provided living cells of *L. plantarum* AP1 and *L. fermentum* AA0014 that are covered by encapsulate materials. An observation of scanning electron micrographs showed there was no free cells or non-encapsulated bacteria surrounding encapsulated bacterial powders indicating that coconut milk as encapsulate material covered cells of *Lactobacillus*. By visual observation, the size of the spray dried capsules produced was varied and estimated relatively small at the scale of micrometer (Fig. 1).

It was reported that an improved stability of spray dried *Lactobacillus rhamnosus* GG (LGG) microcapsules during drying and storage by the mixture of encapsulating materials containing glucose [20]. As well, the cell viability of *L. plantarum* WCFS1 could be retained by formulations during spray drying and storage. In the process of carbohydrate protein-oil encapsulation, soluble and heat stable protein is preferably during the Maillard reaction. The emulsion of a film forming protein, carbohydrate and oil is useful for oxygen sensitive probiotic. The product of Maillard reaction in the protein-carbohydrate mixture is sufficient to provide oxygen scavenging activity during storage [33]. In coconut milk, protein that consisted in natural protein-carbohydrate-oil may act in the protecting against oxidative damage.

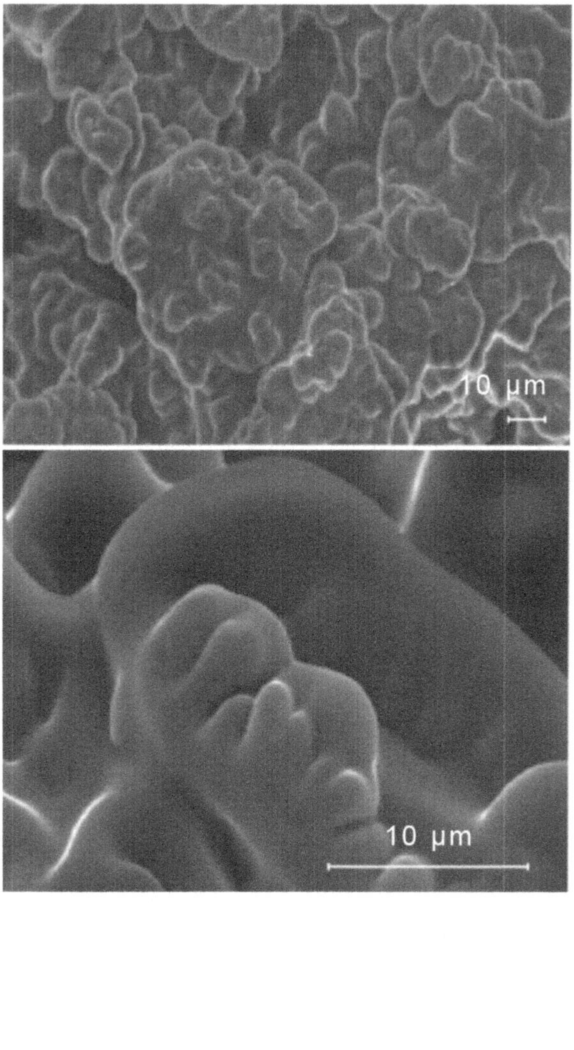

Fig. 1 Micrographs of the coconut milk-based encapsulated cells of *L. plantarum* AP1: magnification of 1000x (top) and 5000x (bottom). Bar represents the length of 10 μm

Coconut protein including prolamine, glutelin-1, and glutelin-2 all were demonstrated to exhibit good radical scavenging activity [34]. Sugars with a reducing group such as monosaccharide (glucose, fructose), disaccharide (maltose, lactose), trisaccharide, prebiotic oligosaccharide, and glucose syrups are a referred carbohydrate in the encapsulation [35]. As reported, cell viability could be improved by certain carbohydrate addition in the encapsulating material, such as trehalose, sucrose, lactose, and maltodextrin DE 6 [36].

Our current study did not supplement those sugars since the substance of encapsulation contains natural carbohydrate in coconut milk emulsion. Carbohydrate in coconut milk was suggested to support the stability of encapsulated cells. Although we did not characterize the sugar groups in coconut milk, Seow and Gwee estimated the carbohydrate contents in coconut milk were about 5.5–8.3% [37]. Incorporation of small sugars increased bacterial survival by replacing water that is removed from enzymes/proteins within cells and reduced the membrane phase transition temperature. Thus, it lead to higher survivability of cells [38] that favor cell viability. Addition of *Ficus carica* powder that contain suitable sugars for *L. bulgaricus* in the microcapsule form was reported to increase the viability [39].

In order to understand the stress response to low pH on cells of *L. plantarum* AP1 and *L. fermentum* AA0014, the incubation of cells was conducted in the MRS broth medium that is adjusted to pH 1.2 (low pH) and 6.5 (neutral pH). As compared to cultivation at neutral pH environment, we found that the cell counts of free cell of *L. plantarum* AP1 decreased up to 3.9 log due to acidic environment at pH 1.2, which final count was equal to 1.6×10^{10} CFU mL^{-1}. Higher reduction in the cell counts was observed up to 5.2 log at pH 1.2 as compared to 2.3×10^{11} CFU mL^{-1} at pH 6.5 in non-encapsulated *L. fermentum* AA0014 (Fig. 2). The growth limitation of Lactobacillus by environmental pH was determined, thus cells are necessary to be encapsulated with sufficient substance to solve the problem of survivability due to low pH. In this study, coconut milk-cells encapsulation was able to protect the viability of bacteria during simulated in vitro gastrointestinal trial of low pH exposures.

In this study we confirmed the toxicity of bile salts to cells. A decreased cell number up to 1.4 log was observed in non-encapsulated *L. plantarum* AP1 when cultivating in medium containing 0.3% bile salts. Similar phenomenon was also observed in non-enacapsulated *L. fermentum* AA0014, although only slight reduction of viability as much of 0.6 log was observed comparing to cultured cells without bile salts in medium (Fig. 3). In this study we recognized that *L. fermentum* AA0014 seems to be more tolerant to bile salts than *L. plantarum* AP1, probably related to source of *L. fermentum* AA0014 which is derived from the human gastrointestinal tract. This result was in agreement with previously reported study that bile salts showed toxicity to living cells. In accordance, Bron et al. [40] reported the changes in morphology *L. plantarum* WCFS1 when cells are expossed with 0.1 and 0.15% bile salts. Higher concentrations of bile interfered the bacterial cells to be shrunken and empty appearance as observed by SEM [41].

In current study, bile salts concentration as much of 0.3% reduced viability of free cells both *L. plantarum* AP1 and *L. fermentum* AA0014. Eventhough bile salts is considered toxic to bacterial cells, in which free bile acids disturb membrane

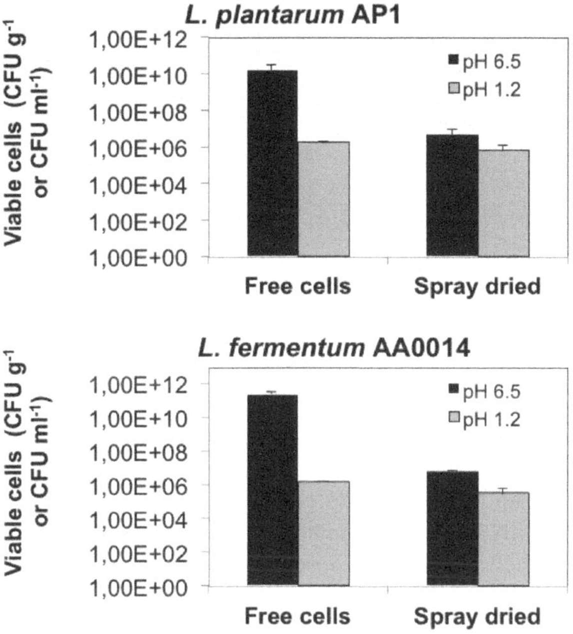

Fig. 2 Cell viability of *L. plantarum* AP1 and *L. fermentum* AA0014 under neutral (pH 6.5) and low pH (pH 1.2) exposures in the form of free cells and spray dried cells

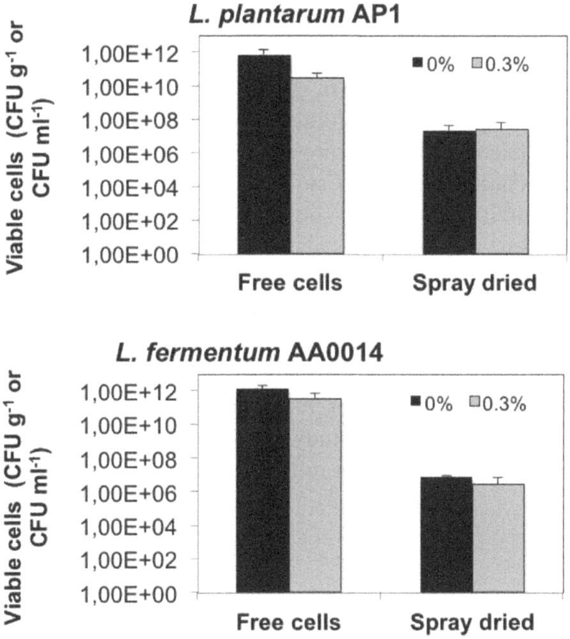

Fig. 3 Cell viability of *L. plantarum* AP1 and *L. fermentum* AA0014 under bile salts (0 and 0.3%) exposure in the form of free cells and spray dried cells

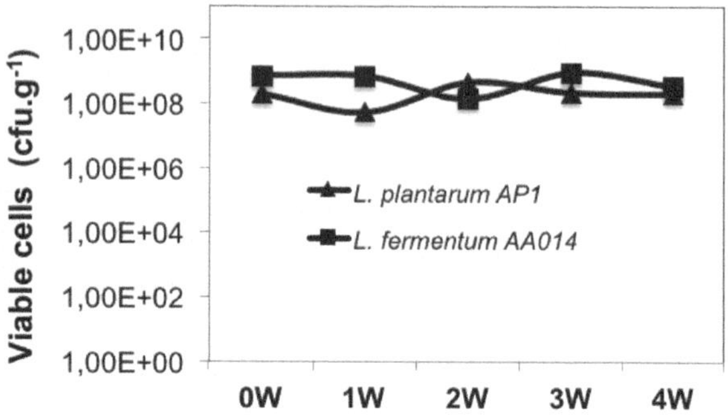

Fig. 4 Viability of spray dried *L. plantarum* AP1 and *L. fermentum* AA0014 cells with coconut milk during storage at room temperature (± 22 °C) for four weeks (0W to 4W)

integrity. As consequence, the proton, potassium ion, and other cellular components are leaked and eventually lead to the cell death [42]. Spray dried encapsulated cells with coconut milk of *L. plantarum* AP1 showed the improved viability against bile salts up to two folds as compared to non-encapsulated cells. Improvement of cell viability by the encapsulation with coconut milk varied amongst the bacterial strain.

The storage of encapsulated *L. plantarum* AP1 and *L. fermentum* AA0014 for four weeks under aerobic conditions reduced viability of the free cells (± 22 °C) was less than 0.4 log cell number (Fig. 4). The total cell number of *L. plantarum* AP1 and *L. fermentum* AA0014 in the spray dried capsules at the end of the storage tests was about 8.32 log CFU g^{-1} and 8.53 log CFU g^{-1}, respectively.

Altogether, both spray dried *L. plantarum* AP1 and *L. fermentum* AA0014 cells showed relatively high in viability against exposures of low acid and bile salts, as well as during long term storage at room temperature (Fig. 5). Those lactobacilli were not inactivated in the presence of oxygen, even after 4-week. This study showed the relationship to the survivability of source of *Lactobacillus* strains, where the sensitivity against bile salts in *L. plantarum* AP1 could be solved by encapsulation within coconut milk as encapsulating substance. In the previous study, a high level of water activity was detrimental to the survival of non-encapsulated probiotic *L. rhamnosus* GG [43]. Altogether, four weeks storage of the encapsulated cells at room temperature as well as other capabilities of protecting cells against low pH and bile acid expossures represented a possible means of supply of the capsules to the food industry. However, future studies such as simulation with intestinal microbiota as model and in vivo study are necessary for supporting the potential application of this developed coconut milk-encapsulated *Lactobacillus* spp.

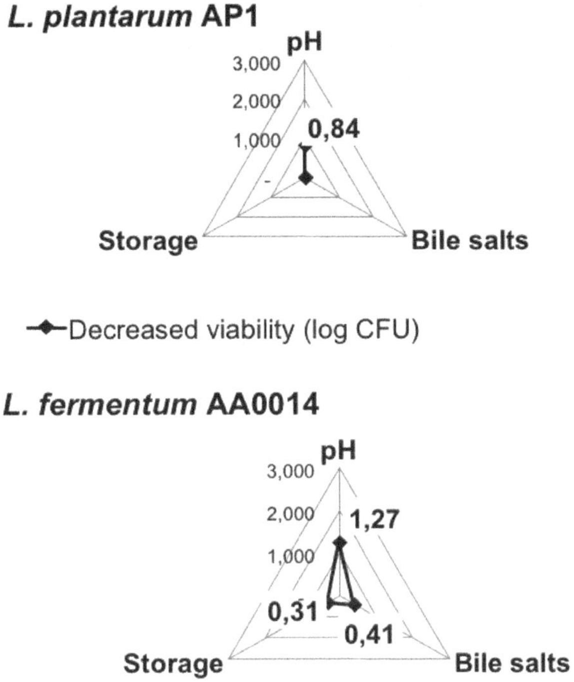

Fig. 5 Decrease in the viability (log) of *L. plantarum* AP1 and *L. fermentum* AA0014 cells encapsulated with coconut milk by spray drying due to encapsulation process, four weeks storage at room temperature, low pH exposure, and bile exposure

4 Conclusion

This study demonstrated the successful encapsulation of *L. plantarum* AP1 and *L. fermentum* AA0014 cells with coconut milk, a natural emulsion of protein-carbohydrate-oil, as novel encapsulating substance through spray drying. Although the technical process of encapsulation affected the cell viability, the coconut milk spray dried cells of assayed *Lactobacillus* spp. survived in acceptable cell number at about 10^8 CFU g^{-1}. The encapsulated cells were relatively more stable in viability than those of non-encapsulated cells under the exposures of low pH and bile salts as well as during storage for four week.

References

1. Putta S, Yarla NS, Lakkappa DB, Imandi SB, Malla RR, Amajala K (2018) Probiotics: supplements, food, pharmaceutical industry. In: Grumezescu AM, Holban AM (eds) Therapeutic, probiotic, and unconventional foods. Elsevier Inc., London, pp 15–25
2. Choi EA, Chang HC (2015) Cholesterol-lowering effects of a putative probiotic strain *Lactobacillus plantarum* EM isolated from kimchi. LWT Food Sci Technol 62:210–217
3. Huang Y, Wang X, Wang J, Wu F, Sui Y, Yang L, Wang Z (2013) *Lactobacillus plantarum* strains as potential probiotic cultures with cholesterol-lowering activity. J Dairy Sci 96:2746–2753

4. Lye H-S, Kato T, Low W-Y, Taylor TD, Prakash T, Lew L-C, Ohno H, Liong M-T (2017) *Lactobacillus fermentum* FTDC 8312 combats hypercholesterolemia via alteration of gut microbiota. J Biotechnol 262:75–83
5. Gbassi GK, Vandamme T, Yolou FS, Marchioni E (2011) In vitro effects of pH, bile salts and enzymes on the release and viability of encapsulated *Lactobacillus plantarum* strains in a gastrointestinal tract model. Int Dairy J 21:97–102
6. Valero-Cases E, Frutos MJ (2015) Effect of different types of encapsulation on the survival of *Lactobacillus plantarum* during storage with inulin and in vitro digestion. LWT Food Sci Technol 64:824–828
7. Martín MJ, Lara-Villosla F, Ruiz MA, Morales ME (2015) Microencapsulation of bacteria: a review of different technologies and their impact on the probiotic effects. Innov Food Sci Emerg Technol 27:15–25
8. O'Riordan K, Andrews D, Buckle K, Conway P (2001) Evaluation of microencapsulation of a *Bifidobacterium* strain with starch as an approach to prolonging viability during storage. J Appl Microbiol 91:1059–1066
9. Krasaekoopt W, Bhandari B, Deeth H (2004) The influence of coating materials on some properties of alginate beads and survivability of microencapsulated probiotic bacteria. Int Dairy J 14:737–743
10. Yari M, Fooladi J, Motlagh MAK (2015) Microencapsulation and fermentation of *Lactobacillus acidophilus* LA-5 and *Bifidobacterium* BB-12. Appl Food Biotechnol 2:27–32
11. Brinques GB, Ayub MAC (2011) Effect of microencapsulation on survival of *Lactobacillus plantarum* in simulated gastrointestinal conditions, refrigeration, and yogurt. J Food Eng 103:123–128
12. Aragón-Rojas S, Quintanilla-Carvajal MX, Hernández-Sánchez H (2018) Multifunctional role of the whey culture medium in the spray drying microencapsulation of lactic acid bacteria. Food Technol Biotechnol 56:381–397
13. Eckert C, Serpa VG, Cristina A, dos Santos CF, da Costa SM, Dalpubel V, Lehn DN, Fernanda C, de Souza V (2017) Microencapsulation of *Lactobacillus plantarum* ATCC 8014 through spray drying and using dairy whey as wall materials. LWT Food Sci Technol 82:176–183
14. McMaster LD, Kokott SA, Slatter P (2005) Micro-encapsulation of *Bifidobacterium lactis* for incorporation into soft foods. World J Microbiol Biotechnol 21:723–728
15. Jasmina H, Mladenovska K, Crcarevska MS, Dimchevska S, Grozdanov A, Popovski E, Chachorovska M, Ivanovska TP, Ugarkovic S, Goracinova K (2017) *Lactobacillus casei* encapsulated in soy protein isolate and alginate microparticles prepared by spray drying. Food Technol Biotechnol 55:173–186
16. Dinoto A, Suksomcheep A, Ishizuka S, Kimura H, Hanada S, Kamagata Y, Asano K, Tomita F, Yokota A (2006) Modulation of rat cecal microbiota by administration of raffinose and encapsulated *Bifidobacterium breve*. Appl Environ Microbiol 72:784–792
17. Marison I, Peters A, Heinzen C (2004) Liquid core caspules for applications in biotechnology. In: Nedovic V, Willaert R (eds) Fundamentals of cell immobilisation biotechnology. Springer Science + Business Media, Dordrecht, pp 185–204
18. Joseph G (2011) Microencapsulation of *Lactobacillus plantarum* (MTCC 5422) by spray-freeze-drying method and evaluation of survival in simulated gastrointestinal conditions. J Microencapsul 28:568–574
19. Rajam R, Anandharamakrishnan C (2014) Microencapsulation of *Lactobacillus plantarum* (MTCC 5422) with fructooligosaccharide as wall material by spray drying. LWT Food Sci Technol 60:773–780
20. Ying D, Sun J, Sanguansri L, Weerakkody R, Augustin MA (2012) Enhanced survival of spray-dried microencapsulated *Lactobacillus rhamnosus* GG in the presence of glucose. J Food Eng 109:597–602
21. Crittenden R, Weerakkody R, Sanguansri L, Augustin M (2006) Synbiotic microcapsules that enhance microbial viability during nonrefrigerated storage and gastrointestinal transit. Appl Environ Microbiol 72:2280–2282

22. Tano-Debrah K, Ohta Y (1997) Aqueous extraction of coconut oil by an enzyme-assisted process. J Sci Food Agric 74:497–502
23. Dinoto A, Rahayu RD, Sulistyo J (2009) Kapasitas *Lactobacillus plantarum* AP1 dalam menghidrolisis isoflavon glikosida kedelai (*Glycine max*). Berkala Penelitian Hayati 3C:63–65 (In Indonesian)
24. Hassan MA (2016) Production of spray-dried coconut milk powder. Pertanika 8:127–130
25. Santana AA, Martin LGP, de Oliveira RA, Kurozawa LE, Park KJ (2017) Spray drying of Babassu coconut milk using different carrier agents. Drying Technol 35:76–87
26. Santana AA, de Oliveira RA, Pinedo AA, Kurozawa LE, Park KJ (2013) Microencapsulation of babassu coconut milk. Food Sci Technol (Campinas) 33:737–744
27. Zafisah NS, Yusof YA, Ali MA, Roslan NS, Tahir A, Aziz MG, Chin NL (2018) Processing of raw coconut milk for its value addition using spray and freeze drying techniques. J Food Process Eng 41:e12602
28. de Man JC, Rogosa M, Sharpe ME (1960) A medium for cultivation of lactobacilli. J Appl Bacteriol 23:130–135
29. Xu M, Gagné-bourque F, Dumont M, Jabaji S (2016) Encapsulation of *Lactobacillus casei* ATCC 393 cells and evaluation of their survival after freeze-drying, storage and under gastrointestinal conditions. J Food Eng 168:52–59
30. Dimitrellou D, Kandylis P, Kourkoutas Y (2016) Effect of cooling rate, freeze-drying, and storage on survival of free and immobilized *Lactobacillus casei* ATCC 393. LWT Food Sci Technol 69:468–473
31. Perdana J, Bereschenko L, Fox MB, Kuperus JH, Kleerebezem M, Boom RM, Schutyser MAI (2013) Dehydration and thermal inactivation of *Lactobacillus plantarum* WCFS1: comparing single droplet drying to spray and freeze drying. Food Res Int 54:1351–1359
32. Kechagia M, Basoulis D, Konstantopoulou S, Dimitriadi D, Gyftopoulou K, Skarmoutsou N, Fakiri EM (2013) Health benefits of probiotics: a review. ISRN Nutrition 13:481651
33. Crittenden R, Sanguansri L, Augustin MA (2005) Probiotic storage and delivery. Patent WO 2005/030229 A1
34. Li Y, Zheng Y, Zhang Y, Xu I, Gao G (2018) Antioxidant activity of coconut (Cocos nucifera L.) protein fractions. Molecules 23:707
35. Augustin MA, Sanguansri L (2003) Encapsulation of food ingredients. Patent US 2003/0185960 A1
36. Perdana J, Fox MB, Siwei C, Boom RM, Schutyser MA (2014) Interactions between formulation and spray drying conditions related to survival of *Lactobacillus plantarum* WCFS1. Food Res Int 56:9–17
37. Seow CC, Gwee CN (1997) Coconut milk: chemistry and technology. Int J Food Sci Technol 32:189–201
38. Castro HP, Teixeira PM, Kirby R (1997) Evidence of membrane damage in *Lactobacillus bulgaricus* following freeze drying. J Appl Microbiol 82:87–94
39. Thakkar U, Preetha R, Nithyalakshmi V (2018) Evaluation of viability of *Lactobacillus bulgaricus* in symbiotic microcapsules: before and after freeze drying. Int Food Res J 25:1642–1646
40. Bron PA, Marco M, Hoffer SM, Van Mullekom E, de Vos WM, Kleerebezem M, Al BET (2004) Genetic characterization of the bile salt response in *Lactobacillus plantarum* and analysis of responsive promoters in vitro and in situ in the gastrointestinal tract. J Bacteriol 186:7829–7835
41. Sarkar A, Ye A, Singh H (2016) On the role of bile salts in the digestion of emulsified lipids. Food Hydrocoll 60:77–84
42. Kurdi P, Kawanishi K, Mizutani K, Yokota A (2006) Mechanism of growth inhibition by free bile acids in lactobacilli and bifidobacteria. J Bacteriol 188:1979–1986
43. Weinbreck F, Bodnár I, Marco ML (2010) Can encapsulation lengthen the shelf-life of probiotic bacteria in dry products? Int J Food Microbiol 136:364–367

The Potential of Kapok/Chitosan/ Cellulose Nanofiber Aerogel as Oil Absorbent

Dian Burhani, Sukma Surya Kusumah, Athanasia Amanda Septevani, Ruby Setiawan, Riska Surya Ningrum, Bernadeta Ayu Widyaningrum, Luthfiah Miftahul Djannah, Muhammad Andrew Putra, and Dewi Sondari

Abstract Kapok/chitosan/cellulose nanofiber aerogel was prepared using a facile method: blending the cellulose nanofiber (CNF) and chitosan suspension with Kapok fiber using Ultra Turrax, followed by freeze-drying. CNF aerogel was fragile and had high porosity; therefore, adding chitosan and Kapok increases the formability of aerogel, which was a good feature for further application of the aerogel for oil spill absorbent. Fourier transform infrared spectroscopy confirmed the materials' blending, and the morphology was analyzed using Field Emission Scanning Electron Microscopy (FE-SEM). The oil absorption capacity of the obtained aerogel for two different oils (used cooking oil and used motor oil) was investigated. FE-SEM analysis displayed a sheet-like structure intertwined with tubular-like fiber, representing the CNF and Kapok. Light film was also observed on the surface of the aerogels, indicating the crosslink of chitosan with CNF. The absorption capacity of the Kapok/CNF aerogels was 24.58 to 26.56 g/g for used cooking oil and 23.15 to 27.3 g/g for used motor oil. The results revealed that the obtained aerogel is a promising candidate for oil absorbent, considering the hydrophilicity state of the obtained aerogel.

Keywords Aerogel · Kapok · CNF · OPEFB · Oil absorbent

D. Burhani (✉) · S. S. Kusumah · R. S. Ningrum · B. A. Widyaningrum · D. Sondari
Research Center for Biomass and Bioproducts, National Research and Innovation Agency, Cibinong, West Java, Indonesia
e-mail: dnburhani@gmail.com

A. A. Septevani
Research Center for Environmental and Clean Technology, National Research and Innovation Agency, Cibinong, West Java, Indonesia

R. Setiawan
Research Center for Biology, National Research and Innovation Agency, Cibinong, West Java, Indonesia

L. M. Djannah · M. A. Putra
Bioprocess Department, Brawijaya University, Malang, East Java, Indonesia

M. A. Rahandi Lubis (ed.), *Proceedings of the 1st International Conference of Lignocellulose*, Green Energy and Technology,
https://doi.org/10.1007/978-981-19-1596-3_2

1 Introduction

Oil spills, one of the major environmental issues in the marine ecosystem, have exposed severe risks to aquatic life and human health. Oil-contaminated water from accidents that might happen during oil production, storage, and transportation needs urgent management. Therefore, developing robust and facile technology to remove the oil pollutant is inevitable [1, 2].

There have been many techniques to clean up the spilled oil, including physical, chemical, and biological. Physical methods such as booms and skimmers, which are commonly used, do not remove oil from the sea surface effectively. Chemical methods involve complicated and expensive processes, including dispersion, in situ, burning, and solidification, which often generate second waste after the oil removal. A biological approach with the use of microorganisms is considered effective. However, it is time-consuming and relies on many parameters, including pH, temperature, dissolved oxygen, and other factors [1].

Natural absorbent materials have been drawn attention from an environmental perspective due to the feasibility of absorbent collection, more effective oil removal, and more cost-efficient. The sorbent disposal is also easier and poses the least environmental hazard than synthetic absorbent [3].

Among other natural absorbents, Kapok (*Ceiba pentandra L.*), a cellulosic fiber with a hollow structure consisting of main components such as cellulose, lignin, polysaccharide, and minor components, including waxy coating. The wax covering the Kapok fiber's surface, allowing Kapok's hydrophobic properties, is a promising potential as an oil absorbent [4, 5].

Cellulose is one of the most abundant and environmentally friendly polymers found in nature [6]. Nanocellulose, the nanoscale version of cellulose, is classified into three types according to their isolation methods, including cellulose nanocrystals (CNC), cellulose nanofiber (CNF), and bacterial cellulose (BC) [7].

The term "aerogel" is defined as lightweight and highly porous material resulting from the sublimation of liquid/solvent to air. With the low density, high porosity, and high specific surface area properties, aerogel is considered an ideal absorbent to absorb more oil and stay floated on water [2]. CNF aerogels have raised attention owing to their excellent absorption performance, environmentally friendly, and sustainability [8].

It is crucial to select a suitable material with particular characteristics as an oil absorbent to improve oil removal from water. Both Kapok and cellulose nanofiber aerogel have shown typical behavior as excellent oil absorbents. Therefore, in this study, with that hypothesis, for the first time, we used Kapok fiber as substrate and cellulose nanofiber as the matrix and fabricated an aerogel using a facile method by freeze-drying. The morphology, mechanical strength, and oil absorption capacity of the obtained aerogel were investigated.

2 Materials and Methods

2.1 Materials

Cellulose nanofiber from a previous study was prepared from oil palm empty fruit bunch by mechanical treatment using SuperMass-Colloider (Model: MKCA6-5J, Masuko Sangyo Co., Ltd, Japan). Kapok provided Kapok.co (Japan). Chitosan 100 mesh (deacetylation degree of 87.5%) and acetic acid were technical grade.

2.2 Methods

A 1.5% w/v cellulose nanofiber and chitosan with a mass ratio of 1:3 and 1:5 was homogenized using Ultra Turrax for 15 min. Then, 1.5% w/v Kapok was blended into the mixture using a laboratory blender (8010 BU, Waring, USA). The CNF/chitosan/ Kapok mixture was placed in a Teflon mould of 17 × 9.5 × 6 cm, then frozen at − 80 °C for at least 8 h, followed by freeze-dried at −50 ± 5 °C, 0.05 mbar for 48 h. The FTIR spectra of Kapok and Kapok/CNF based aerogel were recorded on a Fourier-transform infrared instrument (Perkin Elmer Spectrum Two) with a scan count of 16, resolution of 4 cm^{-1}, and data interval of 1 cm^{-1} in wave number ranging from 400 to 4000 cm^{-1}. The FTIR spectra of all samples were collected using the attenuated total reflection technique (ATR). Morphology and self-assembly behavior of cellulose nanofiber were conducted using Field Emission Scanning Electron Microscope (FE-SEM) Thermo Scientific Quattro S. FE-SEM images were taken at an accelerating voltage of 5 to 7 kV at various magnifications.). The compressive strength of the aerogel was conducted with a Universal Testing Machine (Shimadzu AGS-X series) mounted with a 10 kN load cell and supported with STM-Controller software. The stress-strain curves of a square sample with 48 mm width and 18.55 mm height were obtained at room temperature at a cross-head speed of 1 mm/min.

Meanwhile, the oil absorption test was performed according to ASTM F726. Used cooking oil and motor oil were used for the test. First, a weighed fibrous CNF was put in oil for 10 min to attain its absorption equilibrium and then pulled out for weight measurement. The weight of the fibrous CNF was recorded after the test (W_{final}), and the absorption capacity (g/g), defined as the weight of absorbed substances per unit weight of initial fibrous CNF, was calculated according to the equation below:

$$Q = \frac{W_{final} - W_{initial}}{W_{final}} \tag{1}$$

where W_{final} and $W_{initial}$ are the weights of fibrous CNF after and before adsorption, specific labels were used to simplify the identification. Ka was designated as Kapok aerogel. A015 was defined as CNF aerogel only (1.5% w/v). A015x was CNF aerogel with ratio of chitosan of 1:3. A015y was identified as CNF aerogel with a ratio of chitosan of 1:5. AKax was Kapok/CNF aerogel with the ratio of chitosan of 1:3. 1:3 and AKay were entitled as Kapok/CNF aerogel with the ratio of chitosan 1:5.

3 Results and Discussion

There were three primary materials with a specific role in this study: CNF, chitosan, and Kapok fiber. CNF was acted as a matrix, Kapok fiber as substrate, and chitosan as filler. The CNF was isolated from oil palm empty fruit bunches using mechanical treatment (grinding) with a diameter of 58 nm from the previous study. This study used a facile method to fabricate aerogel without surface modification to alter the hydrophilic aerogel into hydrophobic aerogels. Since Kapok is hydrophobic, we assumed that the introduction of Kapok would affect the surface properties of the aerogels.

The three materials were subsequently homogenized using Ultra Turrax. First, CNF and chitosan (dissolved in acetic acid) were blended, the obtained suspension between CNF and chitosan was white with a slightly gel-like structure. Kapok fiber was then added to the suspension without additional solvent. The suspension was frozen at −80 °C before freeze-drying. The obtained aerogel was slightly white, firm with a rough and uneven surface from the freeze-drying process. The Kapok fibers were still clearly visible since no solvent was used to dissolve the fiber (Fig. 1). The obtained aerogel was lightweight. For comparison, the CNF aerogel from the previous study has a density and porosity of 0.04 g/cm^3 and 97%, respectively.

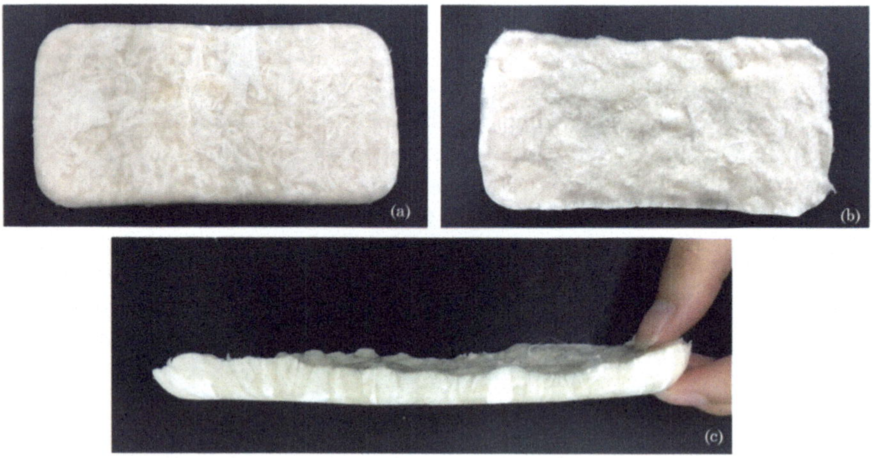

Fig. 1 Visual appearance of Kapok/CNF aerogel **a** top surface **b** bottom surface **c** side view

3.1 FTIR Analysis

FTIR spectra of Kapok fiber and Kapok/CNF aerogel are shown in Fig. 2. Detailed analyses of each peak are revealed in Table 1. The two samples presented two main regions of absorbance; low wavelength (500–1750 cm^{-1}) and higher wavelength (2800–3500 cm^{-1}). Kapok and oil palm empty fruit bunch are lignocellulosic fibers; therefore, similar peaks were observed for both samples.

The peak at 3320 cm^{-1} observed for A015 is attributed to the O-H stretching from the hydroxyl groups of cellulose [5]. Meanwhile, the peaks at 3284–3240 cm^{-1} observed for A015y and AKay are assigned as an overlapping of O–H stretching with N–H stretching and inter-hydrogen bonds of the polysaccharide due to the presence of chitosan [9]. The peaks at 2922–2881 cm^{-1} for all aerogels are attributed to asymmetric and symmetric aliphatic CH_2 and CH_3 stretching. This can be related to plant wax, mostly n-alkanes, and minor components, including alcohols, fatty acids, aldehydes, ketones, and n-alkyl esters [3]. The peaks at 1733–1732 cm^{-1} and 1241 cm^{-1} are associated with the presence of the carbonyl group (C=O) in the ester bonds. The C=O stretching vibrations are assumed to be related to the aliphatic aldehydes, esters, and ketones of Kapok wax which was the characteristic of hemicellulose linkage [3, 5]. Significance differences were observed at peaks 1543–1538 cm^{-1} which is assigned to the amino groups of chitosan in A015y and AKay but were not found in A015 (Fig. 2), indicating the chemical reaction between chitosan and cellulose [10].

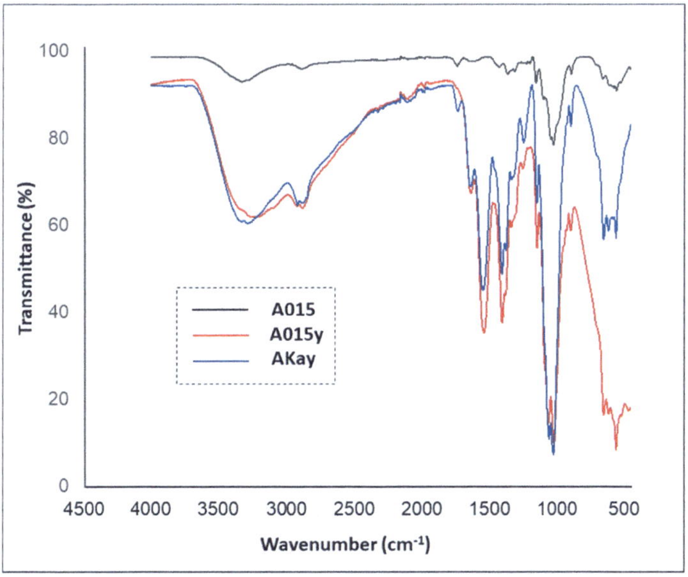

Fig. 2 FTIR spectra of A015 (CNF aerogel), A015y (CNF/chitosan aerogel), AKay (CNF/chitosan/kapok aerogel)

Table 1 Summary of analyzed FTIR Bonds

Peaks (cm^{-1})			Relative bonding	Description
A015	A015y	AKay		
3320	3240	3284	O-H stretching overlapped with N-H stretching	OH group, NH group
2892	2922	2918	C-H stretching of aliphatic	$=CH_2$, and $-CH_3$
–	2881	2881	C-H stretching	$-CH_3$ group
–	2324	2324	Glucose ring stretching	Hemicellulose, cellulose, β-glucoside linkage
–	2115	2114	C-O-C stretching	β-glucoside linkage
–	-	1732	-C=O stretching	Hemicellulose
1634	1632	1636	-OH deformation	H_2O and cellulose
–	1538	1543	N-H bending vibration	-CN (acetamide group)
1425	–	–	C-H deformation	Cellulose and lignin linkage
	1406	1407		
1369	–	1380	C-H deformation	Cellulose, hemicellulose and lignin linkage
1316	1339	1336	-OH in plane deformation	Cellulose
1202	1255	1249	C=O stretching	Hemicellulose and lignin
1159	1151	1152	C-O-C asymmetric stretching	Cellulose, hemicellulose and lignin linkage
–	1065	1060	C-O stretching	Hemicellulose and lignin
1027	1021	1027	C-O-C stretching vibration	Free amino group at C2 of glucosamine
896	898	898	Glucose ring stretching	β-glucoside linkage

Source [3–5, 9, 12–15]

The absence of peaks at 1594–1371 cm^{-1} indicates that all the aerogels contained no more lignin. Although, peak at 1732 cm^{-1} with decreased intensity still appears which might be due to partial removal of hemicellulose. The peaks at 2114 and 896 cm^{-1} were observed in all samples, confirming the typical cellulose structure (due to β-glycosidic linkages of glucose ring of cellulose) [11].

3.2 Morphology Analysis

The morphology of Kapok/CNF aerogel was investigated using FE-SEM. The porous CNF aerogel, self-aggregated into porous and sheet-like structures, was displayed in Fig. 3a. This structure was commonly observed for nanocellulose-based-aerogel fabricated from a freeze-drying method. The freezing process before the freeze-drying significantly affected the microscopic structure of the obtained aerogels. This study used medium freezing speed at −80 °C resulting in bundles of CNF aggregates in the obtained aerogel instead of single CNFs. The aggregates were formed according to the direction of the growing ice crystals. During the drying process, the CNF was entangled and formed hydrogen bonds with each other, leading to sheet-like structure formation [8].

The concentration of suspension also plays a critical role in the structure of aerogels. Here, we used 1.5% w/v initial concentration, which was quite high. In a high concentration, the space between the microfibers was not enough for dispersion, causing tight linking of adjacent microfibers, resulting in a 2D-sheet structure like aerogels [16, 17]. Macropores in the range of μm were also noticed in all aerogels.

Similar sheet-like structures were found for other aerogels Fig. 3b, c, and d. Rough surfaces covered with light film were observed in Fig. 3b, which might be due to the crosslinking reaction of chitosan and cellulose chains. The cross-linking process occurred due to chitosan, which contains the amino group (-NH_2), reacting with the aldehyde group (-CHO) from oxidized cellulose and Schiff base formation. Nanocellulose-based aerogel is fragile. Therefore, the addition of chitosan through the chemical crosslinking improving the integrity and increasing its mechanical properties [10, 18].

Kapok fiber with the distinctive tubular-like fiber was observed intertwining the sheet-like structure of CNF, as shown in Fig. 3c and d. Kapok fiber has a hollow structure which contributed to the additional pores in the aerogel. In addition, CNFs cloaked the Kapok fiber by assembling thin sheets and behaved as a binder and basic skeleton to boost the porous structure, suggesting increasing pore formation from the Kapok fiber [19].

The sheet-like structures were more packed and narrower in Fig. 3c than in d, as the ratio of CNF and chitosan increased from 1:3 to 1:5 and the introduction of Kapok fiber in the aerogel, indicating the increase of density. However, the addition of chitosan and Kapok increases the formability of aerogel, compared to CNF only aerogel, which was a good feature and high porosity for further application of the aerogel for oil spill absorbent.

The stress-strain curve for AKay aerogel was illustrated in Fig. 4. The aerogels displayed maximum compressive stress, 100 kPa at 58% strain. Yang and Cranston [20] explained that the deformation of fiber material is mainly associated with pore bending and the breakdown of pores, which implied that the resistance comes from the hydrogen bonding and physical crosslinking [20, 21].

3.3 Oil Absorption Capacity

The absorption capacity of aerogel can be specified as the weight of the absorbed liquid per weight of the aerogel [22]. Several parameters affect the performance of aerogel to absorb liquid, including porosity and surface properties (hydrophobic/hydrophilic) [2]. This study used hydrophilic aerogels without hydrophobic surface modification. Six aerogels (Ka, A015, A015x, A015y, AKax, and AKay) and two

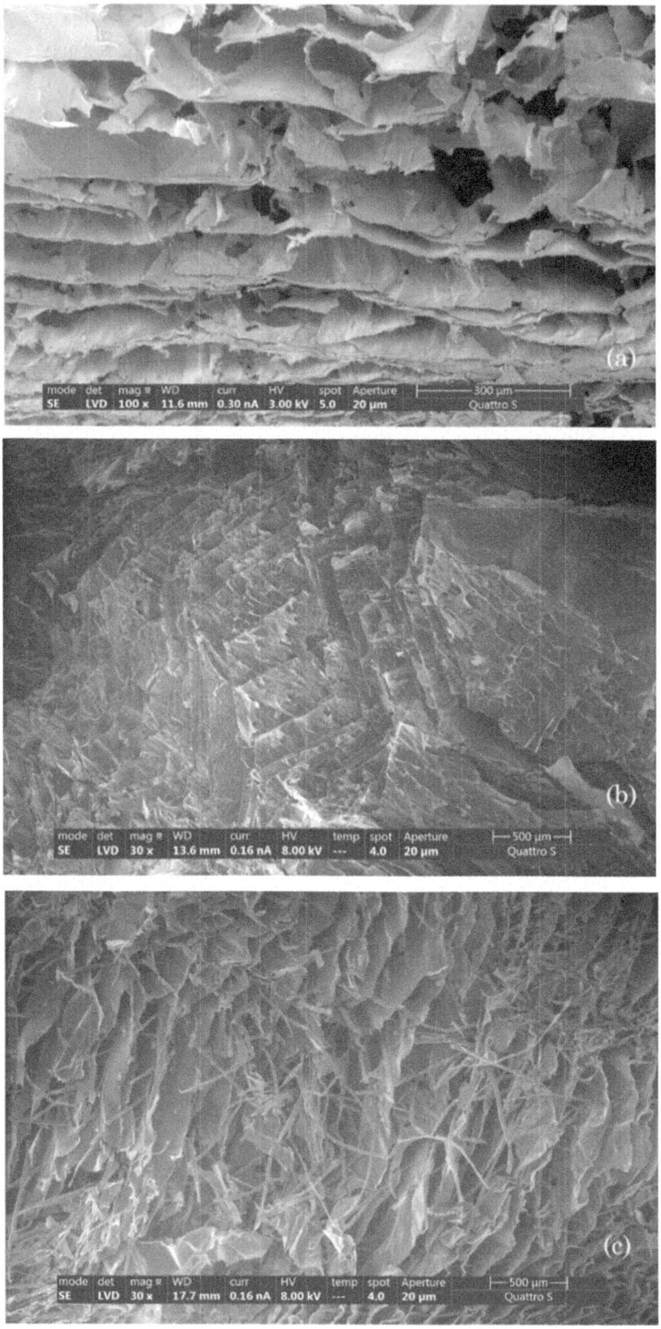

Fig. 3 FE-SEM Images of **a** A015, **b** CNF/Ch aerogel, **c** AKax, and **d** AKay

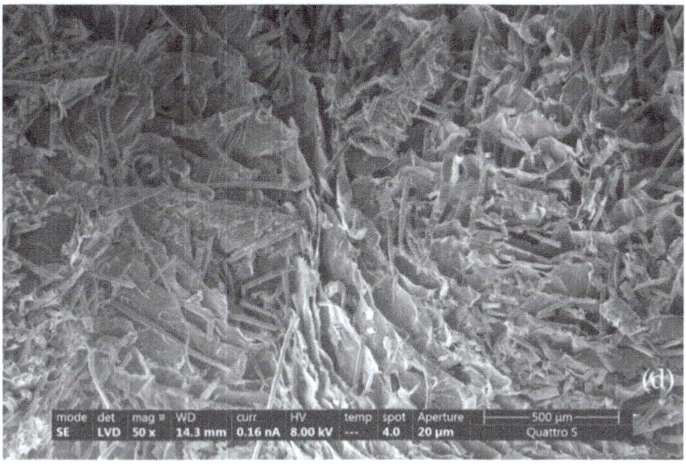

Fig. 3 (continued)

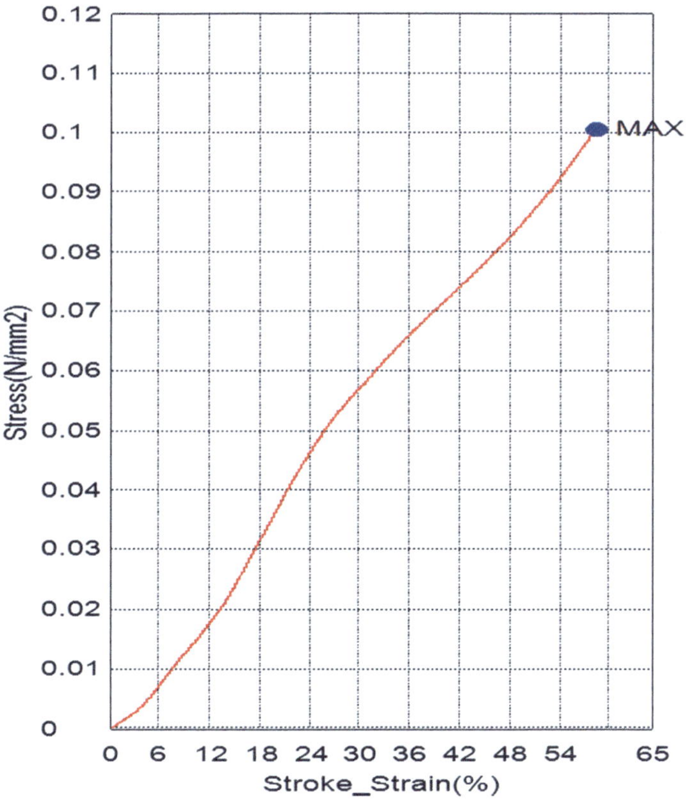

Fig. 4 Stress-strain curve of AKay aerogel

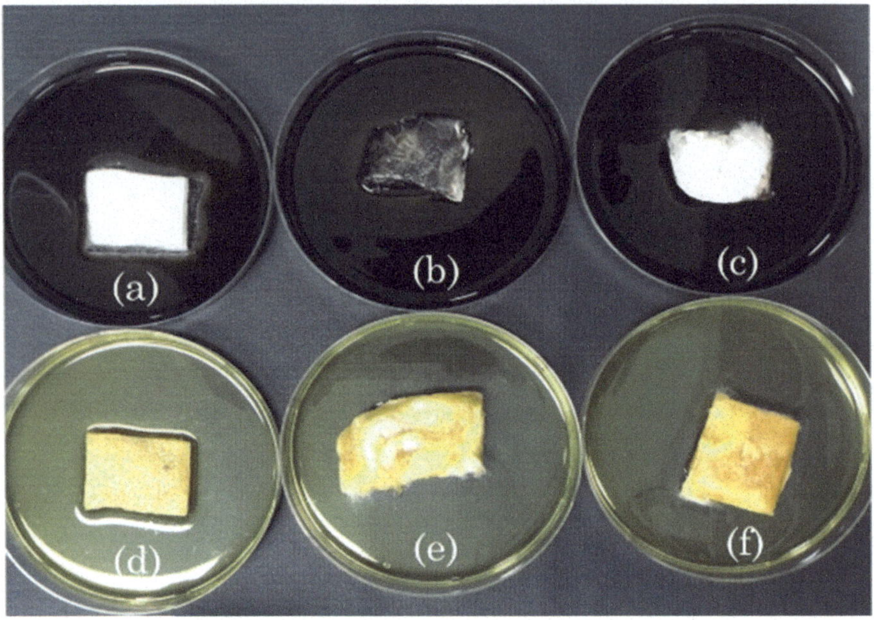

Fig. 5 Oil absorption process **a** Ka in used motor oil, **b** AKax in used motor oil, **c** AKay aerogel in used motor oil, **d** Ka in used cooking oil, **e** AKax in used cooking oil and **f** AKay in used cooking oil

different oils, cooking oil and motor oil, were used to investigate the oil's behavior against the aerogels. Figure 5 shows the final state after the oils' 10-min absorption test.

The absorption capacity of the aerogels was varied from 9.5 to 38.14 g/g. In general, considering the hydrophilicity state of the aerogels in this study, the aerogels showed comparable absorption capacity to hydrophobic aerogels, including hydrophobic cotton aerogels (49.9 g/g for motor oil and 55.8 g/g for sunflower oil) [23] and hydrophobic Kapok/MFC aerogel (130.1 g/g) [19]. However, it showed higher performance than Kapok cellulose aerogel (11.98 g/g) [24].

A015 achieved a maximum absorbency of 38.14 g/g for used cooking oil. Meanwhile, the minimum absorbency of 9.5 g/g was achieved by A015y for used motor oil. However, the absorption capacity of AKax (26.56 g/g for used cooking oil and 27.3 g/g for used motor oil) was higher than AKay (24.58 g/g for used cooking oil and 23.15 g/g for used motor oil). The explanation of this reason is due to the increased concentration of chitosan in AKay (1:5 ratio of CNF: chitosan, meanwhile, AKax contained a 1:3 ratio of CNF: chitosan), making the aerogel more packed, which lowering the porosity of the aerogels (Fig. 6).

The absorption of oil and organic solvents occurred via physical absorption, where the oils and organic solvents are trapped and stored in the pores of the aerogel [22]. A more expansive space between the sheet structure from the aggregates CNF observed

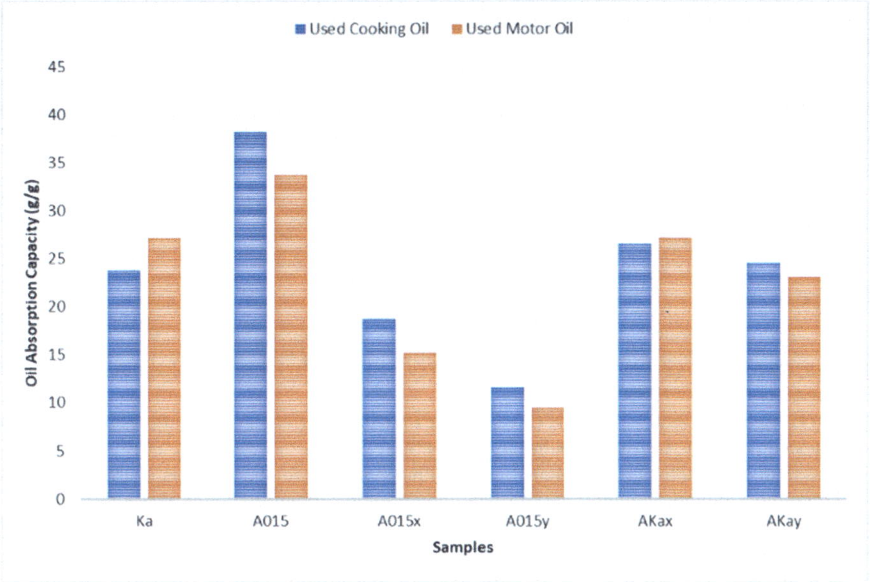

Fig. 6 Oil absorption capacity of the aerogels

in Fig. 3a suggests a higher chance of trap and store oils in the pores. However, with the addition of chitosan, space narrowed, reducing the porosity, therefore, lowering the oil absorption capacity of the aerogels (A015x and A015y). Meanwhile, the addition of Kapok fiber increased the oil absorption capacity. Kapok fiber has a hollow structure and is hydrophobic in nature. Therefore, the introduction of Kapok fiber in the aerogel increased the porosity and the performance of oil absorption of the aerogels.

4 Conclusion

Kapok/chitosan/CNF aerogel was successfully prepared. The obtained aerogel was sturdy, lightweight, with high porosity. The high concentration of CNF affected the microstructure of the aerogel to a sheet-like structure. CNF aerogels were fragile. Therefore, the addition of chitosan through the chemical crosslinking improved the integrity and mechanical properties of the aerogel. However, with the addition of chitosan, space narrowed, reducing the porosity, thus, lowering the oil absorption capacity of the aerogels. In addition, Kapok fiber has a hollow structure which contributed to the additional pores in the aerogel. The absorption capacity of the Kapok/CNF aerogels was in the range of 24.58 to 26.56 g/g for used cooking oil and 23.15 to 27.3 g/g for used motor oil. The results revealed that the obtained aerogel is a promising candidate for oil absorbent, considering the hydrophilicity state of the obtained aerogel.

Acknowledgements This works was supported by CPO Fund of the Ministry of Finance (Badan Pengelola Dana Perkebunan Kelapa Sawit, BPDP-KS) 2021 [grant numbers PRJ-01/DPKS/DIT.IV/ 2021].

References

1. Nguyen ST et al (2013) Cellulose aerogel from paper waste for crude oil spill cleaning. Ind Eng Chem Res 52:18386–18391
2. Liu H, Geng B, Chen Y, Wang H (2017) Review on the aerogel-type oil sorbents derived from nanocellulose. ACS Sustain Chem Eng 5:49–66
3. Abdullah MA, Rahmah AU, Man Z (2010) Physicochemical and sorption characteristics of Malaysian Ceiba pentandra (L.) Gaertn. as a natural oil sorbent. J Hazard Mater 177:683–691
4. Wang J, Zheng Y, Wang A (2012) Effect of kapok fiber treated with various solvents on oil absorbency. Ind Crops Prod 40:178–184
5. Macedo MJP et al (2020) Surface modification of kapok fibers by cold plasma surface treatment. J Mater Res Technol 9:2467–2476
6. Wang S et al (2015) An ultralight, elastic, cost-effective, and highly recyclable superabsorbent from microfibrillated cellulose fibers for oil spillage cleanup. J Mater Chem A 3:8772–8781
7. Zaman A, Huang F, Jiang M, Wei W, Zhou Z (2020) Preparation, properties, and applications of natural cellulosic aerogels: a review. Energy Built Environ 1:60–76
8. Hasan M et al (2019) Robust superhydrophobic cellulose nanofiber aerogel for multifunctional environmental applications. Polymers (Basel) 11:495
9. Kumari S, Rath P, Sri Hari Kumar A, Tiwari TN (2015) Extraction and characterization of chitin and chitosan from fishery waste by chemical method. Environ Technol Innov 3:77–85
10. Yang H, Zhang S, Yan J (2020) Chitosan-reinforced MFC/NFC aerogel and antibacterial property. Adv Polym Technol 2020:1–9
11. Kaushik A, Singh M (2011) Isolation and characterization of cellulose nanofibrils from wheat straw using steam explosion coupled with high shear homogenization. Carbohydr Res 346:76–85
12. Draman SFS, Daik R, Latif FA, El-Sheikh SM (2014) Characterization and thermal decomposition kinetics of kapok (Ceiba pentandra L.)-based cellulose. BioResources 9:8–23
13. Varma R, Vasudevan S (2020) Extraction, characterization, and antimicrobial activity of chitosan from horse mussel Modiolus modiolus. ACS Omega 5:20224–20230
14. Ibitoye EB et al (2018) Extraction and physicochemical characterization of chitin and chitosan isolated from house cricket. Biomed Mater 13(2):025009
15. Oh JW, Chun SC, Chandrasekaran M (2019) Preparation and in vitro characterization of chitosan nanoparticles and their broad-spectrum antifungal action compared to antibacterial activities against phytopathogens of tomato. Agronomy 9(1):21
16. Han J, Zhou C, Wu Y, Liu F, Wu Q (2013) Self-assembling behavior of cellulose nanoparticles during freeze-drying: effect of suspension concentration, particle size, crystal structure, and surface charge. Biomacromol 14:1529–1540
17. Chen W et al (2014) Comparative study of aerogels obtained from differently prepared nanocellulose fibers. Chemsuschem 7:154–161
18. Li AZ, Shao L, Hu W, Zheng T (2018) Excellent reusable chitosan/cellulose aerogel as an oil and organic solvent absorbent. Carbohydr Polym. https://doi.org/10.1016/j.carbpol.2018.03.027

19. Zhang H et al (2020) Ultralitght, hydrophobic, sustainable, cost-effective and floating kapok/ microfibrillated cellulose aerogels as speedy and recyclable oil superadsorbents. J Hazard Mater 406:124758. https://doi.org/10.1016/j.jhazmat.2020.124758
20. Yang X, Cranston ED (2014) Chemically cross-linked cellulose nanocrystal aerogels with shape recovery and superabsorbent properties. Chem Mater 26:6016–6025
21. Rafieian F, Hosseini M, Jonoobi M, Yu Q (2018) Development of hydrophobic nanocellulose-based aerogel via chemical vapor deposition for oil separation for water treatment. Cellulose 25:4695–4710
22. Zhou S et al (2016) Sustainable, reusable, and superhydrophobic aerogels from microfibrillated cellulose for highly effective oil/water separation. ACS Sustain Chem Eng 4:6409–6416
23. Bidgoli H, Mortazavi Y, Khodadadi AA (2019) A functionalized nano-structured cellulosic sorbent aerogel for oil spill cleanup: synthesis and characterization. J Hazard Mater 366:229–239
24. Susilowati T, Muljani S (2020) Synthesis of cellulose aerogels from kapok fiber for cleaning the waste of lubricant oil. Tek Kim 15:2305

Color and Surface Characterictics of Impregnated Wood Melamine Formaldehyde Furfuryl Alcohol and Nano-SiO_2

Istie Rahayu, Shintia Permata Sari, Wayan Darmawan, Akhiruddin Maddu, and Esti Prihatini

Abstract Fast-growing wood such as jabon (*Anthocephalus cadamba*) and ganitri (*Elaeocarpus sphaericus*) which has a less harvest period about 6–7 years, usually has inferior wood characteristics, namely low density, strength and durability. This study aim was to analyze the effect of melamine formaldehyde-furfuryl alcohol (MFFA) and nano-SiO_2 impregnation on color alteration and wettability values of *A. cadamba* and *E. sphaericus* wood. The impregnation solution treatments were consisted of untreated (water treated), MFFA, and MFFA nano-SiO_2 0.5%. The impregnation process was comprised by vacuum of 0.5 bar for 60 min and continued by applying a pressure of 2 bars for 120 min. The results showed that large color change (ΔE) occurred in jabon and ganitri after being treated by 0.5% MFFA and MFFA nano-SiO_2. The content of nano-SiO_2 inside the MFFA solution can increase the color brightness (decrease ΔE) in both wood species. The treatment of MFFA and MFFA nano-SiO_2 0.5% also had a significant effect on the contact angle and he K values. The optimum treatment for the wettability of jabon and ganitri wood was MFFA nano-SiO_2 0.5%.

Keywords Color alteration · Impregnation · Wettability

I. Rahayu (✉) · S. P. Sari · W. Darmawan · E. Prihatini
Department of Forest Products, Faculty of Forestry and Environment, IPB University, Bogor, Indonesia
e-mail: istiesr@apps.ipb.ac.id

A. Maddu
Department of Physics, Faculty of Mathematics and Natural Sciences, IPB University, Bogor, Indonesia

M. A. Rahandi Lubis (ed.), *Proceedings of the 1st International Conference of Lignocellulose*, Green Energy and Technology,
https://doi.org/10.1007/978-981-19-1596-3_3

1 Introduction

The increasing demand for wood raw materials in Indonesia is not in line with the availability of wood produced from natural forests. This gap is addressed by using timber from plantations forest and community forests. Production of logs in 2019 from natural forest was 6.77 million m^3, industrial plantation forests were 36.23 million m^3. This number decreased compared to 2018, where natural forest produced 8.60 million m^3 of logs, and 40.14 million m^3 of industrial plantation forests [1].

Plantation and community forests are usually planted by fast-growing species. Fast growing wood has a less harvest period of about 6–7 years, but has shortages, namely low specific gravity, strength and durability. Jabon and ganitri are widely planted in West Java, Indonesia. Jabon has almost the same quality as sengon, and has a low density [2]. Its heartwood is white to yellowish white. The growth rings are not distinct and texture is rather smooth. The surface is slightly slippery and slightly soft to hard [3]. According to Prihatini et al. [4], ganitri wood has yellow heartwood, the difference is not very clear from white sapwood, the texture is rather smooth, the wood surface is dull, rather light and soft. Ganitri wood can be used for building materials with relatively low durability so that it is susceptible to termites and dry wood beetles [5]. It is necessary to enhance the quality of fast-growing wood, one of which is by modifying the wood.

One of wood modification methods is impregnation. Research on impregnations on fast-growing wood has been carried out, such as on wood polymer nano composite (WPnC) using phenol formaldehyde (PF) and montmorillonite (MMT) [6]. According to Deka and Saikia [7], the impregnation treatment by using melamine formaldehyde caused a significant increase in the surface hardness and modulus of elasticity (MOE) of Jabon wood. Wood modification with nano-based treatment also has the potential to improve wood quality. Nano-SiO_2 is a polymer that is usually used to increase physical and mechanical strength and can be applied to improve the wood qualities that has been furfurized. Impregnated jabon and ganitri wood with melamine formaldehyde-furfuryl alcohol (MFFA) which is capable to enhance the physical properties of wood, including increasing density, weight percent gain (WPG), bulking effect (BE), anti-swelling efficiency (ASE), and decreased water uptake (WU) [8].

The nano-SiO_2 content inside melamine formaldehyde-furfuryl alcohol (MFFA) was able to improve the dimensional stability of *Ficus hispida* wood [9]. The wood experienced a darker color change after furfurylation, so that it became one of the weaknesses in this process [10]. The addition of nano SiO_2 in furfuryl alcohol can make the poplar wood color slightly lighter than using only furfuryl alcohol [11]. The same thing was also stated by Pratama [12], who found a brighter change in the color of sengon wood after being treated with MEG and nano silica. The impregnated wood will eventually be used for a specific purpose, both for exterior and interior use. Therefore, research on color changes and surface characteristics (wettability) of fast-growing wood resulting from MFFA and nano-SiO_2 impregnation is necessary. This study was performed to evaluate the effect of wood impregnation using a mixture of nano-SiO_2 and MFFA on color changes and surface characteristics (wettability) of fast-growing wood.

2 Materials and Method

2.1 Materials and Tools

Six-years old jabon wood (*A. cadamba*) and ganitri wood (*E. sphaericus* Schum) came from a community forest in Sukabumi, West Java, with a branch-free height of about 7–9 m and with a diameter of 28 cm were used in this study. Melamine, formaldehyde, furfuryl alcohol, NaOH, nano-SiO_2 (99.8, white powder), maleic anhydride, PH paper, and distilled water were used.

The instruments such as hot plate, scanner, sonicator, impregnation tube, moisture meter, macro capture microscope, injection, and camera were also utilized.

2.2 Procedure

Preparation of Wood Test Sample. The sample dimension of jabon and ganitri wood were $(6 \times 5 \times 2)$ cm^3 for the color alteration test referring to Cristie [13] and $(10 \times 5 \times 2)$ cm^3 for the wettability test referring to Yuningsih [14]. Each treatment was repeated 5 times, so that the required test samples were 30 for each wood species. The code in this study were as follow:

KJSB = untreated jabon before impregnation
KJSD = untreated jabon after impregnation
MJSB = MFFA treated jabon before impregnation
MJSD = MFFA treated jabon after imprgenation
NJSB = MFFA nano-SiO_2 0.5% treated jabon before impregnation
NJSD = MFFA nano-SiO_2 0.5% treated jabon after impregnation.

KGSB = untreated ganitri before impregnation
KGSD = untreated ganitri after impregnation
MGSB = MFFA treated ganitri before impregnation
MGSD = MFFA treated ganitri after impregnation
NGSB = MFFA nano-SiO_2 0.5% treated ganitri before impregnation
NGSD = MFFA nano-SiO_2 0.5% treated ganitri after impregnation.

Preparation of Impregnation Solution (MFFA Copolymer). The solution preparation method refers to the research of Hazarika and Maji [9] and Yao et al. [15]. Melamine, formaldehyde, furfuryl alcohol (mol ratio 1:3:5) were placed in the flask.

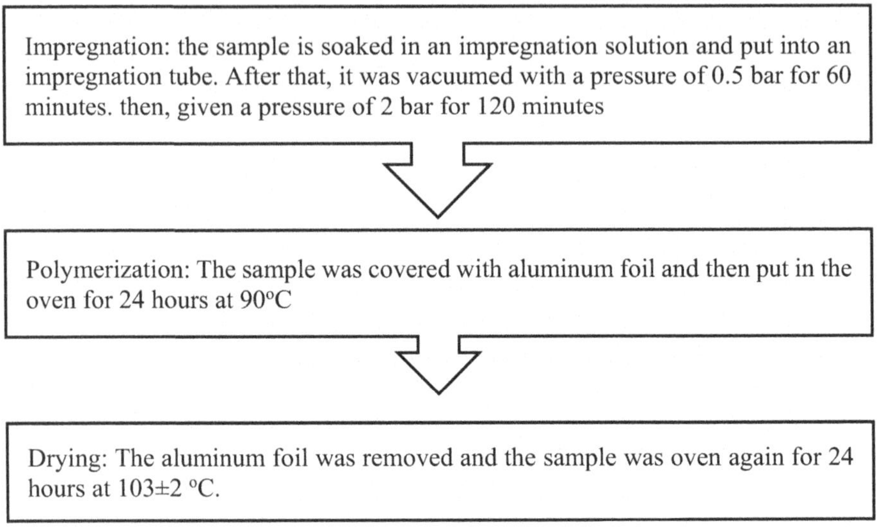

Fig. 1 Wood impregnation process

The pre-reaction of melamine and formaldehyde to hydroxymethylated melamine is the initial step of preparation. The pre-reaction was started when the pH of the medium was changed until the pH ranged from 9.4–9.6 by dripping 10% NaOH solution and the temperature was slowly increased to 98 °C. When the solution looks transparent, then furfuryl alcohol were poured into the solutions. Furthermore, after 10 min, then 1.5% maleic anhydrous catalyst was poured and the solutions was rested for about 10 min. For the impregnation process, the 50% MFFA copolymer was prepared.

The mixing of nano-SiO_2 in the MFFA copolymer was performed according to Hazarika and Maji [9]. Nano-SiO_2 was put into FA solution for 24 h while stirringwith, followed by sonicated for 30 min. Furthermore, the MFFA copolymer was slowly poured into the nano-SiO_2 solution under stirring conditions. Mixture of MFFA and nano SiO_2: FA were 1:1 (v/v). Finally, the mixture solution was sonicated for 15 min. Impregnation solutions were consisted of untreated (water treated), MFFA and MFFA nano SiO_2 0.5%.

Impregnation Process. The impregnation of MFFA and nano SiO_2 was carried out in several stages (Fig. 1). The stages of this process were adapted from Prihatini [8].

2.3 Color and Surface Characteristics Testing

Color Alteration Test. The wood color measurement was carried out by determining the L*a*b* value using the Adobe Photoshop CS 4 software application. The wood test sample is scanned first using a Canon 4400F Scanner, each test sample is marked with five points on the surface to obtain the L*a*b values. L*a*b* values were recorded before and after treatment. The color alteration (ΔE) was formulated according to the CIELab method [13], with the equation:

$$\Delta E = \surd(\Delta L)^2 + (\Delta a)^2 + (\Delta b)^2$$

Information:

ΔE = Color difference
ΔL = Brightness difference L
Δa = Red or green difference
Δb = Yellow or blue difference.

The values of ΔE before and after treatment can be seen in Table 1.

Wettability Measurement. The wettability test was conducted by sizing the contact angle between the water and the wood surface which was adapted from Yuningsih [14] research. The contact angle of each image piece is then measured using Image-J with dropsnake plugin analysis. The constant contact angle value was formulated according to the segmented regression equation between time (x) and contact angle (y) using the PROC NLIN program from SAS. Then the K (K-Value) value was formulated from the S/G model using the XLSTAT 2020 program [17] (Fig. 2).

The S/G model uses the formula:

Table 1 Criteria for the effect of color alteration based on the value of ΔE

Color change (ΔE)	Effect
< 0.2	Not visible
0.2–1.0	Very slight
1.0–3.0	Slight
3.0–6.0	Medium
> 6.0	High

Sumber: Hunter Lab [16]

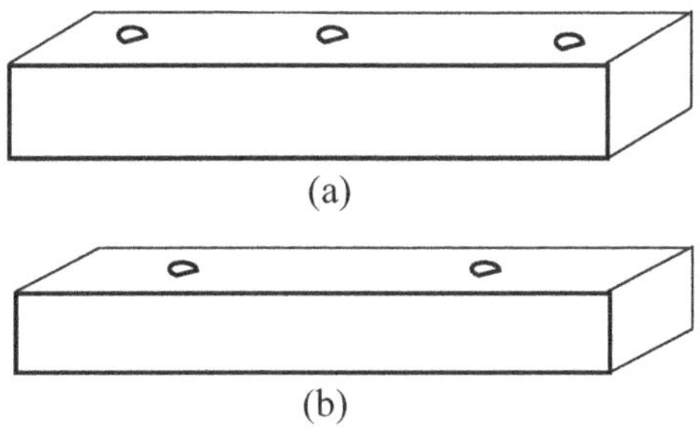

Fig. 2 Illustration of the pattern of liquid dripping on the surface of the top test sample (**a**) the bottom (**b**)

$$\theta = \frac{\theta i.\theta e}{\theta i + (\theta e - \theta i)\exp[K(\theta e\theta e - \theta i)t]}$$

Information:

θ = contact angle at a certain time
θi = initial contact angle
θe = equilibrium contact angle
t = time
K = constant rate of change of contact angle.

2.4 *Data Analysis*

The factorial completely randomized design (CRD) with 2 factors was used in this study. Treatment factor with 3 levels, namely untreated, MFFA, and 0.5% nano-SiO_2 MFFA and wood species factor with 2 levels, namely jabon and ganitri. The data obtained were analyzed by means of variance (ANOVA), if there was a significant difference, it would be continued with Duncan's test.

3 Result and Discussion

3.1 *Color Alteration*

The color of MFFA and MFFA nano-SiO_2 0.5% treated jabon and ganitri wood was darker than untreated wood (Fig. 3). However, the color of MFFA treated was darker than the MFFA nano-SiO_2 0.5%. This brightness is due to an addition of nano-SiO_2.

Figure 3a and b, show the larger and more positive the L* value, the higher the brightness, on the contrary, the lower the L* value, the darker the color [18]. Based on the results of color testing on both woods, the brightness value (L*) decreased after being treated by MFFA and the brightness increased after being treated with MFFA nano-SiO_2 0.5%. Both woods also showed a decrease in yellowish color (a*) when treated with MFFA and an increase in MFFA nano-SiO_2 0.5% treatment (Fig. 3). The same thing also happened to redness value (b*) in the MFFA treatment and an increase in the MFFA nano-SiO_2 0.5% treatment.

The test results after impregnation showed that the highest average brightness value (L*) was jabon wood. In untreated jabon, the brightness (L*) of wood was 72.32 after being treated with MFFA decreased by 46.04 and increased after being treated with MFFA nano-SiO_2 0.5% by 59.20. The brightness (L*) of untreated ganitri wood was 57.32. After being treated with MFFA decreased by 44.24 and increased after being treated with MFFA nano-SiO_2 0.5% by 53.6 (Figs. 4 and 5).

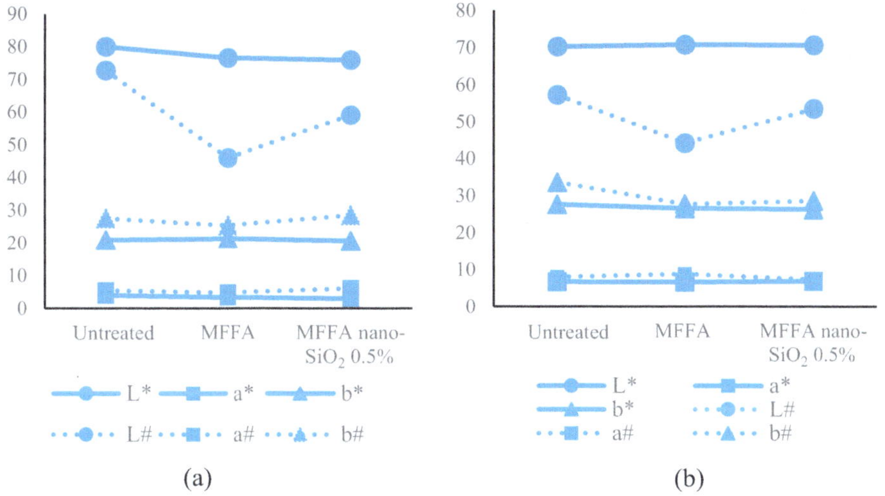

Fig. 3 L*a*b* value (before) and L#a#b# (after) MFFA and nano-SiO_2 impregnation treatment on jabon wood (**a**) and ganitri wood (**b**)

(a) (b) (c)

Fig. 4 The color of untreated (**a**); MFFA treated (**b**) and MFFA nano-SiO_2 0.5% treated jabon wood (**c**)

(a) (b) (c)

Fig. 5 The color of untreated (**a**); MFFA treated (**b**) and MFFA nano-SiO_2 0.5% treated ganitri wood (**c**)

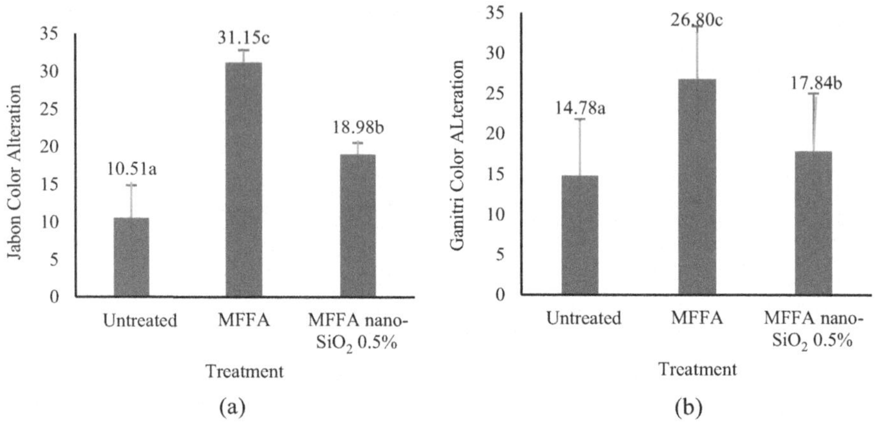

Fig. 6 Color alteration values of various MFFA and nano-SiO_2 impregnation treatments on jabon wood (**a**) and ganitri wood (**b**)

The color alteration (ΔE) is shown in Fig. 6. The color alteration of MFFA and MFFA nano-SiO_2 0.5% were more than 6.0, indicate that the impregnation treatment using MFFA and nano-SiO_2 gave a large color alteration (according to Table 1). MFFA treatment had the highest color alteration in jabon (31.15) and ganitri (26.80). The statistical analysis showed that the wood species did not significantly effect, while the treatment factor had a significant effect on color alteration of both woods. Duncan's test showed that the untreated, MFFA and MFFA nano-SiO_2 0.5% were significantly different.

The color alteration of MFFA treatment in jabon and ganitri wood made the wood color darker, while the MFFA nano-SiO_2 0.5% treatment made the wood color lighter. The dark color produced in the MFFA treatment was caused by the reaction between FA and the components of the wood cell wall [11]. Thus, nano-SiO_2 can increase the value of wood brightness when compared to MFFA treatment. This is due to the nano-SiO_2 confine the MFFA so that it is difficult for MFFA to bond with cell wall components. Impregnated poplar wood using furfuryl alcohol polymer and nano-SiO_2 also showed a brighter color change after the addition of nanoparticles [11].

3.2 Contact Angle

The contact angle is one of the parameter for analyzing the wettability of wood. A contact angle value above 90° refers to poor wettability which made the liquid will be difficult to wet a surface [19]. Alteration in contact angle based on time can be used to analyze the wettability of a material [17].

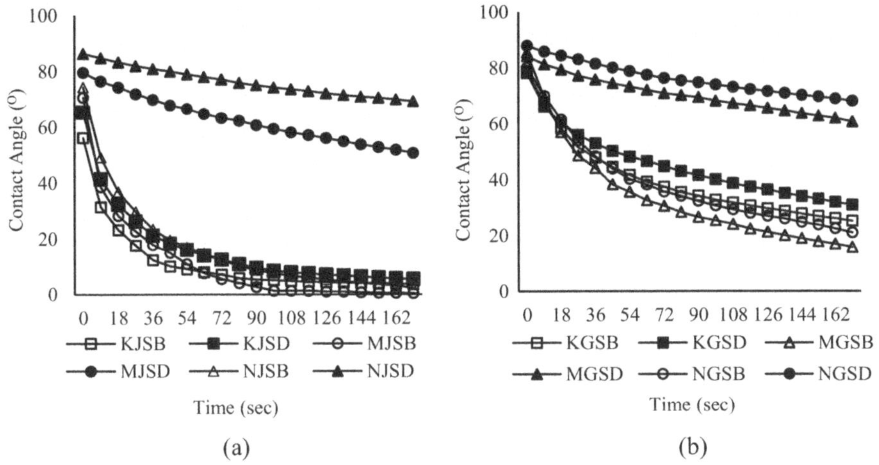

Fig. 7 Relation between contact angle values and time in jabon (**a**) and ganitri wood (**b**) in various treatments

Figure 7 shows that different impregnation solutions can affect contact angle values. The contact angle of untreated woods showed a steeper reduction compared to MFFA and MFFA nano-SiO_2 0.5% impregnated woods. Based on the results of the study, the contact angle of the untreated woods were smaller than contact angle values of MFFA and MFFA nano-SiO_2 0.5%. It was due to the density of both wood species was low so that water molecules tend to easily enter the wood.

Table 2 shows that the value of the constant contact angle of jabon wood tends to be lower than that of ganitri wood in all treatments. However, based on the statistical analysis, wood species has no significant effect. This is due to the density of jabon wood (0.32 g/cm^3) and ganitri wood (0.41 g/cm^3) Prihatini [8] is not much different. The statistical analyses, shows that the treatment has a significant effect on the contact angle. Duncan's test showed that the untreated was significantly different from MFFA nano-SiO_2 0.5% and MFFA treatment on jabon and ganitri wood. While, MFFA treatment was not significantly different from MFFA nano-SiO_2 0.5% treatment for both jabon and ganitri wood.

3.3 Wettability

Wettability is the condition of a surface that allows a liquid to penetrate into the surface. Wettability influences the absorption, adsorption, penetration and spread of adhesives [20]. The wettability value in this study is determined by the K value. The K value determines the wettability of a material. An increase in K value indicates an increase in the wettability of a material (liquid will easily spread and penetrate into the wood surface) [17].

Table 2 Constant contact angle (θe) in each treatment

Samples	θe (^{O})	Samples	θe (^{O})
KJSB	4.322	KGSB	28.077
KJSD	5.906^{a}	KGSD	32.567^{b}
MJSB	1.438	MGSB	30.322
MJSD	44.348^{b}	MGSD	54.926^{b}
NJSB	5.657	NGSB	64.880
NJSD	64.880^{c}	NGSD	65.383^{c}

The significance of letter a,b and c in this table, were resulted from Duncan's test

The K values of untreated, MFFA and MFFA nano-SiO_2 0.5% before and after impregnation on jabon and ganitri wood are shown in Table 3. The results of the calculation of K values in both woods showed a decrease in K values after treatment by MFFA and MFFA nano-SiO_2 0.5%. This indicates that the wettability of jabon and ganitri wood decreased. This could lead to dimensional stability improvement of both woods. This is in line Prihatini [8] which showed that the ASE (Anti-Swelling Efficiency) and WU (Water Uptake) values of jabon and ganitri wood impregnated with MFFA and MFFA nano-SiO_2 0.5% were improved. Based on the improvement in ASE and WU values after impregnation, it was shown that MFFA and the addition of nano-SiO_2 was able to minimize the ability of wood to absorb water because the polymer bonded with cell wall components [8] (Fig. 8).

The percentage reduction of K value in jabon wood after MFFA treatment was 96% and after MFFA nano-SiO_2 0.5% treatment was 98%. Meanwhile, the percentage reduction of K value in ganitri wood after MFFA treatment was 90% and after 0.5% MFFA nano-SiO_2 treatment was 89%. The reduction of K value concludes that the wettability of jabon and ganitri wood after MFFA and MFFA nano-SiO_2 0.5% impregnation treatment become lower. This is presumably because the impregnation material is capable to fill the lumens and pits on the wood cell walls. This is correspond to Prihatini [8] which shows that MFFA nano-SiO_2 and MFFA can enter the pits and fill the pores so that the cell walls and lumens become evenly covered.

Analysis of variance showed that the treatment had a significant effect, while wood species had no significant effect on the K value. This was because the densities

Table 3 Average K values in each treatment

Samples	K values	Samples	K values
KJSB	0.153	KGSB	0.046
KJSD	0.127	KGSD	0.038
MJSB	0.136	MGSB	0.039
MJSD	0.005	MGSD	0.004
NJSB	0.110	NGSB	0.027
NJSD	0.003	NGSD	0.003

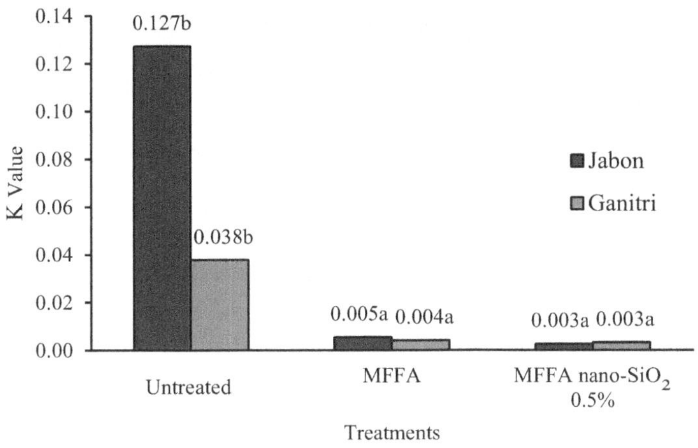

Fig. 8 K values of jabon and ganitri wood after impregnation

of jabon and ganitri wood were not much different, so after being treated the resulting values were also not different. Duncan test showed that the untreated was significantly different from MFFA nano-SiO_2 0.5% and MFFA treatment on jabon and ganitri wood, but the MFFA treatment was not significantly different from MFFA nano-SiO_2 0.5% treatment on both types of wood.

Based on the results of the research above, it can be concluded that there is a larger color change (ΔE) in jabon and ganitri wood cause by MFFA and MFFA nano-SiO_2 0.5% treatment when compared to the untreated. The nano-SiO_2 inside MFFA solution can increase color brightness (decrease ΔE) in both woods. The MFFA nano-SiO_2 0.5% and MFFA treatments also had a significant effect on the contact angle and K values. The most optimum treatment for wood wettability was obtained by MFFA nano-SiO_2 0.5% treatment. Impregnated jabon and ganitri wood by MFFA nano-SiO_2 0.5% had the smallest K value compared to other treatments. A small K value indicates low wettability which can increase the dimensional stability of the two woods. This shows that the good dimensional stability of wood makes the wood tend to be hydrophobic due to the addition of nano-SiO_2. Silica can be used as a support for hydrophobic properties in a layer [21]. This is also in line with Pambudi and Mochamad [22] who found that silica has fairly good material characteristics in hydrophobic stability. The hydrophobicity of impregnated test samples could lead to high dimensional stability and be suitable for exterior use.

4 Conclusion and Suggestion

4.1 Conclusion

Impregnation using MFFA and MFFA nano-SiO_2 0.5% had a significant effect on color alteration and K values in jabon and ganitri wood. Jabon and ganitri wood alter color after being treated with MFFA and MFFA nano-SiO_2 0.5%. The color alteration of the wood became darker after being treated by MFFA, but the brightness increased after the addition of 0.5% nano-SiO_2. Both woods produced lower K values after being treated with MFFA and MFFA nano-SiO_2 0.5%. This indicates that the wettability of jabon and ganitri wood is reduced after treatment. It is necessary to perform further research on measuring contact angles and wettability using other fluids such as paints or adhesives in order to optimize the utilization of impregnated wood. In addition, it is necessary to perform further research related to increase adhesive and paint strength on MFFA and MFFA nano-SiO_2 0.5% impregnated wood surfaces.

References

1. [KLHK] Ministry of Environment and Forestry (2020) Government breakthrough to increase investment and productivity of production forests. http://ppid.menlhk.go.id/siaran_pers/browse/2271. Accessed 15 Dec 2020
2. Widiyanto A, Siarudin M (2016) Physical characteristics of jabon wood (*Anthocephalus cadamba Miq*) in longitudinal and radial directions. Trop For J 4(2):102–108
3. Mandang YJ, Pandit IKN (2002) Guidelines for identification of wood species in the field. Prosea, Bogor (ID)
4. Prihatini E, Maddu A, Rahayu IS, Kurniati M (2020) Basic properties of ganitri wood (Elaeocarpus sphaericus (Gaertn.) K. Schum.) from Sukabumi and potential uses. J For Sci 14:109–118
5. Sani, A (2015) Durability, durability and drying properties of ganitri wood (*Elaeocarpus sphaericus* Schum) from Sukabumi [Skripsi]. Bogor Agricultural University, Bogor (ID)
6. Xue F, Zhao G (2008) Optimum preparation technology for Chinese fir wood/Camontmorillonite (Ca-MMT) composite board. For Stud China 10(3):199–204
7. Deka M, Saikia C (2000) Chemical modification of wood with thermosetting resin: effect on dimensional stability and strength property. Bioresour Technol 73(2):179–181. https://doi.org/10.1016/S0960-8524(99)00167-4
8. Prihatini E (2020) Improved physical properties of fast-growing wood through impregnation of nano-SiO_2 particles [thesis]. Bogor Agricultural University, Bogor (ID)
9. Hazarika A, Maji TK (2013) Properties of softwood polymer composites impregnated with nanoparticles and melamine formaldehyde furfuryl alcohol copolymer. J Polym Eng Sci 54(5):1019–1029. https://doi.org/10.1002/pen.23643
10. Herold N, Dietrich T, Grigsby WJ, Franich RA, Winkler A, Buchelt B, Pfriem A (2013) Effect of maleic anhydride content and ethanol dilution on the polymerization of furfuryl alcohol in wood veneer studied by differential scanning calorimetry. BioResources 8(1):1064–1075
11. Dong Y, Yan Y, Zhang S, Li J (2014) Wood/polymer nanocomposites prepared by impregnation with furfuryl alcohol and nano-SiO_2. BioResources 9(4):6028–6040

12. Pratama A (2020) Physical properties, mechanical properties and durability of sengon wood from nano silica impregnation of bamboo betung leaves [skripsi]. IPB, Bogor (ID)
13. Christie RM (2007) Colour chemistry. The Royal Society of Chemistry Science Park, Cambridge, UK. https://doi.org/10.1039/9781847550590
14. Yuningsih I (2017) Effect of surface roughness and viscosity of acrylic paint on wettability of long and short rotation jato wood [skripsi]. Bogor Agricultural University, Bogor (ID)
15. Yao M, Yang Y, Song J, Yu Y, Jin Y (2017) Melamine formaldehyde modified furfurylation to improve Chinese fir's dimensional stability and mechanical properties. Bioresources 12(2):3057–3066
16. Hunter Lab, Hunter L,a,b Color Scale. http://www.hunterlab.com. Accessed 15 Mar 2018
17. Shi SQ, Gardner DJ (2001) Dynamic adhession wettability of wood. Wood Fiber Sci 33(1):58–68
18. Muflihati M, Nawawi DS, Rahayu IS, Syafii W (2014) Changes in color of jabon wood stained with tanned bark extract (*Syzygium inophyllum*). J Trop Wood Sci Technol 12(1):11–19
19. Yuan Y, Lee TR (2013) Contact angle and wetting properties: Surface science techniques. Springer, pp 3–34. https://doi.org/10.1007/978-3-642-34243-1
20. Marra AA (1992) Technology of wood bonding: principles in practise. Van Nostrand Reinhold, New York, USA
21. Rohmah R, Mochamad Z (2016) The effect of variations in SiO_2 calcination temperature on the wettability of the hydrophobic surface. J Sci Arts 5(2):2337–3520
22. Pambudi DRS, Mochamad Z (2016) Effect of holding time on silica precursor calcination process as hydrophobic coating material. J Sci Arts 5(2):2337–3520

Eco-Friendly Particleboard Made of Alang-Alang/Sorghum Bagasse Mixture Bonded with Citric Acid-Sucrose

Subyakto, Sukma Surya Kusumah, Firda Aulya Syamani, Kurnia Wiji Prasetiyo, Eko Widodo, Lilik Astari, Sasa Sofyan Munawar, and Kenji Umemura

Abstract Alternative raw materials and environmentally friendly adhesives are important for the particleboard industry. The purpose of this experiment is to develop eco-friendly particleboard made of alang-alang (*Imperata cylindrica*) and sorghum (*Sorghum bicolor*) bagasse mixture bonded with citric acid-sucrose. A preliminary experiment was conducted using alang-alang and citric acid only to observe effect of resin content, pressing temperature and pressing time on the board properties. The resin contents of citric acid used were 15, 20, 25, and 30% based on dry weight of the particles. The particles were hot pressed at temperature of 160, 180, and 200 °C and pressure of 25 kgf/cm^2 for 5, 7, and 10 min. Enhancements of the board properties were attempted. The first enhancement properties was conducted by mixing citric acid with sucrose at ratio of 25:75; 20:80; 15:85; 10:90; 5:95; and 0:100. The second attempt was by mixing alang-alang with sorghum bagasse at ratio of 75:25; 50:50; and 25:75. All the boards were made with target density of 0.8 g/cm^3. The Japanese Industrial Standard (JIS) A 5908 was used to evaluate the properties of the boards. The results showed that the alang-alang board properties using 20% citric acid, pressing temperature of 200 °C for 10 min were met the JIS standard for type 8 board. Properties enhancement was achieved at ratio of citric acid and sucrose of 10:90; and ratio of alang-alang and sorghum bagasse of 25:75. It is concluded that alang-alang/sorghum bagasse and citric acid-sucrose are promising materials for preparation of particleboard.

Keywords Particleboard · Alang-alang · Sorghum bagasse · Citric acid · Sucrose

Subyakto (✉) · S. S. Kusumah · F. A. Syamani · K. W. Prasetiyo · E. Widodo · L. Astari · S. S. Munawar
Research Center for Biomass and Bioproducts, National Research and Innovation Agency (BRIN), Jl. Raya Bogor Km. 46, Bogor 16911, Indonesia
e-mail: suby001@brin.go.id

K. Umemura
Research Institute for Sustainable Humanosphere, Kyoto University, Gokasho Uji, Kyoto 611-0011, Japan

M. A. Rahandi Lubis (ed.), *Proceedings of the 1st International Conference of Lignocellulose*, Green Energy and Technology,
https://doi.org/10.1007/978-981-19-1596-3_4

1 Introduction

Alternative of raw materials and environmentally friendly adhesives are important for the particleboard industry. Recently the availability of wood from natural forest is reducing, therefore some other raw materials might be used as substitutes. Researches of non-wood materials for particleboard have been done. Among them are sugarcane bagasse [1], Michantus grass [2], oil palm frond [3], bamboo [4], rapeseed straw [5], salacca frond [6], sorghum bagasse [7], crops residue [8], oil palm trunk [9], and others. Alang-alang (*Imperata cylindrica*) is considered as weed and it grows rapidly. The area of alang-alang grassland in Asia is 35 million hectares where 8.5 million hectares is in Indonesia [10]. The alang-alang has high cellulose and hemicellulose contents [11]. Due to the availability and the chemical components, alang-alang might be used as material of particleboard. Other key factor in manufacturing of particleboard is the adhesive. Synthetic adhesive such as urea formaldehyde, phenol formaldehyde, melamine urea formaldehyde, and isocyanate are commonly used in the industry. These adhesives are non-renewable and contain harmful chemicals to human health. Recently more environmentally friendly adhesive especially citric acid is developed for moulding products and particleboard [12–17]. In addition, citric acid has been used as adhesive for sorghum bagasse particleboard [15, 18–20]. Sucrose was added to citric acid to improve the properties of sorghum bagasse or wood particleboards [16, 21, 22]. Sucrose has derivative compound under heat treatment that contain a hydroxyl group lead to react with carboxyl group of citric acid and other hydroxyl group of lignocellulose material [21]. Citric acid (2-hydroxy-1,2,3-propanetricarboxylic acid) contain three carboxyl groups citrus fruits such as lemon and limes is among the citric acid resources. Citric acid commercially produced by fermenting glucose or glucose and sucrose-containing materials [23, 24]. Sucrose is extremely soluble in water, hence it has high mobility in water and its abundance in hydroxyl groups [25]. Moreover, sucrose is chemically labile and can undergo hydrolysis in the presence of and acid in solution to form fructose and glucose [26]. In addition, sugar modifies its structure when the temperature rises up to 160 °C [27]. During this process, called caramelization, some isomerizations of sucrose are happening and highly reactive furanic derivatives are produced [28]. The studies on utilization of alang-alang as raw material of particleboard are very limited [29–31]. The purpose of this experiment is to develop alang-alang/ sorghum bagasse particleboards using citric acid as adhesive and to improve their properties with addition of sucrose.

Table 1 Chemical components of alang-alang

Chemical component (%)	This experiment	Reference [11]
Cellulose	52.05	40.22
Hemicelluloses	14.87	18.40
Lignin	25.02	31.29
Extractives	2.99	8.09

2 Materials and Methods

2.1 Materials

Particles were prepared by processing alang-alang leaves and sorghum bagasse with a ring flaker (PZ8, Pallmann, Germany). The chemical contents were determined using TAPPI standards for cellulose (TAPPI T 17 wd-97) [32], lignin (TAPPI T 222 om-02) [33], extractives (TAPPI 204 cm-97) in ethanol–benzene [34], and the content of holocellulose, cellulose and hemicelluloses was determined according to Browning [35]. The result is presented in Table 1, and it is comparable with reference [11]. The particles were screened to pass 4 mesh and retained on 14 mesh screens and then dried in an oven at 80 °C for 12 h to get 5% of moisture content before board preparation. Citric acid anhydrous (MERCK) or combination of citric acid and sucrose (MERCK) in powder forms were dissolved in water at concentration of 59% to use as adhesives.

2.2 Preparation of Alang-Alang Particleboard Using Citric Acid

Citric acid solution was applied homogenously to the particles by spray gun. The resin contents were set at 15, 20, 25, and 30% based on dry weight of the particles. The particles contain citric acid were dried in an oven at temperature of 80 °C for 12 h to get dry particles. The particles were matt formed and hot pressed at various temperatures and pressing times. The board size was 250 mm by 250 mm by 9 mm with density target of 0.8 g/cm^3. To observe effect of pressing temperature, the boards were pressed at 160, 180, and 200 °C for 10 min at citric acid content of 20%. While to observe effect of pressing time, following the previous research (Kusumah et al. 2016), the boards were pressed at 200 °C for 5,7 and 10 min at citric acid content of 20%. All boards were hot pressed at pressure of 25 kgf/cm^2.

2.3 Preparation of Alang-Alang Particleboard Using Citric Acid and Sucrose

Adhesives were prepared by mixing citric acid with sucrose at ratio of 25:75; 20:80; 15:85; 10:90; 5:95 and 0:100. The particles were sprayed with the adhesives, dried in an oven at temperature of 80 °C for 12 h, and matt formed. The boards were hot pressed at 200 °C for 10 min at pressure of 25 kgf/cm^2 at adhesive content of 20%. The board size was 250 mm by 250 mm by 9 mm cm with density target of 0.8 g/cm^3.

2.4 Preparation of Alang-Alang and Sorghum Bagasse Particleboard Using Citric Acid

The raw materials were prepared by mixing particles of alang-alang with sorghum bagasse at ratio of 75:25; 50:50 and 25:75. The particles were sprayed with the adhesives, dried in an oven at temperature of 80 °C for 12 h, and matt formed. The boards were pressed at 200 °C for 10 min at pressure of 25 kgf/cm^2 at citric acid content of 20%. The board size was 250 mm by 250 mm by 9 mm with density target of 0.8 g/cm^3.

2.5 Board Testing

The boards were tested for their physical and mechanical properties in accordance with JIS A 5908: Particleboard [36]. Thickness swelling (TS), modulus of rupture (MOR), modulus of elasticity (MOE), internal bonding (IB) strength and screw holding power (SW) were determined. TS was determined by soaked the specimen (50 mm x 50 mm x 9 mm) in water for 24 h, thickness before and after soaking were measured. Bending strength (MOR and MOE), IB and SW were determined using Universal Testing Machine (AG-IS, Shimadzu, Japan). Bending strength with specimen size of 200 mm x 50 mm x 9 mm was measured using three-point bending test at loading speed of 10 mm/minute and span of 150 mm. IB and SW were measured using specimen sizes of 50 mm x 50 mm x 9 mm and 100 mm x 50 mm x 9 mm, respectively.

3 Results and Discussion

3.1 Properties of Alang-Alang Particleboard Using Citric Acid

The effects of citric acid content, pressing temperature and pressing time on the particleboard properties are presented in Tables 2, 3 and 4, respectively.

The thickness swelling of the boards was decreased with increasing of citric acid content, however all values were met the JIS standard for type 8 board. Citric acid as adhesive was performed good physical properties for lingocellulosics due to the presence of ester linkages [13, 14, 18, 21]. Syamani et al. [24] found smaller value (2.04%) of thickness swelling of alang-alang particleboard at citric acid content of 20% and pre-drying of 12 h. The mechanical properties (MOR, MOE, IB) were optimized at citric acid content of 20% where all values were met the JIS A5908

Table 2 Effect of resin content on the alang-alang board properties

Properties	Resin content				JIS A 5908: type 8
	15%	20%	25%	30%	
Thickness swelling (%)	7.14	4.50	3.85	2.95	< 12
Modulus of rupture (MPa)	9.40	10.13	9.78	2.92	> 8
Modulus of elasticity (GPa)	2.85	3.33	3.13	2.14	> 2
Internal bonding strength (MPa)	0.11	0.17	0.12	0.01	> 0.15

Table 3 Effect of pressing temperature on the alang-alang board properties

Properties	Pressing temperature			JIS A 5908: type 8
	160 °C	180 °C	200 °C	
Thickness swelling (%)	34.06	23.93	4.50	< 12
Modulus of rupture (MPa)	5.80	5.98	10.13	> 8
Modulus of elasticity (GPa)	0.93	1.28	3.33	> 2
Internal bonding strength (MPa)	0.05	0.06	0.17	> 0.15

Table 4 Effect of pressing time on the alang-alang board properties

Properties	Pressing time			JIS A 5908: type 8
	5 min	7 min	10 min	
Thickness swelling (%)	22.95	23.93	4.50	< 12
Modulus of rupture (MPa)	6.20	6.91	10.13	> 8
Modulus of elasticity (GPa)	1.05	1.31	3.33	> 2
Internal bonding strength (MPa)	0.05	0.06	0.17	> 0.15

standard type 8 board. The MOR and IB values are higher and MOE is similar when compared with the work by Syamani et al. [24]. All properties were lower compared to sorghum bagasse particleboard with 20% citric acid [15]. This may due to the lower hemicellulose content of alang-alang compare to sorghum bagasse. In addition, the wettability of adhesive might be inhibited by mineral (such as Silica) and wax components of alang-alang.

The optimum pressing temperature of wood or sorghum bagasse particleboard using citric acid is 200 °C [14, 18], the same as this experiment when using alang-alang. In this experiment lower pressing temperatures (160, 180 °C) were also tried, however the properties obtained were not met the JIS A5908 standard. The curing stage of citric acid is not yet achieved at lower temperature, even though the melting point of neat citric acid is 150 °C. Some experiments were tried to use lower the pressing temperature. Wibowo et al. [22] found that addition of $ZnCl_2$ as catalyst could decrease the pressing temperature to obtained good board properties.

The board properties that pressed at 10 min were met the JIS standard, while those of 5 and 7 min were not met the standard. Kusumah et al. [18] found that pressing time of 10 min for sorghum bagasse particleboard using citric acid resulted in the highest IB.

3.2 Properties of Alang-Alang Particle Board Using Citric Acid and Sucrose

The properties of boards at citric acid to sucrose ratio of 25:75; 20:80; 15:85; 10:90; 5:95 and 0:100 were presented in Figs. 1, 2, 3 and 4. Thickness swelling and water absorption were the smallest at ratio of 10:90. The mechanical properties (MOR, MOE, IB, SW) were optimized at ratio of 10:90. At this ratio all of the boards properties were met the JIS standard A5908 type 8 board, except for SW. Addition of sucrose in citric acid as adhesive improve the mechanical properties values by about 200% when compared the properties without sucrose as reported above. Kusumah et al. [16] found the same ratio of citric acid sucrose (10:90) using sorghum bagasse for particleboard. However the properties of using sorghum bagasse is higher compare to alang-alang.

3.3 Properties of Alang-Alang and Sorghum Bagasse Particleboard Using Citric Acid

The properties of boards made from alang-alang and sorghum bagasse at ratio of 75:25; 50:50; and 25:75 are presented in Figs. 5, 6, 7, 8 and 9. The board properties (TS, MOR, MOE, IB, SW) were optimized at ratio of alang-alang to sorghum bagasse of 25:75. The TS, MOR, MOE and IB were met the JIS A5908 type 8 board, except

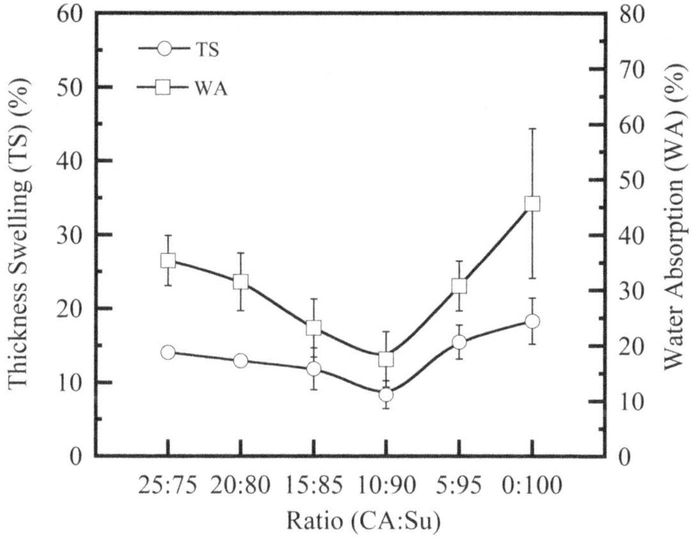

Fig. 1 Thickness swelling of alang-alang board at various citrid acid and sucrose ratio

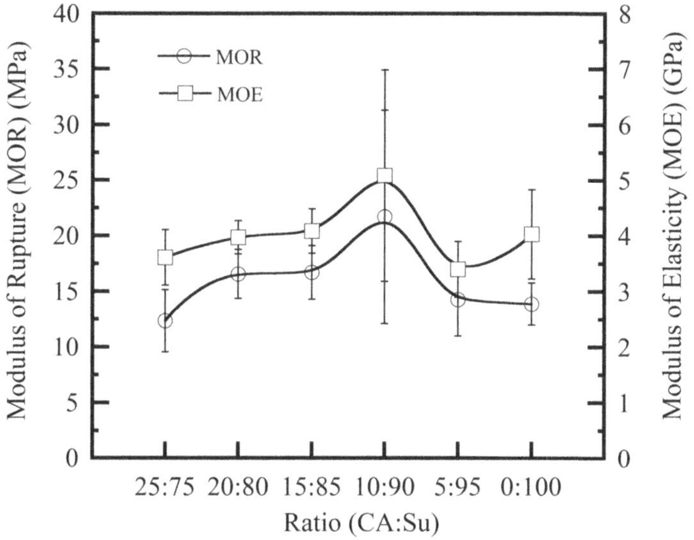

Fig. 2 Bending strength of alang-alang board at various citrid acid and sucrose ratio

for SW. However these properties were lower compared to sorghum bagasse only particleboard [15, 16, 18]. Overall particleboard properties of alang-alang only are lower than those of sorghum bagasse. The properties of origin material (chemical components mainly the minerals and extractives, particle brittleness, or others) may

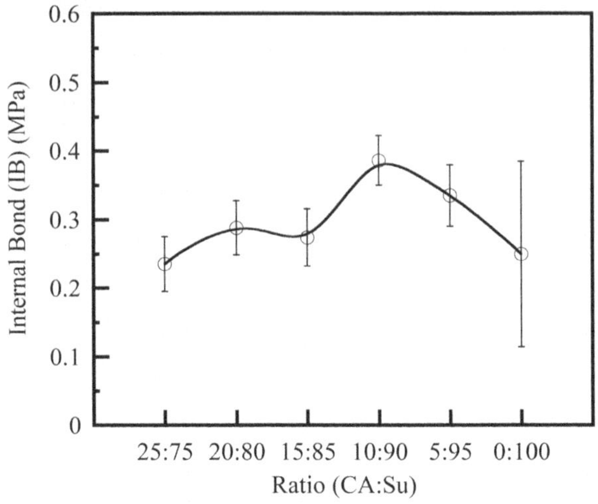

Fig. 3 Internal bonding strength of alang-alang board at various citrid acid and sucrose ratio

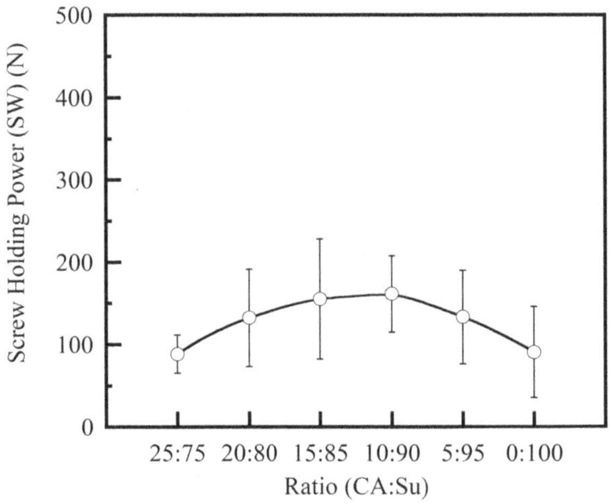

Fig. 4 Screw holding power strength of alang-alang board at various citrid acid and sucrose ratio

influence the board properties. In general wood particles produced higher properties of citric acid particleboard compare to grass particles [14, 21].

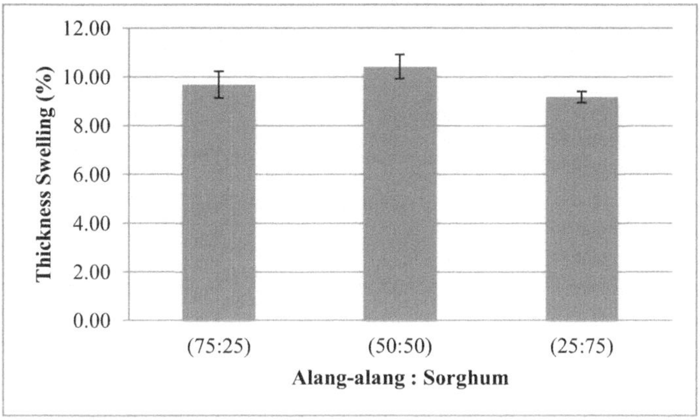

Fig. 5 Thickness swelling of particleboards at various ratio of alang-alang sorghum bagasse

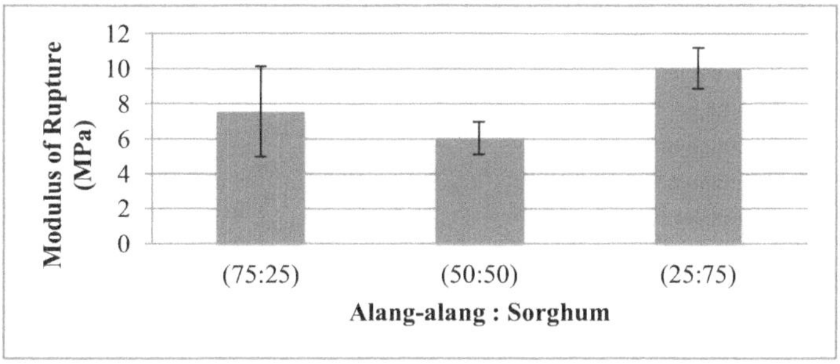

Fig. 6 Modulus of rupture of particleboards at various ratio of alang-alang sorghum bagasse

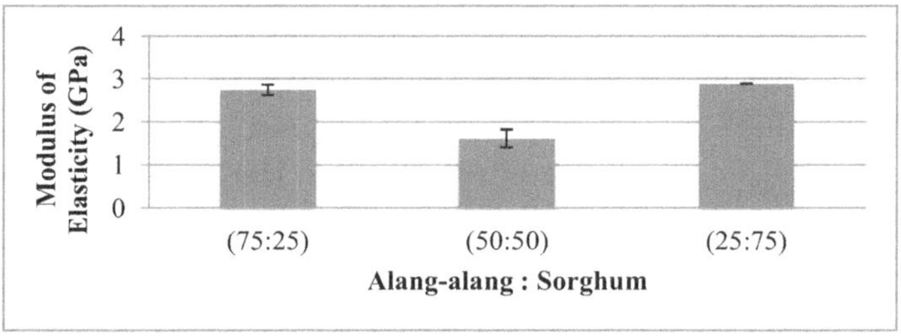

Fig. 7 Modulus of elasticity of particleboards at various ratio of alang-alang sorghum bagasse

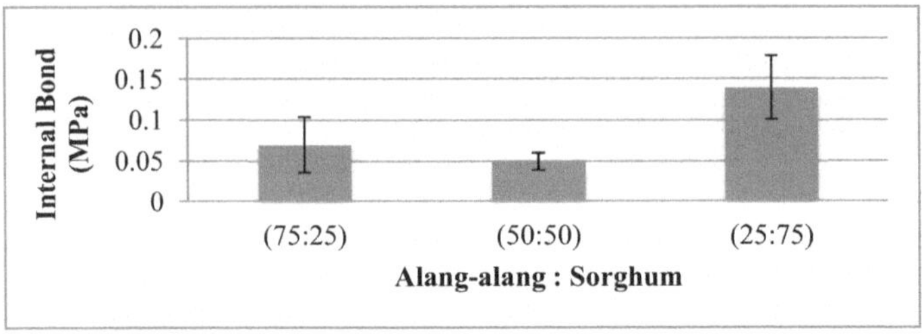

Fig. 8 Internal bonding strength of particleboards at various ratio of alang-alang sorghum bagasse

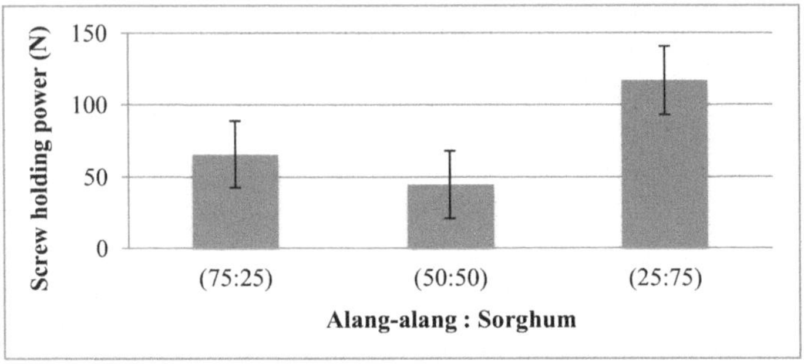

Fig. 9 Screw holding power of particleboards at various ratio of alang-alang sorghum bagasse

4 Conclusion

Alang-alang/sorghum bagasse particleboards were prepared and their properties were determined. The optimum properties of alang-alang particleboard were achieved at 20% of citric acid content, pressing temperature of 200 °C and pressing time of 10 min, which met the JIS standard for type 8 board. Properties improvements were obtained at ratio of citric acid and sucrose of 10:90; and ratio of alang-alang to sorghum bagasse of 25:75. Alang-alang/sorghum bagasse and citric acid-sucrose are promising materials for preparation of particleboard. However, further research on grass particleboard is needed to get better properties.

Acknowledgements Authors wishing to acknowledge the Science and Technology Research Partnership for Sustainable Development (SATREPS) Program on the Project for producing biomass energy and material through revegetation of alang-alang (*Imperata cylindrica*) fields, and Japan Society for Promotion of Sciences (JSPS) for financial support.

References

1. Ghalehno MD, Nazerian M, Bayatkashkooli A (2011) Influence of utilization of bagasse in surface layer on bending strength of three-layer particleboard. Eur J Wood Wood Prod 69:533–535
2. Park HJ, Oh SW, Wen MY (2012) Manufacture and properties of *Micanthus*-wood particle composite boards. J Wood Sci 58:459–464
3. Subyakto S, Zhao Z, Umemura K, Kawai S (2015) Development of particleboard from oil palm frond fibers using citric acid and sucrose as adhesives. In: Proceedings of the international conference on innovation in polymer science and technology. Yogyakarta, pp 166–173
4. Widyorini R, Umemura K, Isnan R, Putra DR, Awaludin A, Prayitno TA (2016) Manufacture and properties of citric acid-bonded particleboard made from bamboo materials. Eur J Wood Wood Prod 74:57–65
5. Dukarska D, Czarnecki R, Dziurka D, Mirski R (2017) Construction particleboards made from rapeseed straw glued with hybrid pMDI/PF resin. Eur J Wood Wood Prod 75:175–184
6. Widyorini R, Umemura K, Septiano A, Soraya DK, Dewi GK, Nugroho WD (2018) Manufacture and properties of citric acid-bonded wood composite board made from Salacca frond: effects of maltodextrin addition, pressing temperature, and pressing method. BioResources 13(4):8662–8676
7. Kusumah SS, Astari LA, Subyakto, Umemura K (2019) Sweet sorghum bagasse as prospective raw material for eco-friendly particleboard. In: Rogers L, Willis M (eds) Sweet sorghum: characteristics, cultivation and uses. Nova Science Publishers, New York, pp 97–131
8. Kariuki SW, Wachira J, Kawira M, Murithi G (2020) Crop residues used as lignocellulose materials for particleboards formulation. Heliyon 6(9):e0525
9. Komariah RN, Miyamoto K, Tanaka S, Prasetiyo KW, Syamani FA, Subyakto, Umezawa T, Kanayama K, Umemura K (2019) High-performance binderless particleboard from the inner part of oil palm trunk by addition of ammonium dihydrogen phosphate. Ind Crop Prod 141:111761
10. Granitty DP, Soekardi M, van Noordwijk M, de la Cruz R, Pathak PS, Gunasena HPM, van So N, Huijun G, Majid NM (1996) The imperata grassland of tropical Asia: area, distribution, and typology. Agrofor Syst 36:3–29
11. Sutiya B, Istikowati WT, Rahmadi A, Sunardi (2012) Chemical content and properties of cogongrass fiber (Imperata cylindrica) as a description of raw materials for pulp and paper. Bioscientiae 9(1):8–19
12. Umemura K, Ueda T, Kawai S (2012) Effects of molding temperature on the physical properties of wood-based moulding bonded with citric acid. For Prod J 62(1):63–68
13. Umemura K, Ueda T, Munawar SS, Kawai S (2012) Application of citric acid as natural adhesive for wood. J Appl Polym Sci 123(4):1991–1996
14. Umemura K, Sugihara O, Kawai S (2014) Investigation of a new natural adhesive composed of citric acid and sucrose for particleboard II: Effect of board density and pressing temperature. J Wood Sci 61(1):40–44
15. Kusumah SS, Umemura K, Yoshioka K, Miyafuji H, Kanayama K (2016) Utilization of sweet sorghum bagasse and citric acid for manufacturing of particleboard I: effects of pre-drying treatment and citric acid content on the board properties. Ind Crops Prod 84:34–42
16. Kusumah SS, Arinana A, Hadi YS, Guswenrivo I, Yoshimura T, Umemura K, Tanaka S, Kanayama K (2017) Utilization of sweet sorghum bagasse and citric acid for manufacturing of particleboard III: influence of adding sucrose on the properties of particleboard. BioResources 12(4):7498–7514
17. Widodo E, Kusumah SS, Subyakto, Umemura K (2020) Development of moulding using sweet sorghum bagasse and citric acid: effects of application method and citric acid content. For Prod J 70(2):151–157
18. Kusumah SS, Umemura K, Guswenrivo I, Yoshimura T, Kanayama K (2016) Utilization of sweet sorghum bagasse and citric acid for manufacturing of particleboard II: influences of pressing temperature and time on particleboard properties. J Wood Sci 63(2):161–172

19. Syamani FS, Ramatia D, Sudarmanto, Akbar F, Kusumah SS, Hermawan D (2020) Combination of citric acid and maltodextrin as bonding agent in sorghum bagasse particleboard. IOP Conf Ser Mater Sci Eng 935:012058 (2020)
20. Kusumah SS, Massijaya SY, Prasetiyo KW, Sutiawan J, Lubis MAR, Hermawan D: Surface modification of eco-friendly particleboard made from sorghum bagasse and citric acid sucrose adhesive. IOP Conf Ser Mater Sci Eng 935:012054
21. Umemura K, Sugihara O, Kawai S (2013) Investigation of a new natural adhesive composed of citric acid and sucrose for particleboard. J Wood Sci 59(3):203–208
22. Wibowo ES, Kusumah SS, Subyakto, Umemura K (2021) Modification of novel bio-based adhesive made from citric acid and sucrose by $ZnCl_2$. Int J Adhes Adhes 108:102866
23. Abou-Zeid A-ZA, Ashy MA (1984) Production of citric acid: a review. Agric Wastes 9:5–76
24. Tsao GT, Cao NJ, Du J, Gong CS (1999) Production of multifunctional organic acids from renewable resources. Adv Biochem Eng/Biotechnol 65:243–278
25. Bock K, Lemieux RU (1982) The conformational properties of sucrose in aqueous solution: intramolecular hydrogen-bonding. Carbohyd Res 100(1):63–74
26. Kwok K, Mauer LJ, Taylor LS (2010) Kinetic of moisture-induced hydrolysis in powder blends stored at and below the deliquescence relative humidity: investigation of sucrose-citric acid mixture. J Agric Food Chem 58:11716–11724
27. Queneau Y, Jarosz S, Lewandowski B, Fitremann J (2007) Sucrose chemistry and applications of sucrochemicals. Adv Carbohydr Chem Biochem 61:217–292
28. Taher AM, Cates DM (1974) A spectrophotometric investigation of the yellow color that accompanies the formation of furan derivatives in degraded-sugar solutions. Carbohyd Res 34(2):249–261
29. Syamani FS, Dalle RH, Subyakto (2017) Characteristics of particle board from cogongrass with variations in particle size and isocyanate adhesive content. In: Proceedings of the XX MAPEKI national seminar, pp 61–73
30. Syamani FA, Kusumah SS, Astari L, Prasetiyo KW, Wibowo ES, Subyakto (2018) Effect of pre-drying time and citric acid content on *Imperata cylindrica* particleboards properties. IOP Conf Ser Earth Environ Sci 209:011001
31. Lubis M, Wijaya S, Ginting MHS, Harahap MB (2020) Effect of cogon grass powder (*Imperata cylindrica* L. Beauv) composition and particle size on physical and mechanical properties of unsaturated polyester resin particle board. AIP Conf Proc 2267:020055
32. TAPPI T 17 wd-97 (1997) Cellulose in wood. Technical Association of the Pulp and Paper Industry, Atlanta
33. TAPPI T 222 om-02 (2002) Acid-insoluble lignin in wood and pulp. Technical Association of the Pulp and Paper Industry, Atlanta
34. TAPPI 204 cm-97 (1997) Solvent extractives of wood and pulp. Technical Association of the Pulp and Paper Industry, Atlanta
35. Browning BL (1967) Methods of wood chemistry. Interscience Publishers, New York
36. JIS A 5908-2003 (2003) Particleboard. Japanese Standards Association, Tokyo

Isolation of Cellulose and Starch from Mango Golek (*Mangifera indica L*) Seed Shells and Its Characterization as Potential New Biocomposites

H. S. Syamsidar, A. Ahmad, S. Fauziah, and D. Sondari

Abstract The increasing use of plastic in life has resulted in pollution. Synthetic plastic is difficult to degrade in nature so that the main ingredients for environmentally friendly plastics are needed from waste, namely cellulose and starch. Cellulose and starch are obtained in mango seeds. These compounds can be used as raw materials or new biocomposite to improve the mechanical and functional properties. Extract mango seed shell was cellulose 35.33%, hemicellulose 9.44% and lignin 6.13% and the inner mango seeds was starch 87.82%. Then analysis of water content 30.94%, ash 0.07%, fat 4.85%, protein content 4.80%, and carbohydrate content 51.96%. Functional group test for mango seed shell cellulose using FT-IR showed absorption peaks are functional group O–H alcohol on 3334.25 cm^{-1}. And the C-H functional group alkane (strain) at 2918.76 cm^{-1}. Mango seed starch while visible aromatic C=C functional group absorption at the 1592.12 cm^{-1} and the C-O group function at 1050.34 cm^{-1}. C-H group function of alkanes (stretch) at the 2917.58 cm^{-1} and C=C group of alkenes at the 1729.50 cm^{-1}. Those are also indicate presence of C-O and C-N single bonds and ester groups at the absorption peak at the 1337.09 cm^{-1}.in mango seed starch. The functional groups of cellulose and starch from mango seed showed FTIR are not differ from standard cellulose and starch..

Keywords Cellulose · Starch · *M indica L* · Biocomposite

H. S. Syamsidar
State Islamic Institute of Religion Bone, Watampone, Indonesia

A. Ahmad (✉) · S. Fauziah
Department of Chemistry, Mathematics and Natural Science Faculty, Hasanuddin University, Makassar, Indonesia
e-mail: ahyarahmad@unhas.ac.id

D. Sondari
Research Center for Biomass and Bioproducts, National Research and Innovation Agency, Cibinong, West Java, Indonesia

M. A. Rahandi Lubis (ed.), *Proceedings of the 1st International Conference of Lignocellulose*, Green Energy and Technology,
https://doi.org/10.1007/978-981-19-1596-3_5

1 Introduction

Currently, plastic is widely used as packaging and is increasing day by day, ranging from household furniture, electronic equipment, food, and beverages [1, 2]. Plastic has the advantages of being light, strong, and affordable, but plastic has the disadvantage of being difficult to degrade so that it has a negative impact and pollutes the environment. Plastics that are mostly derived from synthetic polymers will be degraded for tens or even hundreds of years and if burned the carbon emissions produced can also pollute the environment [3, 4].

The amount of plastic produced annually is 300 million tons [5]. The contribution of plastic waste to the total national waste production reaches 15% with an average growth of 14.7% every year and placing plastic waste as the second largest contributor after organic waste plastic, among others packaging products on the market, one of which is 7.16% low density poluethylene (LDPE) [6]. LDPE plastic is a polymer that cannot be degraded, so it has the potential to cause environmental pollution and threaten ecosystems. In addition, fossil fuel reserves as raw materials for LDPE are dwindling, so that bioplastics are an opportunity to solve problems by utilizing renewable and degradable natural resources [7].

Bioplastic raw materials come from plant compounds such as starch, cellulose, and lignin as well as materials from animals such as; proteins, caseins, lipids [8]. Mango (*Mangifera indica L.*) is one of the fruit plants favored by the people of Indonesia. This fruit can grow well in the tropics and subtropics, including in Indonesia. This fruit contains minerals, vitamin C, and complementary nutrients [9, 10]. However, not only the flesh of the fruit has benefits, but the skin and seeds also have many benefits, such as mango skin which functions to improve blood circulation, rejuvenate the skin, lose weight, and overcome excessive bleeding during pregnancy menstruation [11, 12]. Mango seeds have not been widely used so they only become waste.

Mango production nationally 2,808,939.00 tons [13]. Automatically the waste generated will be the same. Mango seeds as a potential source of starch compared to conventional sources. Starch isolation, structural morphological composition along physicochemical properties various are well discussed. Various modifications to enhance the functionality of mango seed starch (MSS) were comprehensively investigated based on the previous findings. Mango seed starch (MSS) reflects the presence of starch that is more resistant than other conventional starches. The structure of mango seeds starch can be easily manipulated using biological, chemical and physical methods. The utility of mango seeds on a large scale will increase the economic viability of the mango waste treatment industry.

In Indonesia, the use of starch as a raw material for the synthesis of bioplastics has great potential, because many plant are produce starch and cellulose, one of which is the mango seed. Mango seeds are an alternative source of starch and cellulose which are quite potential in Indonesia. These raw materials have the potential to be used as new biocomposite to bioplastics synthezise. Lignocellulosic components are cellulose, hemicellulose and lignin which function to maintain the integrity (support) of the starch structure [14] dan. Starch with the addition of cellulose in the synthesis of bioplastics can improve the mechanical and physical characteristics. The resulting bioplastic still absorbs more water than commercial LDPE plastic. The best bioplastic characteristics were obtained at a starch: cellulose ratio of 6.5: 3.5 (w/w) [15].

2 Materials and Methods

2.1 *Materials*

The materials used in this study are aquadest (H_2O), ethanol (C_2H_5OH), sulfuric acid (H_2SO_4) 97%, PP indicator, filter paper, methanol (CH_3OH), sodium hydroxide (NaOH) pa, sodium hypochlorite (NaOCl) technically, sodium carbonate (Na_2CO_3) pa, sulfuric acid, mango seeds, yodium, Na_2CO_3 5%, NaOH 1 M, H_2O_2 50%.

The tools used in this study include Shimadzu Fourier Transform Infrared (FT-IR) spectrophotometer FTIR is tool used for qualitative analysis of functional groups in a chemical compound, hotplate, magnetic stirrer, blender, analytical balance, autoclave, glassware, thermocouple and magnetic stirrer, thermometer, oven, Erlenmeyer, blender, 100 mesh sieve (25 × 25 × 3 mm), acrylic glass plate, spatula, watch glass, beaker glass, test tube, measuring flask, dropper, volume pipette, measuring pipette, petri dish, porcelain dish, tripod, gauze, lighter fire, oven, balance, Bunsen burner, funnel, filter paper.

2.2 *Methods*

2.2.1 Extraction of Cellulose from Mango Seed Shell

Pretreatment Process

The samples used in this study were mango seed shells from several mango juice sellers in Hertasning street Makassar, Indonesia. The mango seed shell was cleaned and dried in the sun. The dried mango seed shell was crushed and sieved with a 100 mesh size to equalize the sample size. Then the 10 g samples which had been finely mixed with NaOH p.a and 72% H_2SO_4 was stirred and allowed to stand for 1 h. The maceration sample was transferred to a beaker and added 770 mL of water, heated for 2 h and then cooled. Then filtered and the lignin precipitate was washed with hot water (initial lignin).

The mango seed shell was put into an autoclave of 50 g in a 2.5% H_2O_2 solution and 6% NaOH cooked at 121 °C with a sample ratio of 1:15. The residue is separated from cooking [14].

Analysis of Cellulose, Hemicellulose, and Lignin Chesson Method (Datta, 1981)

1 g of dry mango seed shalls was weighed and added 150 mL of H_2O. Refluxed at 100 °C in a water bath for 1 h. One gram of dry sample weight of mango seed shalls powder was weighed, before being mixed with H_2O as a value (a). The filtered residue was washed with aquadest until neutral (b), the volume of distilled water used was 300 ml. The neutral residue was dried in an oven until it was constant. Drying was carried out at 105 °C for 24 h. After constant material is weighed as value (c).

Cellulose Content

The residue was added 10 mL of H_2SO_4 72% at room temperature for 4 h. Added 150 mL of H_2SO_4 1N and refluxed in a water bath for 1.5 h in reverse cooling. The residue was filtered with 400 mL H_2O until neutral, then heated in an oven at 105 °C for 24 h. After 24 h and considered constant the residue was weighed as the value (d).

$$\%Cellulose = \frac{c - d}{a} \times 100\% \quad (1)$$

Hemicellulose Content

The residue was added with H_2SO_4 1N and refluxed with a water bath at 100 °C for 1 h. The residue was filtered with water until neutral as much as 300 ml, then the residue was dried in an oven to a constant weight of 105 °C temperature for 24 h.

$$\%Hemicellulose = \frac{b - c}{a} \times 100\% \quad (2)$$

Lignin Content

The residue is ashed and weighed as value (e). Washing residue using hot distilled water. Total lignin can be calculated as follows:

$$\%Lignin = \frac{d - e}{a} \times 100\% \quad (3)$$

where

a = sample weight (grams)
b = weight of residue on the second weighing (grams)
c = weight of residue on the third weighing (grams)
d = weight of residue on the fourth weighing (grams)
e = weight of ash (grams).

2.2.2 Bleaching

About dry powder mango seed 10 g shell resulting from hydrolysis added 300 ml of NaOCl 10%. The solution was stirred for 3 h at room temperature and then the homogeneous solution was filtered. The residue was washed with distilled water and dried in an oven at 50 °C to constant weight. The resulting cellulose was characterized by FTIR [16].

2.2.3 Extraction of Mango Seed Starch

The dried mango seeds were peeled and then cleaned and then cut into small pieces as much as 100 g. After that, it was mashed using a blender and soaked in 100 ml of hot water at 90 °C for 30 min. Then filtered and allowed to stand for 24 h. A precipitate of mango seed pulp is obtained. Then separated with water using a sieve, then the precipitate was added with water and then re-deposited for 24 h. The precipitate obtained was dried in an oven at 50 °C for 30 min. The dry starch powder was obtained, then the starch powder was sieved through a 100 mesh sieve. FTIR testing was carried out on mango seed starch [17].

Proximate Analysis of Mango Seed Starch

Ash content 1 g of sample flour on a porcelain dish. Then put in the oven for 1 h and weigh the mass. While the water content was weighed 2.5 g of the sample in a porcelain cup and heat in the oven for 24 h. Carbohydrate analysis was also carried out in this experiment, with the method of 2 mL of 3,5 dinitrosalicylate (DNSA) reagent added to 5 mL of the sample (diluted 5 times). Then heated for 5 min in boiling water and then cooled and the sample was measured using a UV–Vis spectrophotometer at maximum 540 nm. For the preparation of the standard curve, 0.1 g of glucose were weighed and dissolved in a 100 mL volumetric flask. Take a solution with a variation of 0; 0.2; 0.4; 0.5; 1; and 1.2 mL and then added water to 5 mL. Then 2 mL of 3.5 DNSA reagent was added and heated in boiling water and then cooled. After that, it was measured with a UV-Vis spectrometer at maximum 540 nm.

To test the protein with Biuret method, mixture sample 0.5 g and add aquadest 5 ml, 0.5 ml of NaOH 1M heated at a temperature of ± 90 °C for 10 min. Then it was filled with distilled water in a 25 mL volumetric flask. After that on a magnetic stir for 15 min and then filtered with filter paper. Take 1 mL of supernatant (blank-aquadest) plus 4 mL of Biuret reagent. The mixture was incubated at room temperature for 15 min and added 20 mL of distilled water. Absorbance was measured at maximum 540 nm.

For the total protein test, 0.5 g of the sample was heated and added 3 ml of H_2SO_4 for 3–5 min and then cooled. Add 10 mL of H_2O_2 50% heated for 3–4 min then cooled. After cooling, filtered and transferred to 100 ml of gray measuring then filled with distilled water. 10 mL sample added 1 ml KOH 12 N, 3 drops of mineral stabilizer shaken. Added 3 drops of Polyvinyl Alcohol (PVA) then shaken and made up to 25 mL with distilled water. Add 1 mL of Nessler's reagent. This research was conducted in duplicate.

Crude fiber testing was 2.5 g of the sample. The sample into a kolf and 50 mL H_2SO_4 was added. The mixture is refluxed for ± 30 min. filtered and rinsed with hot water. The residue was put into a cup, and 50 mL of NaOH was added. The residual mixture was refluxed again for ± 30 min. After that, it was filtered and rinsed with H_2SO_4 1% and 25 mL ethanol 96%. Experimental results were weighed. The method for fat extraction 2.5 g wrapped in filter paper. The sample was put into a Soxhlet and added 150 mL of hexane. And than extracted for 1 h and evaporated in kolf to dry. The result transferred to a 250 mL beaker that has been previously thawed. Baked in the oven overnight at 90 °C and then the beaker and sample is determined.

3 Result and Discussion

3.1 Proximate Analysis of Mango Golek Seed Shells

The measurement results can be seen in Table 1. The results obtained are the cellulose, hemicellulose and lignin content in the mango seed shell obtained as shown in the following table:

Mango seed shell worthy of further research in the manufacture of biocomposites. In accordance with the statement of Dawson and Boophaty (2008), NaOH works in degrading lignin which wrapping cellulose and hemicellulose destroying the lignin that wraps so that the lignin that wraps will be broken or damaged. The process of destroying the structure of the lignin and hemicellulose bonds can be an increase the amount of free cellulose present in the material. In bagasse solution containing NaOH (strong base), energy can be spread through ionic conduction which causes heating [18]. The pretreatment, alkali can cause the polymer to decompose in the cell wall by breaking hydrogen and covalent bonds. Therefore, degradation of the lignin produced is increase, heating treatment was added using an autoclave treatment process. Cellulose content analysis shows the feasibility for further research in the manufacture of new biocomposites to synthesize bioplastics.

Table 1 Average content of mango golek seed shells

Composition	Contents (%)	Average (%)
Cellulose	33.45 37.22	35.33
Hemicellulose	8.36 10.52	9.44
Lignin	4.77 7.49	6.13

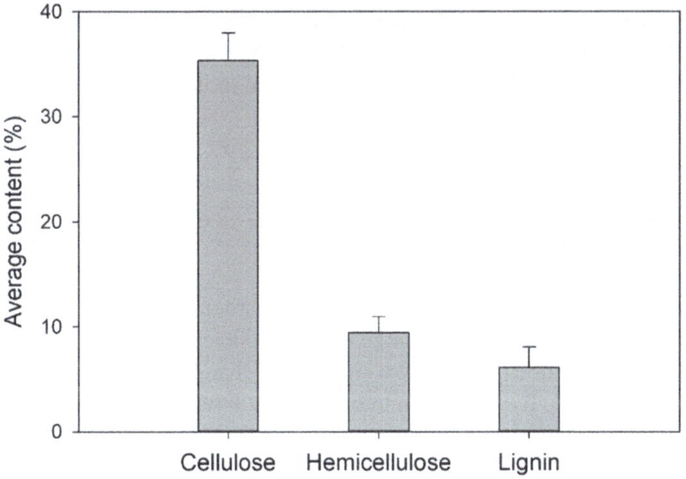

Fig. 1 Average contents of mango seed shalls

The addition of H_2SO_4 will increase the level of decomposed lignin which means reducing the lignin contained in the in the sample, decreased levels of lignin is directly proportional to the increase in H_2SO_4 concentration [19] (Fig. 1).

The results showed a lot of cellulose content was 35.33%. The increase in cellulose content occurred because the lignin content decreased 6.13%.

3.2 Bleaching

Cellulose produced from the extraction of the mango seed shell is in the form of an odorless powder with a color comparison before and after the bleaching process can be seen in Fig. 2.

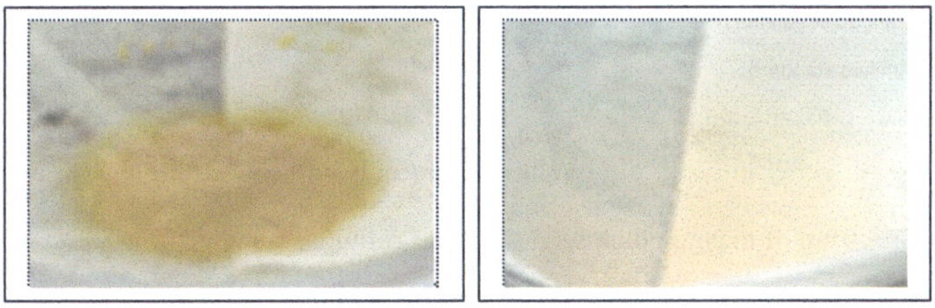

(a) (b)

Fig. 2 Differences in the color of cellulose powder (**a**) before and (**b**) after bleaching

Sodium hypochlorite (NaOCl) causes a decrease in the weight of the powder obtained. This occurs possibly due to the increasing number of lignin and hemicellulose residues that are separated from the cellulose molecules during the bleaching process [20].

3.3 The Functional Group's Analysis

The functional groups of a compound were also characterized by Fourier Transform Infrared (FTIR) with a wavelength range of 500–4000 cm^{-1}. FTIR spectrum results on Mango Seed Shell does not show a significant change in the chemical structure with standard cellulose. It can be seen in Fig. 3 that there are absorption of functional group O-H alcohol (H-bound) on) wavenumber 3334.25 cm^{-1}. and the C-H. functional group alkane (strain) at wavenumber 2918.76 cm^{-1}. The absorption in this area are functional groups from cellulose. While visible aromatic C=C functional group absorption at the1592.12 cm^{-1} wave number and the group ether C-O function at 1050.34 cm^{-1}, the absorption of the functional group present on lignin [21].

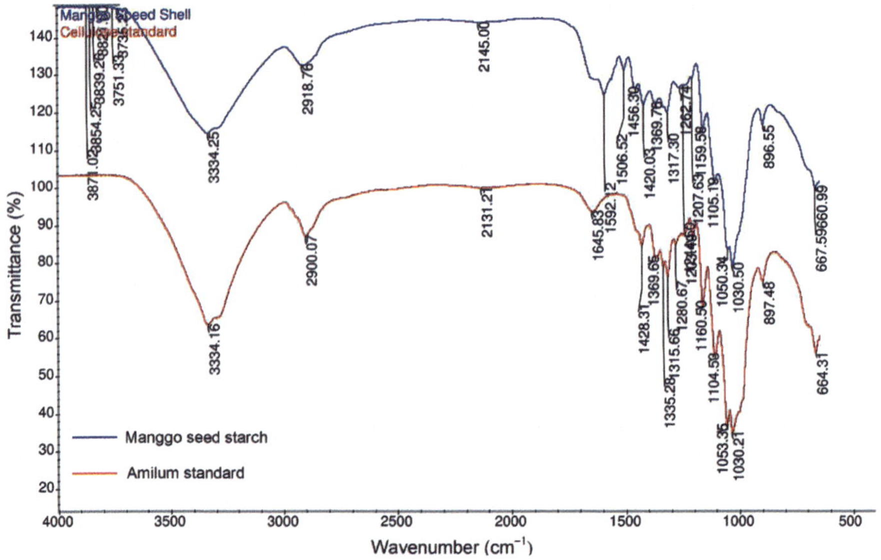

Fig. 3 FTIR Spectrum of mango golek seed shell and cellulose standard

Table 2 Proxymate analysis of mango seed starch

No	Parameter	Result (%)
1	Water content	30.94
2	Ash	0.07
3	Fat	4.85
4	Protein content	4.80
5	Carbohydrate	51.96

3.4 Extraction of Mango Seed Starch Result

Observations on starch yield mango seeds, indicating that the interaction between treatments (precipitation), will affect the color of the starch but has no significant effect on the starch yield of mango golek seeds are 87.82%.

$$\%Yield = \frac{final\ weight\ sample}{first\ weight\ sample} \times 100\% \qquad (4)$$

3.4.1 Mango Seed Starch Proximate Analysis

Based on the results of the proximate analysis of mango seed starch, it is shown in the following Table 2. According to Winarno [22], the content of foodstuffs with a moisture content of less than 15% is classified as dry foodstuffs. Water content in this study is higher. The ash, fat and protein contents in the sample are non-volatile inorganic compounds in mango seed starch is lowest after decomposition at high temperature and associated with mineral elements contained in the material, and then quite high in carbohydrate [22].

3.4.2 The Functional Groups

The functional groups of a compound were also characterized by Fourier Transform Infrared Spectrophotometer (FTIR) with a wavelength range of 500–4000 cm^{-1}. Characterization using FT-IR was carried out the bonds in the mango seed starch produced. Result was shown with the absorbance in the wavenumber (cm^{-1}). The observation of the IR spectra in Fig. 4, several peaks are showed the presence of O-H alcohol groups, C-H alkanes, C=O and C-O ether groups in seed starch mango. The results of the FT-IR analysis of mango seed starch showed the starch content of the mango seeds. Base on literature starch content which is composed of amylose and amylopectin and reducing glucose ($C_6H_{10}O_5$)n [23].

From the results of the FT-IR of mango seed starch compounds, it can be seen the appearance of absorption peaks at 3298.79 cm^{-1} which indicates the presence

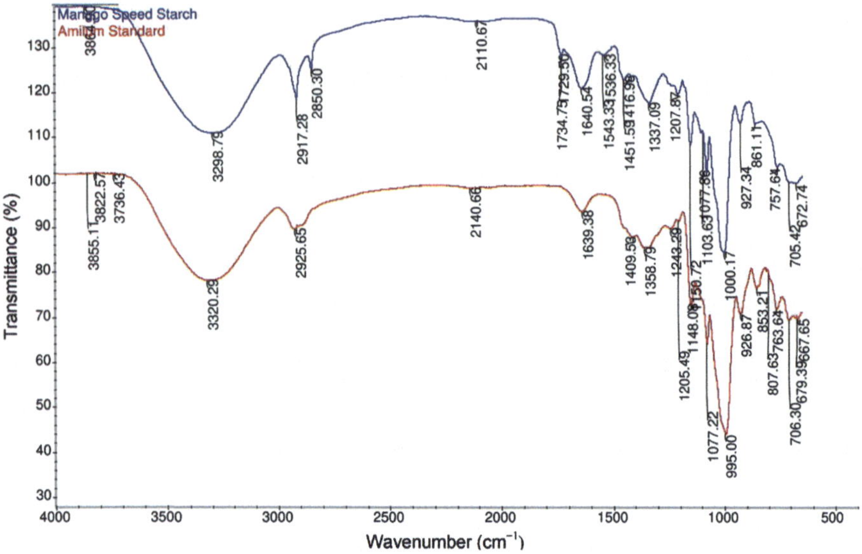

Fig. 4 FTIR spectrum starch of mango golek seed and amilum standard

of an alcohol OH group (H *bonded*). In addition, there is an absorption peak at 2917.58 cm^{-1} which is the presence of the C-H group of alkanes (stretch). The presence of an absorption peak at wavenumber 1729.50 cm^{-1} indicates the presence of the C=C group of alkenes. It is also seen that the absorption peak at from 1337.09 cm^{-1} indicates the presence of C-O and C-N single bonds and ester groups in mango seed starch.

Starch has other content, namely protein, fat and flavonoids in addition to amylose and amylopectin which is the main ingredient. This protein content suspected to be the origin of obtaining the C-N group contained in the starch. Protein was a polymer of amino acid monomers. Then from the FT-IR analysis of seed starch, Mango also has absorption indicating the presence of C-O ester groups. ester group thought to be derived from fatty compounds. Hydrolysis of fats produces fatty acids, where fatty acids are carboxylic acids.

4 Conclusion

Samples from Golek mango seeds in the process of extracting cellulose and starch were obtained using autoclave and maceration. Cellulose from mango seed shell extract was 35.33% and starch from the mango seeds shell was 87.82%. The high content of these two compounds has the potential to be combined into new biocomposites in the manufacture of bioplastics.

References

1. Ariani A (2009) Design of plastic packaging and forming materials. J Dimens Art Des 6(2):191–206
2. Karuniastuti N (2013) The dangers of plastic to health and the environment. Swara Patra 3(1):6–14
3. Kamsiati E, Herawati H, Purwani EY (2017) Potential development of biodegradable plastics based on sago starch and cassava in Indonesia. J Agric Res Dev 36(2):67–76
4. Hasanah YR, Haryanto H (2017) Effect of addition of calcium carbonate ($CaCO_3$) and clay filler on mechanical properties and biodegradable plastics from tapioca waste. Techno (J Fac Eng Univ Muhammadiyah Purwok) 18(2):99–110
5. Novena M (2020) Tahun 2040, 1,3 Miliar ton sampah plastik akan tenggelamkan bumi. *Kompas.com* (online), diakses 12 Maret 2021
6. Masyruroh A, Rahmawati I (2021) Pembuatan recycle plastik HDPE Sederhana menjadi Asbak. abdikarya: Jurnal Pengabdian Dan Pemberdayaan Masyarakat 3(1):53–63
7. Cahyaningtyas AA, Ermawati R, Supeni G, Syamani FA, Masruchin N, Kusumaningrum WB et al (2019) Hydrolysis of modification and characterization of oil palm stem starch as raw material for bioplastics. J Chem Packag 41(1):37–44
8. Shun X (2013) Handbook of biopolymers and biodegradable plastics: overview of plant polymers-resources, demands, and sustainability. PDL Hanbook Series Elsevier, USA
9. Anonymous (2006) Bioplastic Magazine, vol 1. Shopping bags. http://www.bioplasticsmagazine.net. Accessed 27 Oct 2012
10. Halik A, Yudi A, Rosandi (2019) Processing potential of mango to make mango peel chips. Abdikarya J J Lect Stud Serv 3(3):297–299
11. Fitriana F, Setiawan OD (2021) Entrepreneurship training utilization of mango peel waste into various flavored chips snacks as an effort to diversify processed foods that are healthy and rich in benefits. J Sci Technol Community Serv 7(1):16–22
12. Fitranto R, Wahyono ND, Wibisono Y (2020) Strategi Pengembangan Pemasaran Buah Mangga Arumanis 143 PT. Trigatra Rajasa Situbondo Jawa Timur. Jurnal Agribisnis Indonesia (Journal of Indonesian Agribusiness) 8(1):58–68
13. Bangar SP, Kumar M, Whiteside WS (2021) Mango seed starch: a sustainable and eco-friendly alternative to increasing industrial requirements. Int J Biol Macromol 183:1807–1817
14. Septiosari A, Latifah L, Kusumastuti E (2014) Manufacture and characterization of mango seed waste bioplastic with the addition of cellulose and glycerol. Indones J Chem Sci 3(2):157–162
15. Hariyani S (2016) Utilization of activated charcoal from mango golek seed shell (Mangifera Indica) as an adsorbent for lead (Ii) and cadmium (II) ions with Na_2CO_3, pp 1–8
16. Gibson LJ (2012) The hierarchical structure and mechanic soft plant materials. J R Soc Interface 9:2749–2766
17. Darni Y, Sitorus TM, Hanif M (2014) Thermoplastic production of bioplastics from sorghum and cellulose. J Chem Environ Eng 10(2):55–62
18. Kallel F, Bettaieb F, Khiari R, Garcia A, Bras J, Chaabouni SE (2016) Isolation and structural characterization of cellulose nanocrystals extracted from garlic straw residues. Ind Crop Prod 87:287–296
19. Permatasari HR, Gulo F, Lesmini B (2014) Pengaruh konsentrasi H_2SO_4 dan NaOH terhadap delignifikasi serbuk bambu (Gigantochloa apus). Jurnal Penelitian Pendidikan Kimia: Kajian Hasil Penelitian Pendidikan Kimia 1(2):131–140
20. Puspita EI, Ardhyananta H (2013) Effect of heating temperature on the synthesis of carbon black from ori bamboo (Bambusaarundinaceae) and petung bamboo (Dendrocalamusasper). J Pomits Tech 2(1):45–50
21. Amalia YR, Suparto IH, Pramasari DA, Kusumah SS (2019) Chemical components of Andong Bamboo (*Gigantochloa pseudoarundinacea*) based on axial position as raw material for laminated bamboo without adhesive. In: National proceedings of lignocellulose. Biomaterials Research Center, Indonesian Institute of Sciences, Cibinong, pp 73–80

22. Winarno FG (1989) Food chemistry and nutrition. PT Gramedia, Jakarta
23. Tesanika A (2018) Making hybrid biocomposites from mango (*Mangifera indica*) seed starch filled with zinc oxide (ZnO) and clay nanoparticles with ethylene glycol plasticizer. Department of Chemical Engineering, Faculty of Engineering, University of North Sumatra, pp 6–7

The Enhancement of Ultra Violet (UV) Absorption of Mulch Film Based on Linear Low Density Poly-Ethylene (LLDPE)/Cornhusk Cellulose

Fathianissa Agnifa Hernawan, Firda Aulya Syamani, and Mersi Kurniati

Abstract A mulch is a layer of material applied to the surface of soil. The purpose of applying mulch were included conservation of soil moisture and fertility, reducing weed growth and enhancing the visual appeal of the area. In this research, mulch was made by a matrix of Linear Low Density Polyethylene (LLDPE) and a filler of corn husks cellulose. The cellulose was extracted from corn husks, so that can increase the utilization of agricultural waste. In addition, Tinuvin P also be added into LLDPE and cellulose compound, as a UV stabilizer. The treatment variables in this study were cellulose as much as 5 and 10% and Tinuvin P as much as 0.1, 0.3, and 0.5%. In order to obtain mulch compound, the ingredients were mixed using a rheomixer at 40 rpm, 130 °C. Subsequently, the compound was forming into a mulch sheet using a hot press machine. The mechanical and optical properties of mulch were analyzed. The results show that with the addition of cellulose, the value of mulchs tensile strength was decreasing. However, the addition of Tinuvin P could slightly increase the mulch's tensile strength. In the optical properties, there was an absorption peak in the UV wavelength range which indicates that Tinuvin P addition successfully acted as UV absorbent.

Keywords Cornhusk cellulose · LLDPE-based film composite · Mulch film · Tensile strength · Ultra violet absorption

F. A. Hernawan · M. Kurniati
Biophysics Division, Department of Physics, IPB University, Bogor, Indonesia

F. A. Syamani (✉)
Research Center for Biomass and Bioproducts, National Research and Innovation Agency, Cibinong, Indonesia
e-mail: fird003@brin.go.id

M. A. Rahandi Lubis (ed.), *Proceedings of the 1st International Conference of Lignocellulose*, Green Energy and Technology,
https://doi.org/10.1007/978-981-19-1596-3_6

1 Introduction

One of the problems of horticultural cultivation that is often found in dry land is the limitation and availability of water because horticultural crops are usually developed in dry land. Dry land in Indonesia, generally has low water storage capacity. If there is a lack of water, the plant will be thin, wither and even die. However, if the amount of water is excessive, it can cause high soil moisture and stimulate the emergence of fungal diseases. Adjustment of soil moisture to support plant growth can be done by using mulch [1, 2]. The use of mulch in Indonesia has been widely applied to various vegetable crops and is very well used in areas that have a tropical climate. Mulch can help prevent water loss during the dry season and prevent water accumulation during the rainy season, suppress weed growth, improve soil structure, increase soil capacity to hold water and keep soil temperature stable. In addition, mulch can control solar radiation on soil surface, subsequently reducing water evaporation from soil surface. Soil evaporation is a process that determines the availability of water in dry land agriculture [3].

Mulch is divided into two types based on the material, namely organic mulch and inorganic mulch. Organic mulch can be in the form of crop waste such as straw and leaf litter. While inorganic mulch comes from synthetic materials, namely plastic mulch. Plastic mulch material is very effective in controlling groundwater evaporation and can control heat loss at night, so it is widely used in agricultural systems. Plastics that are generally used in mulching technology are Linear Low-Density Polyethylene (LLDPE) plastics. The use of LLDPE as a mulch matrix has the advantages of being flexible, waterproof, translucent and easy to print into thin sheets.

Many studies have been conducted to explore natural resources of non-wood ligno-cellulosic fibers as fillers in synthetic polymers which aim to improve the physical and mechanical properties of a plastic and increase the proportion of biodegradable material content. One of the natural fibers that has not been utilized optimally and has the potential to be developed as a filler for biodegradable plastic materials is corn husk which is still considered as waste [4]. The use of corn husks is only limited to animal feed and food wrapping [5, 6]. Based on data from the Indonesian Statistics Agency (BPS), in 2014 corn production in Indonesia was 19.0 million tons. The production was increased in 2015 to 19.6 million tons. In 2016, the trend of corn production increment continued with a production achievement of 23.6 million tons. In fact, in 2017 corn production reached 28.9 million tons and in 2018 it is predicted that production will continue to increase to reach 30.1 million. Such conditions indicate the large role of corn in the growth of the food crop subsector and the national economy in general. About 50% of the total weight of the corn crop is waste left after harvesting. The percentages of each waste included 50% stalks, 20% leaves, 20% corn cobs and 10% corn husks [7]. This shows that the corn husk waste produced can reach 1.6 million tons.

Based on the research shown in Table 1, it is known that the weaknesses of previous studies have led to less optimal results. Therefore, in this study, cellulose extracted

Table 1 Previous study of LLDPE and corn husk mulch

Materials	Methods	Results	References
LLDPE, corn husk (60 mesh)	Extrusion, then hotpressing	Corn husk powder as filler	[4]
LLDPE, corn husk (200 mesh)	Extrusion, then calandering	The mechanical properties were low	[11]
LLDPE, corn husk (200 mesh), Tinuvin P	Extrusion, then calandering	The mechanical properties were low and Tinuvin were not well dispersed	[12]

from corn husks was used as a biodegradable filler material in the LLDPE polymer matrix. Corn husk cellulose extraction was carried out using the alkaline method [8]. Mulch made from LLDPE, able to retain heat and radiation transmission of sunlight compared to other plastics. However, LLDPE mulch was more easily damaged by exposure to UV radiation from the sun [9]. Moreover, Indonesia has tropical climate, lead the higher intensity of sunlight. The addition of Tinuvin P in this study functions as an anti-UV and stabilizer which is expected to control the degradation process caused by exposure to sunlight [10]. Based on above background, the purpose of this study was to analyze the effect of corn husks cellulose and Tinuvin P addition into LLDPE, on the mechanical and optical properties of the mulch.

2 Materials and Method

2.1 Materials

Linear Low-Density Polyethylene (LLDPE) Asrene UF1810T produced by PT Chandra Asri Petrochemical Tbk, with density of 0.921 g/cm^3, melting temperature of 122 °C. Corn husks were taken from 3rd layer to 10th layer of *Zea mays*. Benzotriazole UV absorber (Tinuvin P), produced by BASF with density of 1.38 g/cm^3, melting temperature of 128–132 °C. Reagens were technical grade natrium hydroxide (NaOH), technical grade natrium hypochlorite (NaOCl), acetone (Pro Analys MERCK).

2.2 Cellulose Extraction

A total of 1.5 kg of corn husks that have been dried and cut into smaller sizes, are put into an autoclave, then processed at 80 °C for 12 h. Afterward, the process was continued by adding 2% NaOH as much as 25 L, at a temperature of 150 °C for 1 h and using a rotational speed of 50 rpm. When the delignification process is complete, the sample was rinsed using tap water until neutralized. Subsequently, bleaching process was carried out using 30 L of NaOCl at room temperature for 30 min with a rotating speed of 100 rpm. After that, the sample was rinsed until the pH was neutral. The extraction results were then tested for fiber content and analyzed for functional groups. The cellulose was stored in a refrigerator before being used for further processing.

2.3 Cellulose Pretreatment

The extracted corn husk cellulose was pre-treated in the form of simple solvent exchange using an organic acetone solution. The cellulose was squeezed to separate the water content, then immersed in acetone solution (1:10 w/v) at room temperature while stirring using an overhead stirrer with a rotating speed of 120 rpm for 2 h. After that, the cellulose was squeezed and then soaked again using a different acetone. This process is repeated 3 times for 1 cycle.

2.4 Materials Compounding and Mulch Forming

Materials compounding was carried out using HAAKE™ Rheomix PolyLab OS Mixers. In this process the ingredients mixed are LLDPE plastic resin, corn husk cellulose with variations of 5% and 10%, and Tinuvin P with variations of 0.1%, 0.3% and 0.5%. These materials were weighed beforehand according to Table 2 with the total mass of 55 g. Then the material was put into the rheomixer slowly with a rotating speed of 40 rpm at a temperature of 130 °C for 10 min. After thoroughly mixed, the sample was then removed from the rheomixer.

Mulch forming was conducting using a hot press machine at a temperature of 130 °C. A total of 3 g of compound were placed between two aluminum plates covered with teflon sheet (20 × 20 cm). Firstly, the compound was melted in a hot press without pressure for 15 min. Secondly, the compression pressure was increased to 10 MPa and the compound was pressed for 15 min. Finally, the compression pressure was increased until it reached a pressure of 40 MPa and pressed for 5 min. The total time required for mulch forming was 35 min. After hot pressing process, the sample was taken out from hot press machine and keep it until reach room temperature before removed the teflon sheet.

Table 2 Formulation of LLDPE and corn husk mulch

Sampel code	LLDPE		Corn husk cellulose		Tinuvin P	
	%wt	weight (g)	%wt	weight (g)	%wt	weight (g)
C1	99.9	54.95	0	0	0.1	0.055
C3	99.7	54.84	0	0	0.3	0.165
C5	99.5	54.73	0	0	0.5	0.274
S5T1	94.9	52.33	5	2.67	0.1	0.003
S5T3	94.7	52.23	5	2.76	0.3	0.008
S5T5	94.5	52.13	5	2.85	0.5	0.014
S10T1	89.9	49.95	10	5.04	0.1	0.005
S10T3	89.7	49.86	10	5.12	0.3	0.015
S10T5	89.5	49.77	10	5.20	0.5	0.026

2.5 Mechanical Properties Analysis

Analysis of mechanical properties was conducted based on ASTM D882 "Standard Test Method for Tensile Properties of Thin Plastic Sheeting", using Instron 5900 Series Universal Testing Systems up to 50 kN. The mechanical properties analysis was carried out at a speed of 10 mm/min and a load cell of 5 kN. Each sample was taken three specimens for testing.

2.6 Optical Property Analysis

Optical properties analysis was carried out by testing UV-Vis absorbance using UV-Vis Ocean Optical UV-Vis spectroscopy USB4000FL series. This test was carried out in the range of wave numbers 200–800 nm. This test was conducted to determine the effect of Tinuvin P addition on the mulch formulation.

3 Results and Discussion

3.1 Corn Husks Cellulose Characterization

One of the efforts to maximize the uses of abundant corn husk waste is to extract the cellulose content in it. The extraction was using alkaline (NaOH), which aims to remove lignin and hemicellulose, so that the cellulose will be obtained. Cellulose content testing was carried using Van Soest analysis. This Van Soest analysis aims to determine the changes in the degree of Acid Detergent Fiber (ADF), lignin, and

Table 3 Corn husk's chemical components

Corn husks	ADF (%)	Lignin (%)	Cellulose (%)
Before extraction	39.77	3.68	35.36
After extraction	75.99	1.80	74.84

cellulose in corn husks before and after extraction by estimating the fiber content in the material and its fractions into certain groups based on their affinity with detergent anions or cations. ADF is a food substance that is insoluble in acidic detergents consisting of cellulose, lignin, and silica. ADF is used as a step to determine lignin so that hemicellulose can be estimated from differences in ADF cell wall structure. This ADF was determined using an acidic detergent solution where the residue consisted of cellulose and lignin [13].

The results of fiber content analysis in corn husks before and after extraction are presented in Table 3. The delignification process using NaOH caused a decrease in lignin content from 3.68 to 1.80%. On the other hand, the ADF content was increase from 39.77 to 75.99%. The cellulose content was also increased along with the increase in the ADF content, which was from 35.36 to 74.84%. The decrease in lignin content was caused by the OH- ion present in the alkali breaking the ether bond between lignin and cellulose and producing phenolic hydroxyl groups [14]. Hydroxyl ions can break the bonds of lignin molecules so that the lignin in the material will dissolve [15].

Functional group analysis was also carried out on corn husks cellulose after extraction. The spectrum of the extracted sample is presented in Fig. 1. Based on the FTIR spectrum, at a wave number of 3394 cm^{-1}, there is an intramolecular stretching of O-H bond from cellulose. The peak at a wave number of 2901 cm^{-1} was associated with the bending of C-H bond, which is identified in the functional groups of alkanes and carboxylic acids [16]. The peak at wavenumber of 1643 cm^{-1} indicates the O-H bond of water absorption [17]. In the range of wavenumber 1600–1500 cm^{-1}, no peak was detected. These wave numbers indicate the presence of C=C bonds in the aromatic ring which is a typical functional group of lignin [16]. This is in accordance with the results of the fiber content analysis, which where reveal that after the extraction process, the lignin content decreased due to the delignification process. Furthermore, at the wave number 1165 cm^{-1} there is a stretching of C-O-C bond. This bonding occurs due to the alkalization process in which associated with reaction of NaOH solution and lignin. The OH- ion reacts with the H group in the lignin to form H_2O [18]. The wave numbers 895 cm^{-1} and 1034 cm^{-1} were associated with C-O bonds streching and C-H vibrations in cellulose. The presence of an absorption peak at the wave number is a marker of the cellulose component, this is in accordance with research conducted by Johar et al. [17]. However, according to Gian [18], the presence of an absorption peak at a wave number of 895 cm^{-1} was a C-H deformation from the bleaching process.

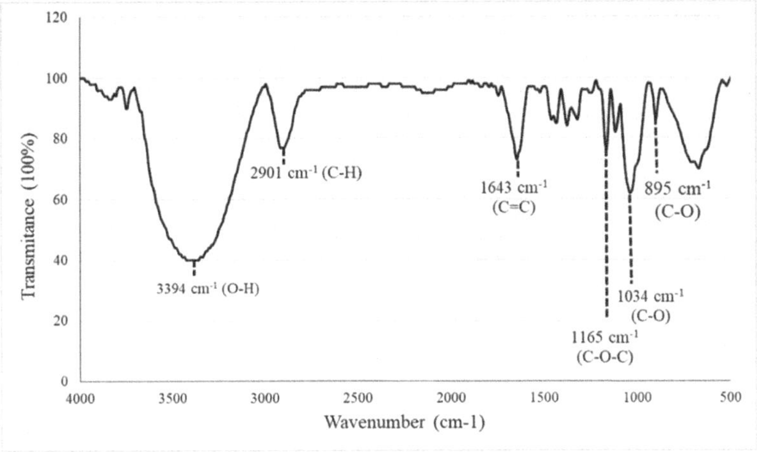

Fig. 1 FTIR spectrogram of corn husks cellulose

3.2 LLDPE/Corn Husks Cellulose Mulch Sheet

The mulch sheets that were successfully formed had a yellowish color, due to the presence of cellulose, meanwhile the control mulch sheet was transparent as shown in Fig. 2. The difference in mulch sheet color was influenced by the constituent materials. The yellowish color from cellulose was caused by the degradation of cellulose during hotpressing. The more cellulose amount, the darker the color of the mulch sheet. Meanwhile, the color of mulch sheet control is more transparent due to the absence of cellulose content in the sheet. The addition of Tinuvin P did not affect the color of the mulch sheet. The mulch sheet thickness ranges from 0.15 to 0.26 mm. The mulch sheet produced in this study were thicker than commercial mulch, which commercial mulch has a thickness of 0.08 mm.

3.3 Mulch Mechanical Properties

The mechanical properties analysis were carried out with a load cell speed of 10 mm/min and a load cell of 5 kN. The data from the tensile test results of mulch sheet are presented in Table 4. There was a decrease in the tensile strength value with the addition of cellulose. This was because the mixture between the materials is not evenly distributed both mechanically and chemically. They cannot be mixed properly because LLDPE is hydrophobic while cellulose has a hydrophilic side so that the two polymers are not compatible [8].

However, before the mixing process using rheomix, the cellulose has been pre-treated by solvent exchange method using acetone. This step was carried out in order to reduce the hydrophilic nature of cellulose. Solvent exchange method was using

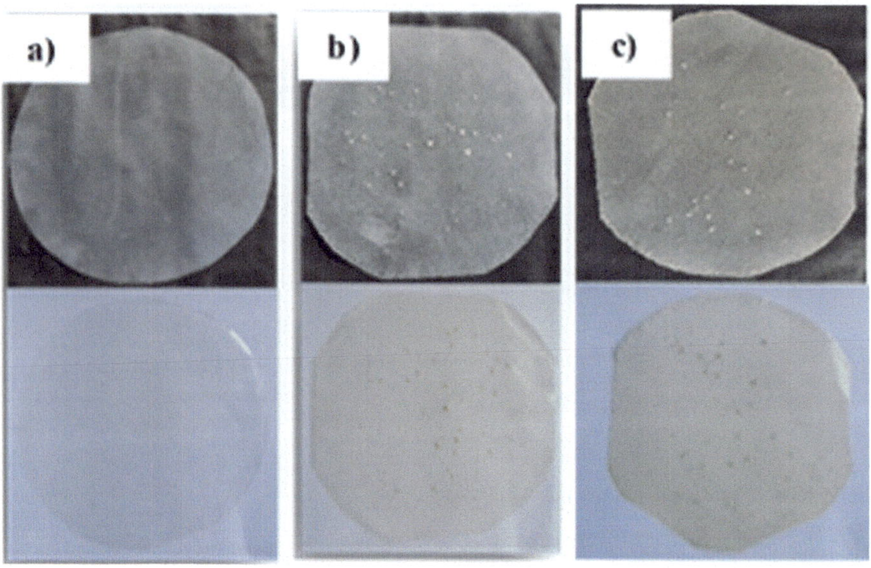

Fig. 2 Mulch with addition of Tinuvin P 5%, **a** Control, **b** Cellulose 5%, **c** Cellulose 10%

Table 4 Corn husk cellulose fibers analysis

Sampel code	Tensile strength (N/mm^2)	Elongation (%)
C1	12.49 ± 0.85	520 ± 4.60
C3	10.62 ± 0.17	244 ± 10.06
C5	10.32 ± 0.16	69 ± 2.18
S5T1	9.16 ± 1.42	23 ± 1.08
S5T3	9.29 ± 0.37	56 ± 4.78
S5T5	9.38 ± 0.05	59 ± 1.66
S10T1	8.13 ± 0.21	37 ± 4.65
S10T3	8.61 ± 0.05	21 ± 1.07
S10T5	8.80 ± 0.90	16 ± 3.72

acetone and causes cellulose to turn into methyl cellulose. Methyl cellulose is often used as an emulsifier between two liquids of different polarity. With the addition of a solvent such as acetone, it can increase the value of the degree of substitution which can make methyl cellulose an emulsifier and more accessible [19]. Later, the methyl group in acetone will replace the hydroxyl group in cellulose by reducing the number of hydrogen bonds in the outer chain so that it can become a bridge between LLDPE and cellulose. With a decrease in the value of tensile strength even though the cellulose has been pre-treated beforehand, it indicates that the solvent exchange process has not been maximized.

Analysis of tensile strength based on the cellulose variation group, as shown in Fig. 3a, at 5% cellulose (S5) and 10% cellulose (S10), the tensile strength value will increase with the increase in Tinuvin P content although it is not significant. This was because with addition of Tinuvin P content, the cellulose content increases proportionally. In addition, the addition of Tinuvin P causes the formation of interactions between anti-UV polymer chains and cellulose in the form of hydrogen bonds [20]. Therefore, in the control mulch sheet, the tensile strength value decreased with the addition of Tinuvin P because no hydrogen bonds were formed due to the absence of cellulose in the control sample. This hydrogen bond will affect the tensile strength of the film [20].

Based on data released by Prospector®, the tensile strength for pure LLDPE has a range of 6.69–31.37 MPa. The tensile strength range is still in accordance with the tensile strength in this study which ranges from 10.32–12.49 N/mm^2. When compared with Dirgantara research [4] which produced mulch made from corn husks powder and LLDPE, the tensile strength value of this study was lower. This was because in Dirgantara study, corn husks powder did not go through the delignification process first, so the natural fibers used still contain lignin. Lignin is more naturally hydrophobic [21], which makes it more compatible with LLDPE. In addition, the use of oleic acid can function as a compatibilizer between LLDPE and corn husk powder [4].

The measurement of tensile strength is usually followed by the result of % elongation. Elongation is the maximum change in length experienced by a material when it is pulled to break. The percent elongation value shows the elasticity of a material. The higher the elongation value, the more elastic the material. The elongation value is presented in Table 4. The highest elongation value is in the control mulch sheet, but the higher the Tinuvin content, the lower the elongation value. The elongation value for the control mulch sheet with the addition of Tinuvin P 0.1% was 520%, with the addition of Tinuvin P 0.3% was 244% and with the addition of Tinuvin P 0.5% was 69%. All % elongation values in the control mulch sheet samples fulfilled the range of elongation values standard for LLDPE issued by Prospector®, namely 47–1000%.

The elongation value in the mulch sheet made by LLDPE and 5% cellulose with the addition of Tinuvin P 0.3% and 0.5% also fulfilled the standard value. However, the elongation value in the mulch sheet made by LLDPE and 5% cellulose was inversely proportional to the control mulch sheet made by LLDPE, along with the addition of Tinuvin P levels.

In the mulch sheet made by LLDPE and 10% cellulose, experienced a decrease in elongation value along with the addition of Tinuvin P content. The decrease in % elongation was thought to be due to a mixture of LLDPE and cellulose. The addition of the number of polymers will make the material stiffer so that the elongation value will be lower [22].

3.4 Mulch Optical Properties

Mulch sheet's optical properties analysis due to the addition of Tinuvin P was carried out by analyze the absorbance of UV–VIS. Measurements were made using a UV–Vis spectrophotometer at a wavelength range of 200–800 nm. The addition of Tinuvin P in polymer, refers to BASF (2010) in the range of 0.1–0.5% [23]. The result of this measurement is a curve of the relationship between absorbance and wavelength. From Fig. 3, the mulch sheet absorbs UV light in the range of 200–400 nm and absorbs visible light in the range of 435–480 nm which is the wavelength of blue visible light. This indicates that the mulch sheet has a complementary color of yellow which corresponds to the color of the mulch sheet with addition of corn husks cellulose, which was yellowish white. The higher the cellulose content, the mulch sheet will be more yellowish in color. The control mulch sheet with the highest Tinuvin P addition (0.5%), has a peak at a wavelength of 245 nm in the UV range.

In the mulch sheet with addition of 5% cellulose (S5), absorption peaks were detected in all samples. The addition of Tinuvin P, effect the increasing of UV absorbance, as shown by the UV absorbance value of S5T3 and S5T5 mulch sheet. This is in accordance with the function of Tinuvin P as a UV absorber [23]. In the mulch sheet with addition of 10% cellulose (S10), absorption peaks were detected around the UV wavelength, but with the addition of 0.5% Tinuvin P the UV absorbance decreased. The curve confirmed that Tinuvin has been dispersed evenly in mulch sheet with addition of 5% cellulose when used at level of 0.3 and 0.5%. Meanwhile, 0.5% Tinuvin could not dispersed evenly in mulch sheet made from LLDPE and 10% cellulose, so that reduce the effectiveness of Tinuvin P as UV absorber.

Tinuvin P has the ability to absorb maximum UV light at a wavelength of 300–400 nm and minimal in the visible light region [24]. Tinuvin is a heat stabilizer and UV stabilizer. The addition of Tinuvin P into LLDPE, can reduce the UV rays that are absorbed so that the mulch is not easily damaged by sunlight radiation exposure.

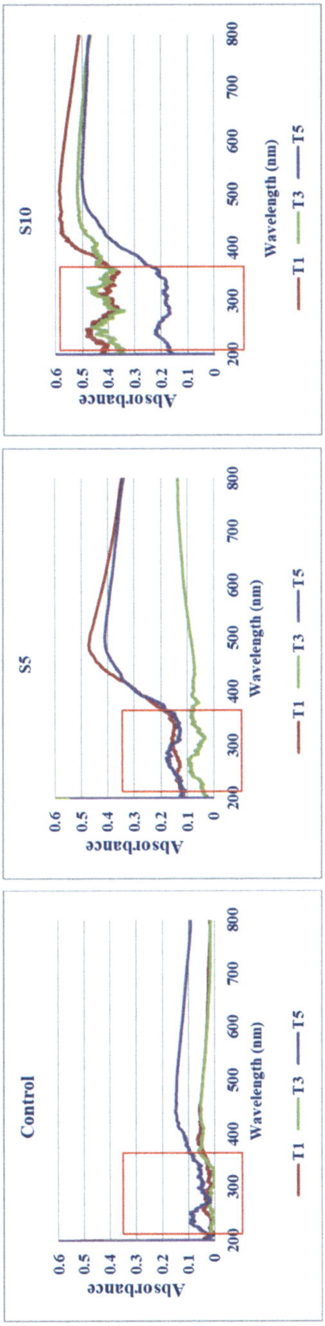

Fig. 3 Mulch UV-Vis absorbance curve **a** control, **b** addition 5% cellulose, **c** addition 10% cellulose

4 Conclusion

Biomulch was successfully made with a mixture of LLDPE, corn husk cellulose and Tinuvin P. Based on mechanical properties analysis, the addition of cellulose reduce the tensile strength of mulch. However, when Tinuvin P was added, the tensile strength of the mulch tends to increase, although not significantly. The mulch with the best mechanical properties was mulch made from LLDPE with the addition of 5% cellulose and 0.5% Tinuvin P (S5T5). Based on the UV-Vis absorbance analysis, Tinuvin P can absorb UV radiation that reach the mulch sheet, so that the mulch will not being easily damaged by sunlight.

Acknowledgements We thank the Japan Society for the Promotion of Science (JSPS) for partial funding of the publication fee.

References

1. Aditya A, Hendarto K, Pangaribuan D, Hidayat KF (2013) Pengaruh Penggunaan Mulsa Plastik Hitam Perak dan Jerami Padi terhadap Pertumbuhan dan Produksi Tanaman Cabai Merah (*Capsicum annum*) di Dataran Tinggi. Jurnal Agrotek Tropika 1(2):147–152
2. Sintim HY, Bandopadhyay S, English ME, Barya AI, DeBruyn JM, Schaeffer SM, Miles CA, Reganold JP, Flury M (2019) Impacts of biodegradable plastic mulches on soil health. Agric Ecosyst Environ 273:36–49
3. Heryani N, Kartiwa B, Sugiarto Y, Handayani T (2013) Pemberian Mulsa dalam Budidaya Cabai Rawit di Lahan Kering: Dampaknya terhadap Hasil Tanaman dan Aliran Permukaan. Jurnal Agronomi Indonesia 41(2):147–153
4. Dirgantara, M, Kurniati M (2013) Effect of corn husk and LLDPE ratio on the properties by thermo-pressing. In: Proceedings the international conference on innovation in polymer science and technology. Indonesian Polymer Association, Yogyakarta, pp 77–88. http://hpi-polimer.org/images/proceeding_IPST2013/pp.87-98_PPA16.pdf
5. Faesal (2013) Pengolahan Limbah Tanaman Jagung untuk Pakan Ternak Sapi Potong. Prosiding Seminar Nasional Inovasi Teknologi Pertanian, pp 181–190. Badan Litbang Pertanian, Jakarta
6. http://kalsel.litbang.pertanian.go.id/ind/images/pdf/prosiding/18%20faesal.pdf
7. Chafid M (2016) Outlook for agricultural commodities of corn food crops. Center for Agricultural Data and Information Systems, Ministry of Indonesia, Jakarta
8. Sulityo HM, Ismiyati (2012) Effect of cassava starch–cellulose formulation on mechanical properties and hydrophobicity in the manufacture of bioplastics. Conversion 1(2):23–30
9. Huda S, Yang Y (2008) Chemically extracted cornhusk fibers as reinforcement in light-weight poly(propylene) composites. Macromol Mater Eng 293:235–243
10. Kasirajan S, Ngouajio M (2012) Polyethylene and biodegradable mulches for agricultural applications: a review. Agron Sustain Dev 32:501–529
11. Nurdayat S, Kemala T, Sudirman, Karo-Karo A (2004) Effect of adding anti UV Tinuvin 783 on the characteristics of Polyvinyl Chloride. Indones Mater Sci 6(1):32–39
12. Indriani A (2016) Effect of variation in composition of LLDPE and corn husk on mechanical characteristics and biodegradation of bioplastic mulch [Thesis]. Bogor Agricultural University, Bogor
13. Ramdhiani F (2016) Effect of variation in composition of corn husk and anti-UV on biomulsa [Thesis]. Bogor Agricultural University, Bogor

14. Van Soest PJ, Robertson JB, Lewis BA (1991) Methods for dietary fiber, neutral detergent fiber, and nonstarch polysaccharides in relation to animal nutrition. J Dairy Sci 74:3583–3597. https://www.journalofdairyscience.org/article/S0022-0302(91)78551-2/pdf
15. Safaria S, Idiawati N, Zaharah TA (2013) Effectiveness of a mixture of cellulase enzymes from aspergillus ni-ger and trichoderma reesei in hydrolyzing coconut fiber substrate. J Chem Equat 2(1):46–51. https://jurnal.untan.ac.id/index.php/jkkmipa/article/view/1779/1727
16. Simatupang H, Nata A, Herlina N (2012) Study of isolation and lignin yield from empty oil palm bunches (TKKS). USU Chem Eng J 1(1):20–24. https://talenta.usu.ac.id/jtk/article/view/1401/888
17. Kampeerapappun (2015) Extraction and characterization of cellulose nanocrystals produced by acid hydrolysis from corn husk. J Metals Mater Miner 25(1):19–26
18. Johar N, Ahmad I, Dufresne A (2012) Extraction, preparation and characterization of cellulose fibres and nanocrystals from rice husk. Ind Crops Prod 37:93–99
19. Gian A, Farid M, Ardhyananta H (2017) Isolation of cellulose from oil palm empty fruit bunch fiber for sound absorption composite nano filler: FTIR analysis. ITS Eng J 6(2):228–231
20. Rahmidar L, Nurilah I, Suadiarty T (2018) Characterization of methyl cellulose synthesized from corn husk (Zea mays). J Sci Educ 2(1):117–122
21. Kaijanen L, Paakkunainen M, Pietarinen S, Jernströml E, Reinikaine S (2015) Ultraviolet detection of monosaccharides: multiple wavelength strategy to evaluate results after capillary zone electrophoretic separation. Int J Electrochem Sci 10(4):2950–2961. http://www.electrochemsci.org/papers/vol10/100402950.pdf
22. Suryanto H (2016) Review of natural fibers: composition, structure, and mechanical properties [review]. State University of Malang, pp 1–14
23. Lismeri L, Irmalinda G, Darini Y, Herdiana N (2018) Application of cellulose fiber from cassava stem waste as low density polyethylene-based composite. In: Proceedings of the 7th national seminar on leather, rubber and plastics, pp 69–82. http://prosiding.bbkkp.go.id/index.php/SKKP/article/view/396
24. [Ciba] Ciba Specialty Chemical (1998) Tinuvin P Benzotriazole UV Absorber. http://www.polivinilplastik.com/urunler/tinp.pdf

Optimization of Pressing Temperature and Pressing Time Sweet Sorghum Bagasse Moulding Bonded by Sucrose-Citric Acid Binder

Sukma Surya Kusumah, Ismadi, Dwi Jaksana Megah Santosa, Dede Hermawan, Jajang Sutiawan, Deni Purnomo, Sudarmanto, Ahmad Syahrir, and Fazhar Akbar

Abstract Agricultural residues such as sweet sorghum bagasse are potentially used for mouldings materials. This study analysed the optimum pressing temperature and time for manufacturing sweet sorghum bagasse mouldings (SSB-moulding). The binder was prepared by mixing sucrose with citric acid with a composition of 90:10 (wt.%), and the sucrose-citric acid content used in this study was 20% of the total air-dried weight of the sorghum bagasse powder. The dumbbell-shaped moulding was manufactured with three different pressing temperatures: 180, 200, 220 °C, and different pressing times of 8, 10, and 12 min. The pressing time in this study did not significantly affect SSB-moulding's physical and mechanical properties. Meanwhile, the pressing temperature affected the properties of SSB-moulding. The mechanical properties of SSB-moulding increased together with increasing pressing temperature. The optimum pressing temperature and time were obtained in the SSB-moulding pressed at 200 °C for 12 min based on the physical and mechanical properties.

Keywords Mouldings · Pressing temperature · Pressing time · Sorghum · Sucrose-citric acid

---

S. S. Kusumah (✉) · Ismadi · J. Sutiawan · D. Purnomo · Sudarmanto · A. Syahrir · F. Akbar
Research Center for Biomass and Bioproduct, National Research and Innovation Agency, Cibinong, Indonesia
e-mail: sukm002@brin.go.id

D. J. M. Santosa · D. Hermawan
Forest Products Department, Faculty of Forestry and Environment, IPB University, Bogor, Indonesia

J. Sutiawan
Forest Product Department, Faculty of Forestry, Universitas Sumatera Utara, Medan, Indonesia

M. A. Rahandi Lubis (ed.), *Proceedings of the 1st International Conference of Lignocellulose*, Green Energy and Technology,
https://doi.org/10.1007/978-981-19-1596-3_7

1 Introduction

Sorghum is one of the agricultural wastes whose availability is relatively abundant. The average biomass of sorghum stalks and sorghum bagasse reach 23.06 tons/ha and 4.89 tons/ha, respectively [1]. In the world, the total land area of sorghum in 2020 is 40.81 million ha [2]. Sorghum bagasse waste was successfully utilized as raw material for particleboard [3–6] and moulding [7].

Widodo et al. [7] state that mouldings-citric acid as a binder was potentially substituted for conventional wood plastic composite products. Widodo et al. [7] research showed that the moulding quality of sorghum bagasse was influenced by the citric acid binder used with an optimal binder content of 20%. Binder powder obtains strength and resistance to water with lower binder content than liquid binders. The high content of citric acid as an acid compound causes the brittleness of composite [4, 8]. The addition of sucrose can reduce the content of citric acid [9]. The hydroxyl group of the derivative of sucrose can react with the carboxyl group of citric acid. This helps improve the brittleness and the particleboard's mechanical and physical properties [9].

The effect of adding sucrose to the manufacture of composite moulding has not been carried out yet. Moulding quality is influenced by binder curing. The temperature and pressing time parameters was effect by the curing mechanism of the binder. The use of too high a temperature and time causes the binder to be over curing and low-temperature results in a partially cured binder that reduces the quality of the board [4]. Therefore, this study investigates the optimum temperature and pressing time in composite sweet sorghum bagasse moulding (SSB-moulding) with sucrose-citric acid binders.

2 Materials and Methods

2.1 Materials

This study used sweet sorghum bagasse (Sorghum bicolor (L.) Moench), citric acid, and sucrose from Nacalai Tesque, Inc. (Kyoto, Japan) as materials.

2.2 Material Preparation

Preparation of raw materials involved preparing powders and binders. First, sweet sorghum bagasse powder was sieved using a 60 mesh. The filtered powder was dried at 60 °C for 15 h to a final moisture content of less than 5%. Next, the binder was prepared by mixing sucrose with citric acid in a 90:10 (wt.%) composition, according

Table 1 SSB-moulding variation in this study

Groups	Pressing temperature (°C)	Pressing time (min)
Group 1	180	8
		10
		12
Group 2	200	8
		10
		12
Group 3	220	8
		10
		12

to Kusumah et al. [9]. The sucrose-citric acid content used in this study was 20% of the total air-dried weight of the sorghum bagasse powder [10].

2.3 Moulding Production

The raw material for sweet sorghum bagasse powder was mixed with a sucrose-citric acid binder using a homogenizer. After thoroughly mixed, the mixture was dried at 60 °C for 15 h [7]. Furthermore, the mixture was put into a dumbbell-shaped mould of the Japanese Industrial Standard K 7139 [11], hot-pressed with temperature variations of 180, 200, and 220 °C and pressing time variations for 8, 10, and 12 min pressure of 4 MPa (Table 1). All of the variations of SSB-moulding were performed in six replications.

2.4 Moulding Conditioning and Cutting

The conditioning of the sample in this study refers to the research of Widodo et al. [7] by inserting the sample into a desiccator containing silica gel. The test samples that have been conditioned were cut to the length and width of sample 80 × 10 mm for testing density, modulus of elasticity, and modulus of rupture. Measures with a length and width of 20 × 20 mm at both ends of the dumbbell for moisture content and repeated boiling treatment.

2.5 Physical and Mechanical Properties Testing of Moulding

Physical properties were tested in density, moisture content, and repeated boiling treatment. Repeated boiling testing in this study refers to the research of Widodo et al. [7]. The test sample was dried at 60 °C for 15 h before treatment (Original). After that, the treatment conditions were boiling water immersion for 4 h (Boiling), drying at a temperature of 60° C for 20 h in the oven (Oven) and then boiling water immersion for 4 h (2nd Boiling), and vacuum-drying at 60 °C for 15 h (Vacuum oven). The changes in thickness and weight of the SSB-moulding were measured in each treatment. Meanwhile, the mechanical properties tests were Modulus of elasticity (MOE) and Modulus of rupture (MOR) with a loading speed and span 5 mm/min and 50 mm, respectively.

2.6 Data Analysis

The completely randomized was used to analyse the pressing temperature factor with 180, 200, 220 °C, and the pressing time factor with levels of 8, 10, and 12 min. The influence of factors on the response was seen using analysis of variance at the p-value < 0.05 confidence level. In addition, Duncan's further test was carried out to determine the differences between the levels of the influencing factors. Data analysis in this study used the SPSS.

3 Results and Discussion

3.1 Physical Properties of Moulding

The SSB-moulding in this study had a characteristic dark brown colour without delamination (Fig. 1). Figure 1 shows that the resulting SSB-moulding is darker as the temperature and pressing time increase, especially at 220 °C. Following Winarno [12] Statement: Heated sucrose beyond its melting point (160 °C) will caramelize, causing the material to turn brown. The resulting SSB-moulding colour is similar to the research conducted by Widodo et al. [7].

The density test results can be seen in Fig. 2. The density of SSB-moulding obtained in this study (0.75–0.88) g/cm^3. Furthermore, the pressing temperature significantly affected the density of SSB-moulding ($p < 0.05$), while the pressing time and interaction between the two had no significant effect ($p > 0.05$). Therefore, the density SSB-moulding obtained in this study is relatively uniform.

The density of SSB moulding decreases with increasing temperature and pressing time. This is because, at lower temperatures and shorter pressing times, less water

Fig. 1 Sample for bending test of SSB-moulding

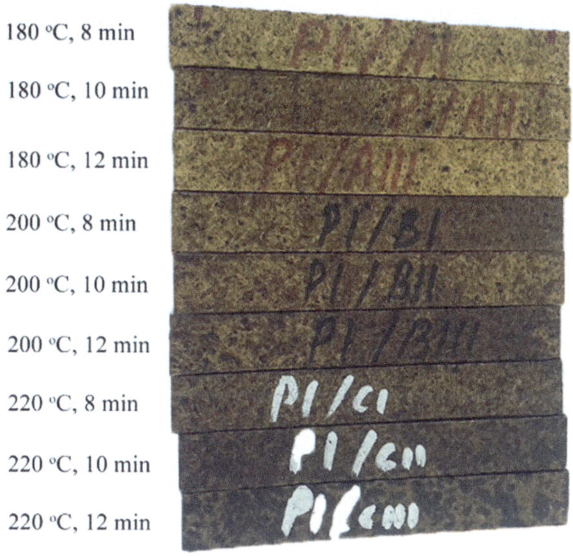

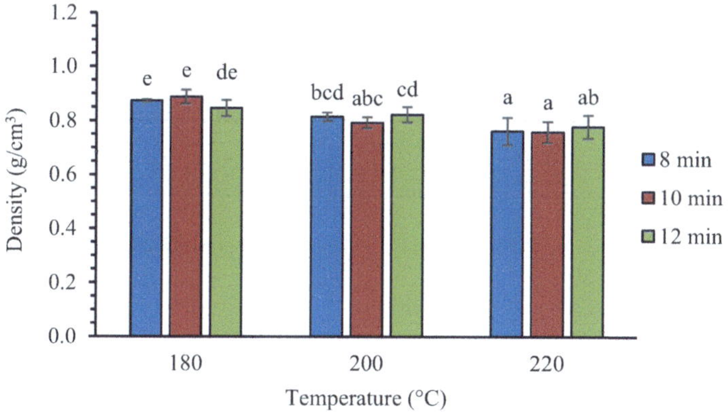

Fig. 2 The density of SSB-moulding at various temperatures and pressing times

evaporates during the pressing process. In addition, less volatile materials are released at lower temperatures and shorter pressing times.

Figure 3 shows that the moisture content of SSB-moulding. The moisture content of SSB-moulding ranged from 2.66 to 8.97%. Thus, the pressing temperature, pressing time, and the interaction of the two significantly affected the moisture content of the SSB-moulding ($p < 0.05$). The highest moisture content, SSB-moulding, was found at pressing conditions of 180 °C for 8 min (8.97%). Meanwhile, the lowest moisture content was found at a temperature of 220 °C for 12 min (2.66%). The higher temperature and the length of the pressing time produced low moisture content of SSB-moulding. Iswanto et al. [13] state that increasing temperature and the length of time was decreased moisture content.

The thickness change that occurred in SSB-moulding repeated boiling treatment is shown in Fig. 4. The SSB-moulding with a pressing temperature of 180 °C decomposed entirely in the first boiling treatment. This was due to the incurring of the sucrose binder so that it had poor resistance to repeated boiling. The pressing temperature at 180 °C was resulted not provide water resistance for the sucrose binder.

Changes in thickness in the SSB-moulding temperature of 200 °C are higher than the SSB-moulding temperature of 220 °C. The first boiling (33.9%) and final dried (−8.82%) treatment was affecting on thickness change SSB-moulding at pressing

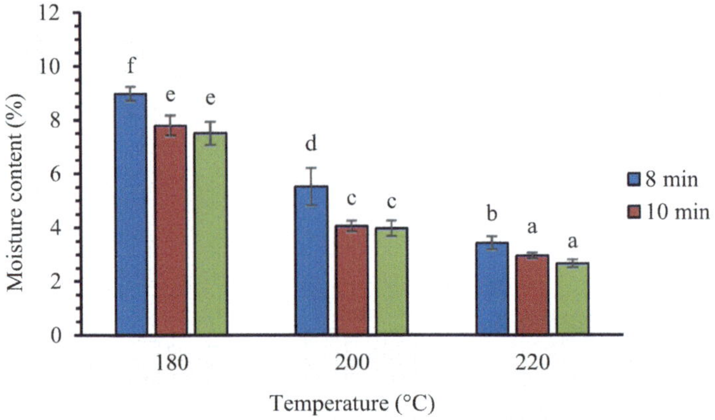

Fig. 3 The moisture content of SSB-moulding at various temperatures and pressing times

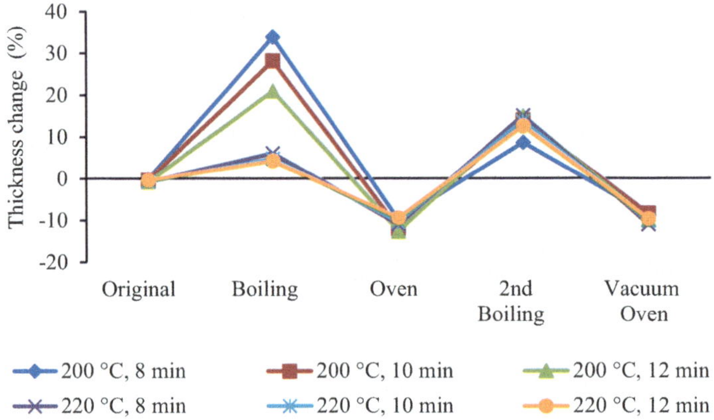

Fig. 4 Thickness change during repeated boiling treatment of SSB-moulding at various temperatures and pressing times

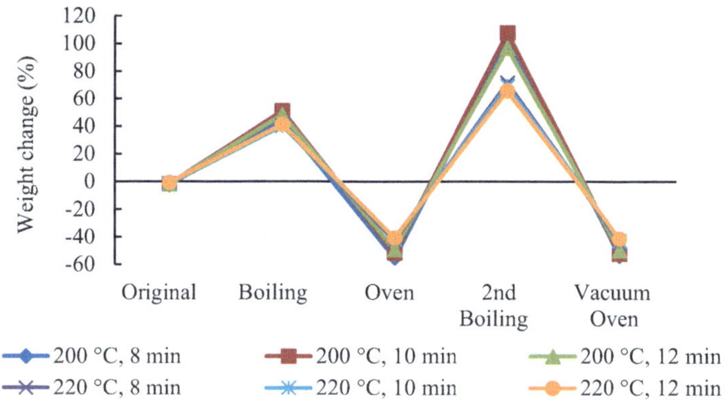

Fig. 5 Weight change during repeated boiling treatment of SSB-moulding at various temperatures and pressing times

conditions 200 °C for 8 min. Meanwhile, the minor thickness change in the first boiling treatment was found in the SSB-moulding pressed at 220 °C for 12 min (4.33%), and thickness change after being finally dried (−9.66%). The reason for this seems to be that at a temperature of 200 °C, sucrose becomes a material with high solubility so that it has low dimensional stability [14].

The weight change of SSB-moulding in repeated boiling treatment is shown in Fig. 5. The weight change of SSB-moulding decreases with increasing pressing temperature and time, especially during boiling treatment. The weight change at 200 °C is more significant than at 220 °C. The biggest weight change in the first boiling treatment was found in SSB-moulding under pressing conditions at 200 °C for 10 min (55.1%), while the smallest weight change in the first boiling treatment was found in SSB-moulding pressed at 220 °C for 10 min (40.3%). However, after the final drying treatment, the most significant weight changed at 220 °C for 8 min (−53.76%), and the smallest at 220 °C for 12 min (−41.13%). This indicates the presence of elution or soluble components in the SSB-moulding during treatment. Based on Figs. 6 and 7, water resistance increases at temperatures above 180 °C, which can increase with increasing pressing temperature and time.

3.2 *Mechanical Properties of Moulding*

The modulus of elasticity (MOE) SSB-moulding obtained in this study ranged from 0.43 to 3.33 GPa (Fig. 6). The pressing temperature significantly affected the MOE of SSB-moulding ($p < 0.05$), while the time and interaction between the two had no significant effect ($p < 0.05$). Furthermore, the MOE SSB-moulding significantly increased at a pressing temperature of 200 °C compared to 180 °C. Meanwhile, the MOE SSB-moulding at 200 and 220 °C is relatively the same.

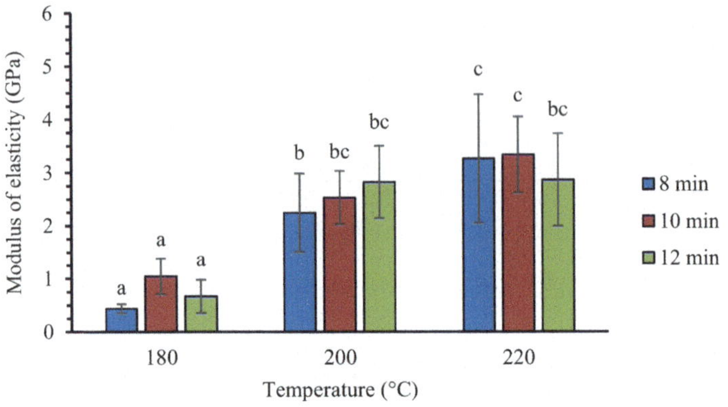

Fig. 6 The modulus of elasticity of SSB-moulding at various temperatures and pressing times

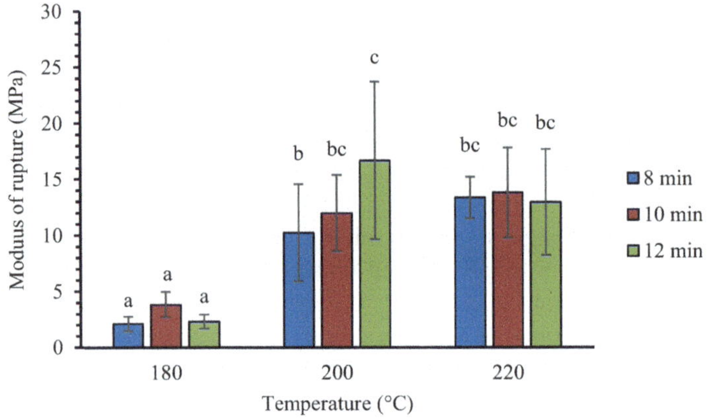

Fig. 7 The Modulus of rupture of SSB-moulding at various pressing temperatures and times

Based on the study results, the higher the pressing temperature resulted, the higher the MOE SSB-moulding. These phenomena are due to an ester bond between citric acid, sweet sorghum bagasse powder, and sucrose. Umemura et al. [15] stated that the ester bond formed increases at high pressing temperatures. Furthermore, good adhesion and physical and mechanical properties were found due to the formation of ester linkages [16]. However, the longer time in hot pressing made the MOE SSB-moulding decreased, as shown in Fig. 6 at 180 and 220 °C. This is because presumably of the degradation of the bonding as hot pressing time was longer.

Figure 7 shows the modulus of rupture (MOR) of SSB-moulding. The MOR of SSB-moulding obtained in this study ranged from 2.11 to 16.70 MPa. The pressing temperature was affecting the MOR of SSB-moulding ($p < 0.05$). Meanwhile, the pressing time and interaction between the two had no significant effect ($p > 0.05$). The highest MOR of SSB-moulding was obtained at 200 °C, while at 180 °C, a lower

MOR was obtained. Umemura et al. [17] stated that the bond between the binder and the raw material is affected by the melting point. The small MOR of SSB-moulding at 180 °C is due to the not melting point of the sucrose binder, which is around 183 °C [18].

4 Conclusions

The quality of SSB-moulding was affecting by pressing temperature and time. Temperature, time, and the interaction of the two only affect the moisture content of SSB-moulding. Meanwhile, the modulus of elasticity and modulus of rupture is only affected by the pressing temperature. As the pressing temperature increases, the modulus of elasticity and modulus of rupture of SSB-moulding has increased. The optimum pressing temperature and time were obtained in the SSB-moulding pressed at 200 °C with 12 min based on the quality of SSB-moulding.

Acknowledgements Author would like to thank to JSPS/JAAI/JASTIP for publication fee and related parties who have helped carry out this research.

References

1. Pabendon MB, Sarungallo R, Mas'ud S (2012) Pemanfaatan nira batang, bagas, dan biji sorgum manis sebagai bahan baku bioetanol. Penelit. Pertan. Tanam. Pangan. 31:180–187. https://doi.org/10.21082/jpptp.v31n3.2012.p%25p
2. Foreign Agricultural Service (FAS): World agricultural production. Ekon. APK. 59–65 (2019). https://doi.org/10.32317/2221-1055.201907059
3. Kusumah SS, Umemura K, Yoshioka K, Miyafuji H, Kanayama K (2016) Utilization of sweet sorghum bagasse and citric acid for manufacturing of particleboard I: effects of pre-drying treatment and citric acid content on the board properties. Ind Crops Prod 84:34–42. https://doi.org/10.1016/j.indcrop.2016.01.042
4. Kusumah SS, Umemura K, Guswenrivo I, Yoshimura T, Kanayama K (2017) Utilization of sweet sorghum bagasse and citric acid for manufacturing of particleboard II: influences of pressing temperature and time on particleboard properties. J Wood Sci 63:161–172. https://doi.org/10.1007/s10086-016-1605-0
5. Sutiawan J, Hermawan D, Kusumah SS, Ayu B, Sukara E (2020) Pemanfaatan Maltodextrin Singkong untuk Perekat Ramah Lingkungan dalam Pembuatan Papan Partikel dari Bagas Sorgum. J Sylva Lestari 8:144–154
6. Sutiawan J, Mardhatillah S, Hermawan D, Syamani FA (2020) Karakteristik papan partikel campuran perekat asam sitrat menggunakan perekat asam sitrat. J Penelit Has Hutan 38:139–150
7. Widodo E, Kusumah SS, Umemura K (2021) Development of moulding using sweet sorghum bagasse and citric acid : effects of application method and citric acid content. For Prod J 70:151–157. https://doi.org/10.13073/FPJ-D-19-00060
8. Sutiawan J, Hermawan D, Massijaya MY, Kusumah SS, Lubis MAR, Marlina R, Purnomo D, Sulastiningsih IM (2021) Influence of different hot-pressing conditions on the performance

of eco-friendly jabon plywood bonded with citric acid adhesive. Wood Mater Sci Eng, 1–10. https://doi.org/10.1080/17480272.2021.1884898
9. Kusumah SS, Arinana A, Hadi YS, Guswenrivo I, Yoshimura T, Umemura K, Tanaka S, Kanayama K (2017) Utilization of sweet sorghum bagasse and citric acid in the manufacturing of particleboard. III: influence of adding sucrose on the properties of particleboard. BioResources 12:7498–7514. https://doi.org/10.15376/biores.12.4.7498-7514
10. Kusumah SS, Massijaya SY, Prasetyo KW, Sutiawan J, Lubis MAR, Hermawan D (2020) Surface modification of eco-friendly particleboard made from sorghum bagasse and citric acid sucrose adhesive. IOP Conf Ser Mater Sci Eng 935:012054. https://doi.org/10.1088/1757-899X/935/1/012054
11. Japanese Standards Association (JSA): Plastics-Test specimens. Japan Industrial Standard K 7139. Tokyo (2009)
12. Winarno FG (2010) Enzim Pangan. M-Brio Press, Bogor
13. Iswanto AH, Azhar I, Supriyanto, Susilowati A (2014) Effect of resin type, pressing temperature and time on particleboard properties made from sorghum bagasse. Agric For Fish 3:62. https://doi.org/10.11648/j.aff.20140302.12
14. Umemura K, Sugihara O, Kawai S (2014) Investigation of a new natural adhesive composed of citric acid and sucrose for particleboard II: effects of board density and pressing temperature. J Wood Sci 61:40–44. https://doi.org/10.1007/s10086-014-1437-8
15. Umemura K, Ueda T, Kawai S (2012) Characterization of wood-based molding bonded with citric acid. J Wood Sci 58:38–45. https://doi.org/10.1007/s10086-011-1214-x
16. Umemura K, Sugihara O, Kawai S (2013) Investigation of a new natural adhesive composed of citric acid and sucrose for particleboard. J Wood Sci 59:203–208. https://doi.org/10.1007/s10086-013-1326-6
17. Santoso M, Widyorini R, Prayitno TA, Sulistyo J, Hamidah N (2020) Effect of pressing temperatures on bonding properties of sucrose-citric acid for nipa palm fronds particleboard. Wood Res 65:747–756. https://doi.org/10.37763/WR.1336-4561/65.5.747756
18. Widyorini R, Nugraha PA, Zakky M, Rahman, Arief PTA (2016) Bonding ability of a new adhesive composed of citric. Bioresour Technol 11:4526–4535

Versatile Access to 1,3,5-Trisubstituted 5'-Hydroxyimidazolidine-2,4-diones (Hydantoins) from α-Keto Acids and Carbodiimides: Towards the Synthesis of Marine Thermochromic Polyandrocarpamide D

Yudhi Dwi Kurniawan and Patrick Perlmutter

Abstract An efficient and straightforward one-step access to 1,3,5-trisubstituted 5'-hydroxyimidazolidine-2,4-diones (hydantoins) is described. Under simple heating, this reaction proceeded through a sequential condensation/intramolecular nucleophilic addition/O → N acyl migration between α-keto acids and carbodiimides. A series of 1,3,5-trisubstituted 5'-hydroxyhydantoins, including protected polyandrocarpamide D analogues, has been successfully synthesized in moderate to excellent yields.

Keywords Hydantoin · α-keto acid · Carbodiimide · Thermochromic

1 Introduction

5-Monosubstituted imidazolidine-2,4-diones (hydantoins) are biologically relevant heterocycles present as structural units in several natural products [1–4]. These frameworks also function as prominent enzyme substrates in the construction of optically active amino acids [5–8] as well as acting as the basis of numerous pharmaceuticals [9, 10].

Y. D. Kurniawan (✉)
Research Center for Pharmaceutical Ingredient and Traditional Medicine, National Research and Innovation Agency, Tangerang Selatan, Indonesia
e-mail: yudh010@brin.go.id

School of Chemistry, Monash University, Clayton, VIC, Australia

P. Perlmutter
La Trobe Institute of Molecular Science, La Trobe University, Bundoora, VIC, Australia

M. A. Rahandi Lubis (ed.), *Proceedings of the 1st International Conference of Lignocellulose*, Green Energy and Technology,
https://doi.org/10.1007/978-981-19-1596-3_8

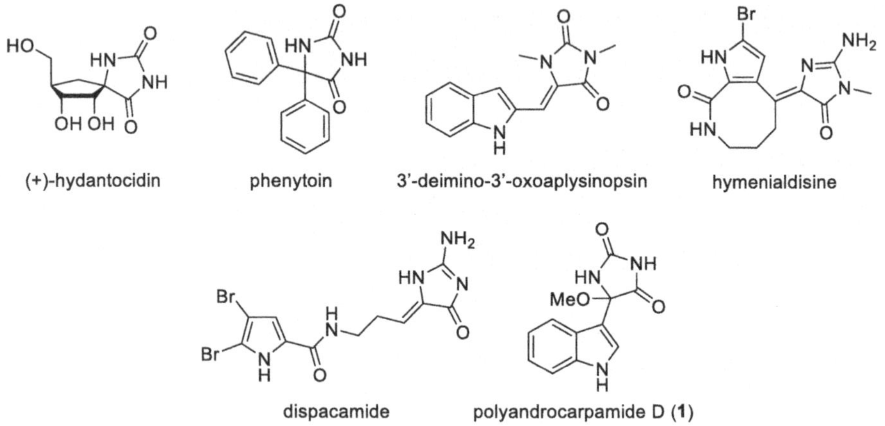

Fig. 1 Substituted hydantoins with various biological activities

Hydantoins bearing a quaternary center at position 5 display activity as exemplified by herbicidal (+)-hydantocidin [4] and anticonvulsant phenytoin (Fig. 1) [11–13] but are more commonly employed to access several 4-aminoimidazolone-based natural products with crucial physiological activity, including cytotoxic 3'-deimino-3'-oxoaplysinopsin [14], anti-inflammatory hymenialdisine [15], and antihistamine dispacamide [16] amongst numerous others. Moreover, the first thermochromic system originated from marine sponge *Zyzza massalis*, polyandrocarpamide D (**1**), also contains this moiety [17].

Numerous methods have been reported in literature for the preparation of substituted hydantoins and notable reviews on this area are available [10, 18]. Volonterio et al. [19] developed an innovative novel route to 1-alkyl/aryl-3,5-disubstituted hydantoins from α-bromoaryl acetic acids or α,β-unsaturated carboxylic acids and *N*-*tert*-butyl-carbodiimides or *N*-trityl-carbodiimides through a domino condensation/nucleophilic attack/O → N acyl migration (Fig. 2). Hydantoins with both *tert*-butyl- and trityl groups at N-3 position were obtained with total regioselectivity, which, after deprotection, provided access to 1,5-disubstituted products.

In the course of the investigation of Steglich esterification as part of a zaragozic acid synthetic program, we discovered a concise and novel route to 1,3,5-trisubstituted 5'-hydroxyhydantoins from α-keto acids and carbodiimides under simple heating (Fig. 2). This reaction could enrich the existing synthetic toolbox in accessing fully substituted 5'-hydroxyhydantoins and offer a more efficient alternative to synthesize this class of compounds. In our study, this reaction was envisaged as the key strategy to achieve **1**, which was not previously synthesised.

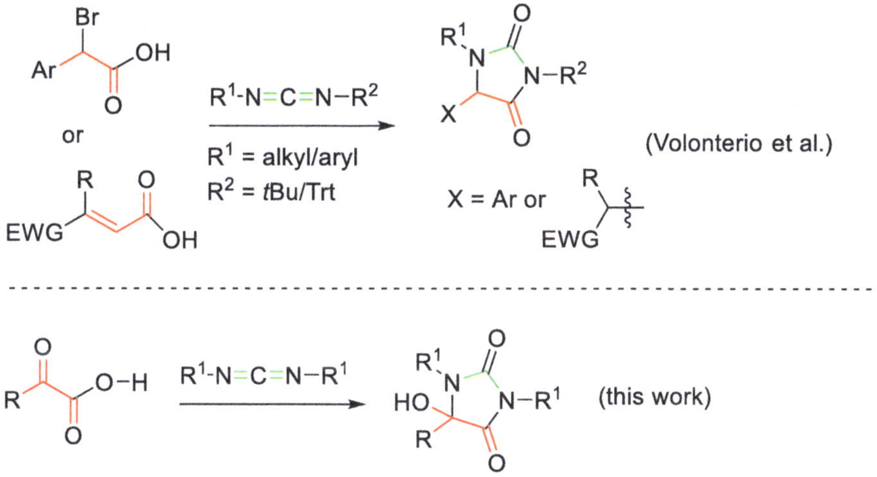

Fig. 2 Substituted hydantoins with various biological activities

2 Experimental Procedure

Bis(2,4-dimethoxybenzyl)methanediimine (3e). To a stirred solution of 1-(azidomethyl)-2,4-dimethoxybenzene (509.3 mg, 2.6 mmol) in CH_2Cl_2 (25 mL) was added triphenylphosphine (691.4 mg, 2.6 mmol) at room temperature, and stirring was maintained for 19 h. Solution of 1-(isocyanatomethyl)-2,4-dimethoxybenzene (506.7 mg, 2.6 mmol) in dry CH_2Cl_2 (28 mL) was added to the mixture and left to stir for 24 h at room temperature. After completion, the solvent was evaporated under vacuum and purified using column chromatography providing bis-2,4-dimethoxybenzyl carbodiimide as a pale green oily liquid (711.8 mg, 79% yield, $R_f = 0.4$ (EtOAc:petroleum benzine 1:3)). ^{1}H NMR (400 MHz, $CDCl_3$, δ ppm): 3.77 (s, 6 H, OCH_3), 3.80 (s, 6 H, OCH_3), 4.24 (s, 4 H, CH_2Ph), 6.39–6.42 (m, 4 H, CH_{Ar}), 7.04–7.06 (d, 2 H, J 8.9, CH_{Ar}). ^{13}C NMR (100 MHz, $CDCl_3$, δ ppm): 45.3 (CH$_2$Ph), 55.3 (OCH$_3$), 55.3 (OCH$_3$), 98.4 (CH$_{Ar}$), 103.9 (CH$_{Ar}$), 119.5 ($C_{Ar\,quat}$), 129.7 (CH$_{Ar}$), 142.7 ($N = C = N$), 158.0 ($C_{Ar\,quat}$), 160.3 ($C_{Ar\,quat}$). HRMS: calculated [M + Na] m/z 365.1477, found [M + Na] 365.1476. IR (neat, cm^{-1}): 3000, 2936, 2835, 2111, 1610, 1587, 1505, 1457, 1438, 1285, 1264, 1204, 1154, 1127, 1031, 965, 922, 916, 822, 781. **Preparation of 1-(azidomethyl)-2,4-dimethoxybenzene**. To a stirred solution of 2,4-dimethoxybenzyl alcohol (423.5 mg, 2.5 mmol) in dry toluene (4.5 mL) DPPA (651 μL, 3.0 mmol) and DBU (451 μL, 3.0 mmol) were added at 0 °C under nitrogen atmosphere. The reaction mixture was then stirred at that temperature for 1 h. After completion, diethyl ether (10 mL) was added to the mixture followed by saturated brine solution (20 mL). The organic phase was collected through separating funnel and the water layer was washed (3 × 10 mL) with diethyl ether. The combined organic layer was then dried over Na_2SO_4, evaporated, and dried under high vacuum. Flash chromatography (EtOAc:petroleum benzine 1:5) of the crude residue provided the desired product (437.3 mg, 90%) as a colourless oil ($R_f = 0.42$). ^{1}H NMR (400 MHz, $CDCl_3$, δ ppm): 3.82 (s, 3 H, OCH_3), 3.84 (s, 3 H, OCH_3), 4.28 (s, 2 H, CH_2Ph), 6.46–6.50 (m, 2 H, CH_{Ar}), 7.14–7.16 (d, 1 H, J 8.2, CH_{Ar}). Data is in accordance with literature values [20]. **Preparation of 1-(isocyanatomethyl)-2,4-dimethoxybenzene**. To a stirred solution of 2,4-dimethoxybenzylamine (601.8 mg, 3.6 mmol) in CH_2Cl_2 (41 mL) a saturated aqueous solution of $NaHCO_3$ (24 mL) was added. Triphosgene (1068.1 mg, 3.6 mmol) was then added to this mixture at 0 °C and allowed to stir for 1 h. After the reaction completion, diethyl ether (30 mL) was added to the mixture followed by a saturated brine solution (30 mL). The organic phase was collected through separating funnel and the water layer was washed (3 × 30 mL) with diethyl ether. The combined organic layer was then dried over Na_2SO_4, evaporated, and dried under high vacuum. Flash chromatography (EtOAc:petroleum benzine 1:5) of the crude residue provided the desired product (600.4 mg, 87.5%) as a colourless oil ($R_f = 0.42$). ^{1}H NMR (400 MHz, $CDCl_3$, δ ppm): 3.82 (s, 3 H, OCH_3), 3.85 (s, 3 H, OCH_3), 4.34 (s, 2 H, CH_2Ph), 6.45–6.48 (m, 2 H, CH_{Ar}), 7.15–7.17 (d, 1 H, J 8.2, CH_{Ar}). Data is in accordance with literature values [21].

Dibenzylmethanediimine (3f). To a stirred solution of 1,3-dibenzyl thiourea (825.7 mg, 3.2 mmol) in dry CH_2Cl_2 (16 mL) was added dry triethylamine (945 μL, 6.8 mmol) and 2-chloro-1-methylpyridin-1-ium iodide (Mukaiyama's reagent) (987.4 mg, 3.9 mmol) at room temperature. After being stirred for 15 min at the same temperature, water (20 mL) was added to the reaction and then diluted with EtOAc (30 mL). The organic phase was collected through separating funnel and the water layer was extracted with EtOAc (3 × 20 mL). The combined organic phase was dried over Na_2SO_4 and evaporated under reduced pressure. Column chromatography purification using EtOAc/hexane 1/5 as the eluent gave dibenzylmethanediimine as a yellow oil (242.3 mg, 34%, $R_f = 0.58$ (EtOAc:hexane 1:5). ^{1}H NMR (400 MHz, $CDCl_3$, δ ppm): 4.38 (s, 4 H, CH_2Ph), 7.28–7.29 (m, 4 H, CH_{Ar}), 7.36–7.40 (m, 6 H, CH_{Ar}). ^{13}C NMR (100 MHz, $CDCl_3$, δ ppm): 50.2 (CH_2Ph), 127.3 (CH_{Ar}), 127.4 (CH_{Ar}), 128.5 (CH_{Ar}), 138.2 ($CH_{Ar\ quat}$), 141.1 (N = C = N). Data is in accordance with literature values [22]. **Preparation of 1,3-dibenzylthiourea**. To a stirred solution of (isothiocyanatomethyl)benzene (1.1746 g, 7.9 mmol) in hexane (30 mL) benzylamine (860 μL, 7.9 mmol) was added at room temperature. A turbid solution was immediately formed in this addition. The mixture was stirred at the same temperature for 1.5 h. After reaction completion, the precipitate was filtered with diethyl ether and dried under reduced pressure providing 1,3-dibenzyl thiourea as a white solid (1.9130 g, 95%, MP 146 °C, (lit. [23] 146–147 °C)). ^{1}H NMR (400 MHz, $CDCl_3$, δ ppm): 4.60 (s, 2 × 2 H, CH_2Ph), 6.2 (s, 2 H, NH), 7.21–7.23 (m, 4 H, CH_{Ar}), 7.26–7.33 (m, 6 H, CH_{Ar}). ^{13}C NMR (100 MHz, $CDCl_3$, δ ppm): 48.5 (CH_2Ph), 127.5 (CH_{Ar}), 127.9 (CH_{Ar}), 128.9 (CH_{Ar}), 136.6 ($C_{Ar\ quat}$), 182.0 (C = S). Data is in accordance with literature values [23]. **Preparation of (isothiocyanatomethyl)benzene**. To a stirred solution of benzylamine (1 mL, 9.2 mmol) in abs. ethanol (10 mL) carbon disulfide (5.5 mL, 91.4 mmol) was added followed by dry Et_3N (1.3 mL, 9.3 mL) at room temperature. After 30 min stirring, the mixture was cooled to 0 °C and then added with solution of Boc_2O (1.98 g, 9.1 mmol) in abs. ethanol (2 mL). Subsequently, solution of DMAP (33.6 mg, 0.28 mmol) in abs. ethanol (2 mL) was added. The mixture was kept at that temperature for 5 min before warming up to room temperature for 15 min. After reaction completion, the mixture was filtered over a short plug of silica gel and the filtrate was concentrated and subjected to column chromatography purification to obtain (isothiocyanatomethyl)benzene as a colorless oil (1.1746 g, 86%, $R_f = 0.45$ (EtOAc/hexane 1/49)). ^{1}H NMR (400 MHz, $CDCl_3$, δ ppm): 4.71 (s, 2 H, CH_2Ph), 7.32–7.42 (m, 5 H, CH_{Ar}). ^{13}C NMR (100 MHz, $CDCl_3$, δ ppm): 48.6 (CH_2Ph), 126.8 (CH_{Ar}), 128.3 ($C_{Ar\,quat}$), 128.9 (CH_{Ar}), 134.1 (N = *C* = S). Data is in accordance with literature values [24].

5-(*Tert*-butyl)-5-hydroxy-1,3-diisopropylimidazolidine-2,4-dione (4a). To a stirred solution of trimethylpyruvic acid **2a** (0.24 mL, 1.92 mmol) in dry DMF (8 mL) at rt was added *N*,*N*′-di-*iso*-propylcarbodiimide **3a** (0.30 mL, 1.92 mmol). The solution was raised to 110 °C and stirred overnight, after which time the reaction was cooled to room temperature and concentrated *in vacuo*. The crude oil was purified by flash chromatography (20% EtOAc in petroleum benzine) to yield the title compound **4a** as a white amorphous solid (0.464 g, 95%, R_f 0.44, 20% EtOAc in petroleum benzine) which was recrystallized from boiling CH_2Cl_2 to yield the title compound **4a** as a crystalline solid (MP 121–123 °C). ^{1}H NMR (400 MHz, $CDCl_3$, δ ppm): 1.03 (s, 9 H, -C(CH_3)$_3$), 1.31–1.34 (m, 9 H, N3-CH(CH_3)$_2$, N1-CH(CH_3)(CH_3)), 1.51–1.53 (d, 3 H, *J* 7.0, N1-CH(CH_3)(CH_3)), 3.65–3.71 (hep, 1 H, *J* 6.8, N1-*C*H(CH_3)$_2$), 3.93 (s, 1 H, O*H*), 4.18–4.25 (hep, 1 H, *J* 6.9, N3-C*H*(CH_3)$_2$). ^{13}C NMR (100 MHz, $CDCl_3$, δ ppm): 19.2 (N3-CH(CH_3)$_2$), 19.7 (N3-CH(CH_3)$_2$), 20.6 (N1-CH(CH_3)$_2$), 20.9 (N1-CH(CH_3)$_2$), 24.4 (-C(CH_3)$_3$), 39.5 (C5-*C*(CH_3)$_3$), 43.4 (N3-*C*H(CH_3)$_2$), 46.1 (N1-*C*H(CH_3)$_2$), 89.7 (C_{quat}, C5), 155.2 (C2, *C* = O), 174.2 (C4, *C* = O). IR (neat) ν_{max}: 3373 (OH, H-bonded), 2975 (CH_3 stretch), 2878 (CH_3 stretch), 1762 (imide), 1688 (imide). HRMS: Calculated ($C_{13}H_{24}N_2O_3H^+$) *m/z* 257.1865, found $[M + H]^+$ 257.1854, [$C_{13}H_{23}N_2O_2{}^{\cdot}$] 239.1749.

1,3-Di-*tert*-butyl-5-(furan-2-yl)-5-hydroxyimidazolidine-2,4-dione (4b). To a stirred solution of α-oxo-2-furanacetic acid **2b** (0.25 g, 1.78 mmol) in dry DMF (7 mL) at rt under inert atmosphere was added *N*,*N*′-di-*tert*-butylcarbodiimide **3b** (0.34 mL, 1.78 mmol). The suspension was raised to 110 °C and stirred overnight, after which time the reaction was cooled to room temperature and concentrated *in vacuo*. The crude oil was purified by flash chromatography (20% EtOAc in petroleum benzine) to yield the title compound **4b** as a pale tan amorphous solid (0.462 g, 88%, R_f 0.31, 20% EtOAc in petroleum benzine) which was recrystallized from boiling CH_2Cl_2 to yield the title compound **4b** as a pale orange-yellow crystalline solid (MP 111–114 °C). ^{1}H NMR (400 MHz, $CDCl_3$, δ ppm): 1.37 (s, 9 H, -C(CH_3)), 1.60 (s, 9 H, -C(CH_3)), 4.17 (s, 1 H, -O*H*), 6.39–6.41 (dd, 1 H, *J* 1.8, 3.3, -C*H*, C2'), 6.52–6.53 (dd, 1 H, J 1.0, 3.3, -C*H*, C3'), 7.386–7.388 (dd, 1 H, J 0.7, 1.9, -C*H*, C4'). ^{13}C NMR (100 MHz, $CDCl_3$, δ ppm): 28.3 (-C(CH_3)$_3$), 28.7 (-C(CH_3)$_3$), 56.7 (-*C*(CH_3)$_3$), 58.4 (-*C*(CH_3)$_3$), 83.6 (C_{quat}, C-5), 109.4 (-*C*H, C4'), 111.2 (-*C*H, C3'), 142.6 (-*C*H, C5'), 150.5 (C_{quat}, C2'), 155.2 (*C* = O, C2), 171.9 (*C* = O, C4). IR (neat) ν_{max}: 3353 (OH, H-bonded), 2978 (CH_3 stretch), 2941 (CH_3 stretch), 1765 (imide), 1700 (imide) HRMS: Calculated ($C_{15}H_{22}N_2O_4H^+$) *m/z* 295.1658, found [M + H]$^+$ 295.1653, [$C_{11}H_{12}N_2O_3$ + H]$^+$ 221.0920.

1,3-Di-*tert*-butyl-5-hydroxy-5-methylimidazolidine-2,4-dione (4c). To a stirred solution of pyruvic acid **2c** (0.25 mL, 2.84 mmol) in dry DMF (12 mL) at rt was added *N*,*N*′-di-*tert*-butylcarbodiimide **3b** (0.30 mL, 1.56 mmol). The solution was raised to 110 °C and stirred overnight, after which time the reaction was cooled to room temperature and concentrated *in vacuo*. The crude oil was purified by flash chromatography (19:1 to 9:1, Petroleum benzine:EtOAc) to yield the title compound **4c** as a white amorphous solid (0.352 g, 93%, R_f 0.16, 10% EtOAc in petroleum benzine) which was recrystallized from boiling EtOAc to yield the title compound 4e as a crystalline solid (MP 98–106 °C). ^{1}H NMR (400 MHz, $CDCl_3$, δ ppm): 1.270–1.273 (d, 3 H, *J* 1.4, C5-CH_3), 1.524–1.527 (d, 9 H, *J* 1.2, -C(C*H*$_3$)$_3$), 1.546–1.550 (d, 9 H, *J* 1.4, -C(C*H*$_3$)$_3$), 1.651–1.654 (d, 3 H, *J* 1.2, C5-CH_3, from its enantiomer probably), 4.08–4.27 (brds, 2 H, O*H*). ^{13}C NMR (100 MHz, $CDCl_3$, δ ppm): 25.4 (OH-C5-CCH_3), 28.7 (-C(C*H*$_3$)$_3$), 29.0 (-C(C*H*$_3$)$_3$), 29.6 (OH-C5-CCH_3, from its enantiomer probably), 56.4 (-*C*(CH_3)$_3$), 57.8 (-*C*(CH_3)$_3$), 84.9 (C_{quat}, C5), 154.9 (C2, *C* = O), 175.3 (C4, *C* = O). IR (neat) ν_{max}: 3354 (OH, H-bonded), 2967 (CH_3 stretch), 2929 (CH_3 stretch), 1766 (imide), 1697 (imide). HRMS: Calculated ($C_{12}H_{22}N_2O_3H^+$) *m/z* 243.1709, found [M + H$^+$] 243.1760, [$C_{12}H_{21}N_2O_2{}^\cdot$] 225.1599.

1,3-Di-*tert*-butyl-5-hydroxy-5-(1H-indol-3-yl)imidazolidine-2,4-dione (4d). To a stirred solution of 3-indole glyoxylic acid **2d** (0.25 g, 1.32 mmol) in dry DMF (6 mL) containing 4 Å MS at r.t was added *N*,*N*′-di-*tert*-butylcarbodiimide **3b** (0.25 mL, 1.32 mmol). The solution was raised to 110 °C and refluxed overnight, after which time the reaction was cooled to room temperature, filtered and concentrated *in vacuo*. Chloroform (40 mL) was added to the crude oil and the subsequent suspension sonicated for 1 min. The resultant precipitate was collected via vacuum filtration and recrystallized from MeOH to yield the title compound **4d** as an amorphous tan solid (0.267 g, 59%). The filtrate was concentrated and further purified by flash chromatography (9:1 to 5:1 to 5:2, petroleum benzine:EtOAc) to yield a second crop of title compound **4d** as a white amorphous solid. Combined product was recrystallised from boiling MeOH to give the title compound as a crystalline solid (0.347 g, 77%, MP > 250 °C, R_f 0.44 (40% EtOAc in petroleum benzine)). ^{1}H NMR (400 MHz, d_6-DMSO, δ ppm): 1.29 (s, 9 H, -C(C*H*$_3$)$_3$), 1.56 (-C(C*H*$_3$)$_3$), 6.97–7.01 (m, 1 H, -CH_{Ar}), 7.06–7.11 (m, 2 H, -CH_{Ar}), 7.24 (brds, 1 H, O*H*), 7.36–7.40 (m, 2 H, -CH_{Ar}), 11.15 (s, 1 H,). ^{13}C NMR (100 MHz, d_6-DMSO, δ ppm): 28.7 (-C(C*H*$_3$)$_3$), 29.2 (-C(C*H*$_3$)$_3$), 56.0 (-*C*(CH_3)$_3$), 57.1 (-*C*(CH_3)$_3$), 85.7 (C_{quat}, C5), 112.4 (-CH_{Ar}), 114.9 (NH*C*H), 118.9 (*C*$_{quat}$, Ar), 119.4 (*C*H_{Ar}), 121.5 (*C*H_{Ar}), 124.3 (*C*$_{quat}$, Ar), 125.3 (*C*$_{quat}$, Ar), 137.1 (*C*H_{Ar}), 155.7 (C2, *C* = O), 173.1 (C4, *C* = O). IR (neat) ν_{max}: 3350 (NH), 2981 (CH_3 stretch), 2942 (CH_3 stretch), 1758 (imide), 1691 (imide). HRMS: Calculated ($C_{19}H_{25}N_3O_3Na^+$) *m/z* 366.1788, found [M + Na$^+$] 366.1791.

1,3-Bis(2,4-dimethoxybenzyl)-5-hydroxy-5-(1*H*-indol-3-yl)imidazolidine-2,4-dione (4g). To a stirred solution of carbodiimide **3e** (3261.6 mg, 9.5 mmol) in dry

DMF (43 mL) was added with 4 Å molecular sieves and indole **2d** (1802.0 mg, 9.5 mmol). The mixture was then heated to 125 °C and left to stir at that temperature for 24 h. After completion, saturated brine was added to the reaction and was diluted with ethyl acetate. The organic phase was collected using separating funnel. The water layer was extracted with ethyl acetate three times and the combined ethyl acetate layers were dried over Na_2SO_4. Column chromatography purification of the residue after evaporation of the solvent under reduced pressure gave **4g** as a foamy pale brown solid (2509.0 mg, 50%, R_f 0.22 (EtOAc:petroleum benzine 5:3), MP 85–89 °C). ^{1}H NMR (400 MHz, d_6-DMSO, δ ppm): 3.63 (s, 3 H, OCH_3), 3.71 (s, 3 H, OCH_3), 3.75 (s, 3 H, OCH_3), 3.76 (s, 3 H, OCH_3), 4.02–4.04 (d, 1 H, *J* 16.7, CH_APh), 4.44–4.48 (d, 1 H, *J* 16.7, CH_APh), 4.69 (s, 2 H, CH_{2B}Ph), 6.39–6.42 (m, 2 H, C$H_{Ar\ benzyl}$), 6.47–6.50 (dd, 1 H, *J* 2.3 8.4, C$H_{Ar\ benzyl}$), 6.58–6.59 (d, 1 H, *J* 2.4, C$H_{Ar\ benzyl}$), 6.94–6.98 (m, 1 H, C$H_{Ar\ indole}$), 7.11–7.15 (m, 2 H, C$H_{Ar\ indole}$), 7.18–7.20 (d, 1 H, *J* 8.4, C$H_{Ar\ benzyl}$), 7.22–7.24 (d, 1 H, *J* 8.2, C$H_{Ar\ benzyl}$), 7.41–7.42 (s, 1 H, O*H*), 7.44 (s, 1 H, C$H_{Ar\ indole}$), 7.49–7.50 (d, 1 H, *J* 2.6, NHCH_{indole}), 11.28–11.29 (d, 1 H, *J* 2.4, N*H*). ^{13}C NMR (100 MHz, d_6-DMSO, δ ppm): 37.2 (*C*H$_2$Ph), 38.1 (*C*H$_2$Ph), 55.6 (O*C*H$_3$), 55.6 (O*C*H$_3$), 55.7 (O*C*H$_3$), 55.9 (O*C*H$_3$), 86.7 (*C*OH$_{quat}$, C4), 98.2 (*C*H$_{Ar\ benzyl}$), 98.8 (*C*H$_{Ar\ benzyl}$), 104.6 (*C*H$_{Ar\ benzyl}$), 104.8 (*C*H$_{Ar\ benzyl}$), 111.2 (*C*$_{quat\ indole}$), 112.5 (*C*H$_{Ar\ indole}$), 116.6 (*C*$_{Ar\ quat\ benzyl}$), 118.3 (*C*$_{Ar\ quat\ benzyl}$), 118.9 (*C*H$_{Ar\ indole}$), 119.0 (*C*H$_{Ar\ indole}$), 121.9 (*C*H$_{Ar\ indole}$), 124.4 (*C*$_{Ar\ quat\ indole}$), 126.7 (NH-*C*H$_{indole}$), 128.7 (*C*H$_{Ar\ benzyl}$), 129.9 (*C*H$_{Ar\ benzyl}$), 137.2 (*C*$_{Ar\ quat\ indole}$), 155.7 (*C* = O, C2), 157.4 (*C*$_{Ar\ quat\ benzyl}$), 158.2 (*C*$_{Ar\ quat\ benzyl}$), 159.9 (*C*$_{Ar\ quat\ benzyl}$), 160.7 (*C*$_{Ar\ quat\ benzyl}$), 172.9 (*C* = O, C5). IR (neat) υ_{max}: 3320, 2935, 2835, 1774, 1702, 1611, 1588, 1507, 1449, 1412, 1337, 1287, 1261, 1206, 1155, 1103,031, 1011, 933, 904, 825, 786, 741. HRMS: Calculated [M + H$^+$] *m/z* 532.2078, found [M + H$^+$] 532.2082.

1,3-Dibenzyl-5-hydroxy-5-(1*H*-indol-3-yl)imidazolidine-2,4-dione (4h). To a stirred solution of carbodiimide **3f** (242.3 mg, 1.1 mmol) in dry DMF (5 mL) was added with 4 Å molecular sieves and indole **2d** (206.2 mg, 1.1 mmol). The mixture was then heated to 125 °C and left to stir at that temperature for 24 h. After completion, saturated brine was added to the reaction and was diluted with ethyl acetate. The organic phase was collected using separating funnel. The water layer was extracted with ethyl acetate three times and the combined ethyl acetate layers were dried over Na_2SO_4. Column chromatography purification of the residue after evaporation of the solvent under reduced pressure gave **4h** as a foamy pale white solid (232.7 mg, 52%, R_f 0.5 (EtOAc:petroleum benzine 1/1), MP 73–74 °C). ^{1}H NMR (400 MHz, $CDCl_3$, δ ppm): 4.09–4.13 (d, 1 H, *J* 15.4, CH_APh), 4.58–4.62 (d, 1 H, *J* 15.6, CH_APh), 4.70–4.78 (dd, 2 H, *J* 3.4, 32.3, CH_{2B}Ph), 5.07 (s, 1 H, O*H*), 6.85–6.89 (m, 1 H, C$H_{Ar\ indole}$), 6.97–6.99 (d, 1 H, *J* 8.0, C$H_{Ar\ indole}$), 7.05–7.13 (m, 4 H, C$H_{Ar\ indole}$, CH_{Ar}), 7.19–7.29 (m, 7 H, CH_{Ar}, C$H_{Ar\ indole}$, NHCH_{indole}), 7.42–7.45 (m, 2 H, CH_{Ar}), 9.02–9.03 (d, 1 H, *J* 1.3, NH_{indole}). ^{13}C NMR (100 MHz, $CDCl_3$, δ ppm): 42.4 (*C*H$_{2A}$Ph), 43.6 (*C*H$_{2B}$Ph), 86.5 (*C*OH$_{quat}$, C4), 110.3 (*C*$_{quat\ indole}$), 111.6 (*C*H$_{Ar\ indole}$), 118.3 (*C*H$_{Ar\ indole}$), 120.1 (*C*H$_{Ar\ indole}$), 122.0 (*C*H$_{Ar\ indole}$), 123.6 (*C*$_{quat\ indole}$), 125.6 (NH*C*H$_{indole}$), 126.9 (*C*H$_{Ar}$), 127.7 (*C*H$_{Ar}$), 127.9 (*C*H$_{Ar}$), 128.1

(CH_{Ar}), 128.4 (CH_{Ar}), 128.6 (CH_{Ar}), 135.7 ($C_{Ar\,quat}$), 136.7 ($C_{quat\,indole}$), 137.3 ($C_{Ar\,quat}$), 155.2 (C = O, C2), 172.5 (C = O, C5). HRMS: Calculated [M + H$^+$] m/z 412.1656, found [M + H$^+$] 412.1643. IR (neat) υ_{max}: 3337, 3061, 3032, 2923, 1769, 1693, 1549, 1496, 1444, 1417, 1337, 1248, 1204, 1132, 1104, 1074, 1029, 1011, 992, 905, 894, 861, 822, 740, 697.

***N*-(2,4-Dimethoxybenzyl)-*N*-((2,4-dimethoxybenzyl)carbamoyl)-2-(1*H*-indol-3-yl)-2-oxoacetamide (5).** To a stirred solution of carbodiimide **3e** (208.6 mg, 482.4 μmol) in dry DMF (2.8 mL) was added with 4 Å molecular sieves and indole **2d** (115.3 mg, 609.5 μmol). The mixture was then heated to 110 °C and left to stir at that temperature for 24 h. After completion, saturated brine was added to the reaction and diluted with ethyl acetate. The organic phase was collected using separating funnel. The water layer was extracted with ethyl acetate three times and the combined ethyl acetate was dried over Na_2SO_4. Purification of the residue after evaporation of the solvent under reduced pressure gave **4g** as a foamy pale brown solid (139.2 mg, 43%, R_f 0.22 (EtOAc:petroleum benzine 5:3)) and **5** as a brown viscous liquid (16.3 mg, 5%, R_f = 0.3 (EtOAc:petroleum benzine 5:3)). ^{1}H NMR (400 MHz, d_6-DMSO, δ ppm): 3.43 (s, 3 H, OCH_3), 3.59 (s, 3 H, OCH_3), 3.75 (s, 3 H, OCH_3), 3.76 (s, 3 H, OCH_3), 4.30–4.31 (d, 2 H, *J* 4.1, CH_2Ph), 4.60–4.72 (d, 2 H, *J* 14.8, CH_2Ph), 4.69 (s, 2 H, CH_2Ph), 6.04–6.05 (d, 1 H, *J* 2.3, $CH_{Ar\,benzyl}$), 6.14–6.17 (dd, 1 H, *J* 2.3, 8.4, $CH_{Ar\,benzyl}$), 6.47–6.50 (dd, 1 H, *J* 2.3, 8.4, $CH_{Ar\,benzyl}$), 6.58–6.59 (d, 1 H, *J* 2.4, $CH_{Ar\,benzyl}$), 6.94–6.96 (d, 1 H, *J* 7.9, $CH_{Ar\,indole}$), 7.02–7.04 (d, 1 H, *J* 8.4, $CH_{Ar\,benzyl}$), 7.14–7.18 (m, 1H, $CH_{Ar\,indole}$), 7.19–7.21 (d, 1 H, *J* 8.4, $CH_{Ar\,benzyl}$), 7.35–7.39 (m, 3 H, $CH_{Ar\,indole}$), 7.59–7.63 (m, 2 H, $CH_{Ar\,indole}$), 7. 69 (s, 1 H, O*H*), 8.28–8.30 (m, 2H, $CH_{Ar\,indole}$), 8.42 (s, 1 H, NHCH_{indole}), 12.65 (s, 1 H, N*H*CH_{indole}). Note: **5** is presumably in equilibrium with **4g**. ^{13}C NMR (100 MHz, d_6-DMSO, δ ppm): 37.4 (CH_2Ph), 37. 6 (CH_2Ph), 55.2 (OCH_3), 55.5 (OCH_3), 55.7 (OCH_3), 55.9 (OCH_3), 84.6 (COH_{quat}, C4), 97. 6 ($CH_{Ar\,benzyl}$), 98.7 ($CH_{Ar\,benzyl}$), 104.2 ($CH_{Ar\,benzyl}$), 104.8 ($CH_{Ar\,benzyl}$), 113.1 ($C_{quat\,indole}$), 113.5 ($CH_{Ar\,indole}$), 116.2 ($CH_{Ar\,indole}$), 116.9 ($C_{Ar\,quat\,benzyl}$), 119.6 ($CH_{Ar\,indole}$), 120.4 (NH-CH_{indole}), 121.7 ($CH_{Ar\,indole}$), 123.8 ($CH_{Ar\,indole}$), 124.7 ($CH_{Ar\,indole}$), 124.8 ($CH_{Ar\,indole}$), 125.6 ($C_{quat\,indole}$), 125.8 ($CH_{Ar\,indole}$), 127.0 ($CH_{Ar\,indole}$), 127.3 ($C_{quat\,indole}$), 130.5 ($CH_{Ar\,benzyl}$), 130. 6 ($CH_{Ar\,benzyl}$), 135.8 ($C_{quat\,indole}$), 137.6 ($C_{quat\,indole}$), 139. 6 (NH-CH_{indole}), 155.4 (C = O, C2), 157.7 ($C_{quat\,Ar\,benzyl}$), 158.3 ($C_{quat\,Ar\,benzyl}$), 160.2 ($C_{quat\,Ar\,benzyl}$), 160.8 ($C_{quat\,Ar\,benzyl}$), 164.9 (C = O, C5), 171.5 (C = O, C5), 182.2 (C = O, C4). Note: 9 is presumably in equilibrium with 4i. HRMS: Calculated [M + H$^+$] m/z 532.2078, found [M + H$^+$] 532.2078. IR (neat) υ_{max}: 3316, 2932, 2826, 1774, 1701, 1610, 1587, 1507, 1440, 1341, 1289, 1242, 1206, 1154, 1105, 1028, 1009, 933, 819, 777, 742.

***Tert*-butyl (2-(1*H*-indol-3-yl)-2-oxoacetyl)carbamate (7).** To a stirred solution of **2d** (205.6 mg, 0.74 mmol) and bis-Boc-thiourea (117.3 mg, 0.62 mmol) in dry CH_2Cl_2 (3.2 mL) was added dry triethylamine (280 μL, 2.0 mmol) and Mukaiyama's reagent (190.1 mg, 0.74 mmol) at room temperature. After being stirred for 20 min at the same temperature, water (10 mL) was added to the reaction and then diluted with EtOAc (10 mL). The organic phase was collected through separating funnel and the water layer was extracted with EtOAc (3 × 10 mL). The combined organic phase was dried over Na_2SO_4 and evaporated under reduced pressure. Column chromatography purification using EtOAc/petroleum benzine 1/1 as the eluent gave **7** as a yellowish solid (119.6 mg, 67%, R_f 0.58 (EtOAc:petroleum benzine 1:1), MP 180–182 °C). Treatment of **7** with TFA (86 μL, 1.1 mmol) in dry CH_2Cl_2 (2.6 mL) for 30 min at rt provided known 2-(1*H*-indol-3-yl)-2-oxoacetamide as a brown solid after column chromatography purification (95%, R_f 0.42 (EtOAc:petroleum benzine 5:3). For **7**: ^{1}H NMR (400 MHz, $CDCl_3$, δ ppm): 1.58 (s, 9 H, Boc), 7.29–7.35 (m, 2 H, $CH_{Ar\ indole}$), 7.51–7.53 (m, 1 H, $CH_{Ar\ indole}$), 8.35–8.37 (m, 1 H, $CH_{Ar\ indole}$), 9.16–9.17 (d, 1 H, *J* 3.4, $NHCH_{indole}$), 9.50 (s, 1 H, NH_{amide}), 10.51 (s, 1 H, $NHCH_{indole}$). ^{13}C NMR (100 MHz, $CDCl_3$, δ ppm): 28.0 ($OCCH_3$), 83.4 ($OCCH_3$), 112.0 ($C_{quat\ indole}$), 112.3 ($CH_{Ar\ indole}$), 122.2 ($CH_{Ar\ indole}$), 123.5 ($CH_{Ar\ indole}$), 124.2 ($CH_{Ar\ indole}$), 126.7 ($C_{Ar\ indole}$), 136.2 ($C_{Ar\ indole}$), 140.1 ($NHCH_{indole}$), 149.8 (C = $OCCH_3$), 160.9 (C = ONHBoc), 177.5 (C = OCONHBoc). LRMS: calculated [M + Na] m/z 311.1, found [M + Na] 311.1. IR (neat) υ_{max}: 3323, 3168, 3058, 2976, 2924, 1748, 1687, 1615, 1486, 1431, 1392, 1369, 1315, 1281, 1236, 1122, 1045, 1010, 978, 926, 877, 849, 808, 773, 758, 731, 688. For 2-(1*H*-indol-3-yl)-2-oxoacetamide: ^{1}H NMR (400 MHz, d_6-DMSO, δ ppm): 7.25–7.28 (m, 2 H, $CH_{Ar\ indole}$), 7.53–7.55 (m, 1 H, $CH_{Ar\ indole}$), 7.70 (s, 1 H, N*H*), 8.07 (s, 1 H, N*H*), 8.22–8.24 (m, 1 H, $CH_{Ar\ indole}$), 8.69 (d, 1 H, *J* 1.3, $NHCH_{indole}$), 12.20 (s, 1 H, $NHCH_{indole}$). ^{13}C NMR (100 MHz, d_6-DMSO, δ ppm): 112.6 ($C_{quat\ indole}$), 113.0 ($CH_{Ar\ indole}$), 121.7 ($CH_{Ar\ indole}$), 122.9 ($CH_{Ar\ indole}$), 123.8 ($CH_{Ar\ indole}$), 126.6 ($C_{Ar\ quat\ indole}$), 136.8 ($C_{Ar\ quat\ indole}$), 138.7 $NHCH_{indole}$), 166.5 $CONH_2$), 183.4 ($COCONH_2$). Data is in accordance with literature values [25]. **Preparation of bis-Boc-thiourea**. To a stirred solution of heptane-washed NaH (1182.3 mg, 29.6 mmol) in dry THF (130 mL) was added with thiourea (500.0 mg, 6.6 mmol) at 0 °C. The mixture was stirred at the same temperature for 5 min and then warmed to room temperature and continued to stir for 10 min. Di-*tert*-butyl dicarbonate (3584.0 mg, 16.4 mmol) was added to the mixture at 0 °C and stirred at the same temperature for 30 min. The mixture was warmed to room temperature and continued to stir for 2 h. After completion, saturated aqueous solution of $NaHCO_3$ was added to the mixture followed by saturated brine and ethyl acetate. The organic phase was collected using separating funnel and the remaining water layer was extracted with EtOAc (3 × 40 mL). The combined organic layer was dried over Na_2SO_4 and evaporated at reduced pressure. The residue was the purified using column chromatography providing bis-Boc-thiourea as a pale yellow solid (1271.0 mg, 70%, R_f = 0.42 (EtOAc/hexane 1/5), MP 135–136 °C, (lit. [26] 124–127 °C)). ^{1}H NMR (400 MHz, $CDCl_3$, δ ppm): 1.39 (s, 9 H, $C(CH_3)_3$), 1.43 (s, 9 H, $C(CH_3)_3$). ^{13}C NMR (100 MHz, $CDCl_3$, δ ppm): 27.7 (CH_3), 83.6 ($OCCH_3$), 149.6 (C = O), 177.7 (C = S). Data is in accordance with literature values [26].

1,3-Bis(2,4-dimethoxybenzyl)-5-(1*H*-indol-3-yl)-5-(phenylthio)imidazolidine-2,4-dione (10). To a stirred solution of **4g** (152.2 mg, 286.3 μmol) in CH_2Cl_2 (1.9 mL) thiophenol (35 μL, 340.9 μmol) was added at room temperature. The mixture was cooled to -78 °C and $SnCl_4$ (1 M in CH_2Cl_2, 300 μL, 300.0 mmol) was added. After 10 min stirring at the same temperature, saturated aqueous $NaHCO_3$

was added to the mixture followed by dilution with CH_2Cl_2. The organic phase was collected through a separating funnel and the aqueous phase was extracted three times with 10 mL CH_2Cl_2. The combined organic phase was dried over Na_2SO_4 and evaporated under reduced pressure. Column chromatography purification using EtOAc:petroleum benzine 5:3 of the residue gave **10** as an off-white foamy solid (140.9 mg, 79%, $R_f = 0.58$ (EtOAc:petroleum benzine 5:3), m.p. 88–91 °C). ^{1}H NMR (400 MHz, $CDCl_3$, δ ppm): 3.44 (s, 3 H, OCH_3), 3.61 (s, 3 H, OCH_3), 3.69 (s, 3 H, OCH_3), 3.79 (s, 3H, OCH_3), 4.28–4.31 (d, 1 H, *J* 14.8, CH_APh), 4.37–4.41 (d, 1 H, *J* 14.8, CH_APh), 4.71 (s, 2 H, CH_{2B}Ph), 5.95–5.96 (d, 1 H, *J* 2.4, $CH_{\text{Ar benzyl}}$), 6.04–6.06 (dd, 1 H, *J* 2.4, 8.4, $CH_{\text{Ar benzyl}}$), 6.35–6.38 (dd, 1H, *J* 2.4, 8.4, $CH_{\text{Ar benzyl}}$), 6.41–6.42 (d, 1 H, *J* 2.4, $CH_{\text{Ar benzyl}}$), 6.77–6.79 (m, 1 H, $CH_{\text{Ar indole}}$), 6.89–6.91 (d, 1 H, *J* 8.0, $CH_{\text{Ar indole}}$), 6.94–7.07 (m, 4 H, 2 $CH_{\text{Ar benzyl}}$, 2 $CH_{\text{Ar indole}}$), 7.16–7.17 (d, 1 H, *J* 2.8, $NHCH_{\text{indole}}$), 7.30–7.34 (m, 2 H, $CH_{\text{Ar thiophenol}}$), 7.44–7.46 (m, 3 H, $CH_{\text{Ar thiophenol}}$), 9.06–9.07 (d, 1 H, *J* 2.4, $NHCH_{\text{indole}}$). ^{13}C NMR (100 MHz, $CDCl_3$, δ ppm): 37.3 (CH_2Ph), 37.9 (CH_2Ph), 54.5 (OCH_3), 55.1 (OCH_3), 55.1 (OCH_3), 55.2 (OCH_3), 76.1 ($CSPh_{\text{quat}}$, C4), 97.1 ($CH_{\text{Ar benzyl}}$), 98.1 ($CH_{\text{Ar benzyl}}$), 103.4 ($CH_{\text{Ar benzyl}}$), 103.9 ($CH_{\text{Ar benzyl}}$), 108.4 ($C_{\text{quat indole}}$), 111.4 ($CH_{\text{Ar indole}}$), 116.0 ($C_{\text{quat Ar benzyl}}$), 117.1 ($C_{\text{quat Ar benzyl}}$), 118.8 ($CH_{\text{Ar indole}}$), 119.5, ($CH_{\text{Ar indole}}$), 121.7 ($CH_{\text{Ar indole}}$), 124.3 ($C_{\text{quat indole}}$), 125.3 ($NHCH_{\text{indole}}$), 127.8 ($C_{\text{quat indole}}$), 128.9 ($CH_{\text{Ar thiophenol}}$), 130.2 ($CH_{\text{Ar thiophenol}}$), 130.6 ($CH_{\text{Ar benzyl}}$), 131.1 ($CH_{\text{Ar benzyl}}$), 136.2 ($C_{\text{quat thiophenol}}$), 137.00 ($CH_{\text{Ar thiophenol}}$), 155.1 ($C$ = O, C2), 157.6 ($C_{\text{quat Ar benzyl}}$), 158.0 ($C_{\text{quat Ar benzyl}}$), 159.8 ($C_{\text{quat Ar benzyl}}$), 160.4 ($C_{\text{quat Ar benzyl}}$), 171.4 ($C$ = O, C5). HRMS: Calculated $[M + H]^+$ 624.2168, $[M + H]^+$ 624.2166; calculated $[M + Na]^+$ 646.1982, found 646.1983. IR (neat) υ_{max}: 3301, 2934, 2835, 1769, 1701, 1611, 1588, 1507, 1438, 1409, 1339, 1288, 1264, 1207, 1156, 1109, 1033, 985, 906, 880, 830, 785, 744, 705, 692.

1,3-Dibenzyl-4-(1*H*-indol-3-yl)-2,5-dioxo-2,5-dihydro-1*H*-imidazol-3-ium (12). To a stirred solution of di-Bn hydantoin **4h** (196.6 mg, 477.8 µmol) in abs. ethanol (30 mL) 20 wt% $Pd(OH)_2$/C (87.3 mg, 124 µmol) and ammonium formate (3.0 g, 48 mmol) were added at room temperature. The resulting mixture was heated to reflux and left to stir for 18 h. After reaction completion, the crude reaction was filtered over a short plug of silica gel and the filtrate was evaporated under reduced pressure. The residue was then diluted with EtOAc (20 mL) and water (20 mL). The organic layer was collected, and the aqueous layer was washed with EtOAc (3 × 20 mL). The combined organic phase was dried over Na_2SO_4 and concentrated providing **12** as a white viscous liquid (**4h**:**12** = 1.00:0.42 (^{1}H NMR ratio), $R_f = 0.67$ (EtOAc:petroleum benzine 1:1)). ^{1}H NMR (400 MHz, $CDCl_3$, δ ppm): 3.65–3.69 (d, 1H, *J* 15.1, $H_{\text{A,benzylic}}$), 5.03–5.06 (d, 1H, *J* 15.0, $H_{\text{A,benzylic}}$), 4.78–4.90 (m, 2H, $H_{\text{B,benzylic}}$), 6.72–7.54 (m, $H_{\text{Ar,indole 4h,12}}$, $H_{\text{Ar 4h,12}}$, $NHCH_{\text{indole 4h,12}}$), 8.9 (s, 1H, NH_{indole}). ^{13}C NMR (100 MHz, $CDCl_3$, δ ppm): 42.8 ($C_{\text{A,benzylic}}$), 44.0 ($C_{\text{B,benzylic}}$), 105.3 (C4), 109.9 ($C_{\text{quat,indole}}$), 111.8 ($C_{\text{Ar,indole}}$), 118.1 ($C_{\text{Ar,indole}}$), 120.2 ($C_{\text{Ar,indole}}$), 122.5 ($C_{\text{Ar,indole}}$), 124.5 ($C_{\text{quat,indole}}$), 126.4 ($NHCH_{\text{indole}}$), 127.1 (CH_{Ar}), 127.8 (CH_{Ar}), 127.9 (CH_{Ar}), 128.2 (CH_{Ar}), 128.8 (CH_{Ar}), 129.0 (CH_{Ar}), 135.6 ($C_{\text{Ar quat}}$), 135.7 ($C_{\text{quat indole}}$), 136.6 ($C_{\text{Ar quat}}$), 155.7 (C = O, C2), 172.5(C = O,

C5). HRMS: Calculated $[M]^+$ 394.1550, found $[M]^+$ 394.1545. IR (neat) υ_{max}: 3344, 3062, 3033, 2925, 2853, 1773, 1706, 1546, 1496, 1439, 1415, 1372, 1338, 1314, 1274, 1245, 1205, 1120, 1104, 1074, 1045, 1008, 983, 930, 881, 843, 822, 739, 697.

3 Discussion

We selected dimethylformamide (DMF) as the reaction medium and performed the reaction at an elevated temperature (110 °C). This condition was applied to a range of substrates, which were commercially available α-keto acids **2a–d**, including the derivative **2d**, a potential precursor of polyandrocarpamide D, and symmetrically substituted carbodiimides **3a–f** (Fig. 3). Carbodiimides **3b–f** were selected for their anticipated ease of removal after hydantoin formation. Keto acids **2a–c** reacted with carbodiimides **3a&b** affording hydantoins **4a–c** in high yields. The reaction of indole-derived keto acid **2d** with carbodiimides **3b&f** [27] also provided hydantoins **4d&h**, respectively, in moderate to good yields. Carbodiimide **3e**, prepared in 79% yield through aza-Wittig reaction [28], was effectively reacted with substrate **2d** at 110 ºC in the presence of 4 Å MS to give hydantoin **4g** in a moderate yield, together with a minor amount of urea **5**. This urea became undetectable by TLC upon further heating to 125 °C.

Simple recrystallization from boiling methanol of selected product **4d** followed by vacuum filtration produced a clean mass of X-ray quality crystal confirming the α-hydroxy-α-indole substitutions at C-5 of the newly formed imidazolidine-2,4-dione core (Fig. 4).

In general the behavior described above held with two exceptions, the reactions of 3-indoleglyoxylic acid (**2d**) with bis-TMS-carbodiimide (**3c**) and with *N,N'*-bis(*tert*-butoxycarbonyl) carbodiimide (**3d**). Instead of the expected product **4e**, reaction of **2d** and **3c**, yielded a plethora of products. Purification of this mixture from several reaction batches provided only one isolable product, the unusual cyanimide **6**, whose structure was confirmed by X-ray diffraction.

When highly reactive but non-isolable [29–31], *N,N'*-bis(*tert*-butoxycarbonyl) carbodiimide (**3d**), was synthesized and treated in situ with α-keto acid **2d** the expected compound **4f** was not formed in detectable amounts. Rather Boc-ketoamide **7** was isolated in good yield within less than half an hour.

Having established a useful synthetic protocol for the construction of 1,3,5-trisubstituted 5'-hydroxyhydantoins, we successfully prepared hydantoins **4d**, **4g&h** each bearing a 3-indole moiety with obvious relevance to polyandrocarpamide D (**1**).

There are a number of published examples describing efficient acid-catalyzed removal of *tert*-butyl protecting amido- and/or imido groups in systems relatively similar to hydantoins [19, 32, 33]. Exploration of these conditions through, treatment of, e.g., **4d** with both mild and strongly acidic conditions [19, 32] provided no traces of the desired fully deprotected imidazolidinedione **8** (Table 1). Instead, only degradation of the starting material occurred. We then moved to the hydantoin **4g** bearing 2,4-DMB protecting group. This functionality is often used for maleimide

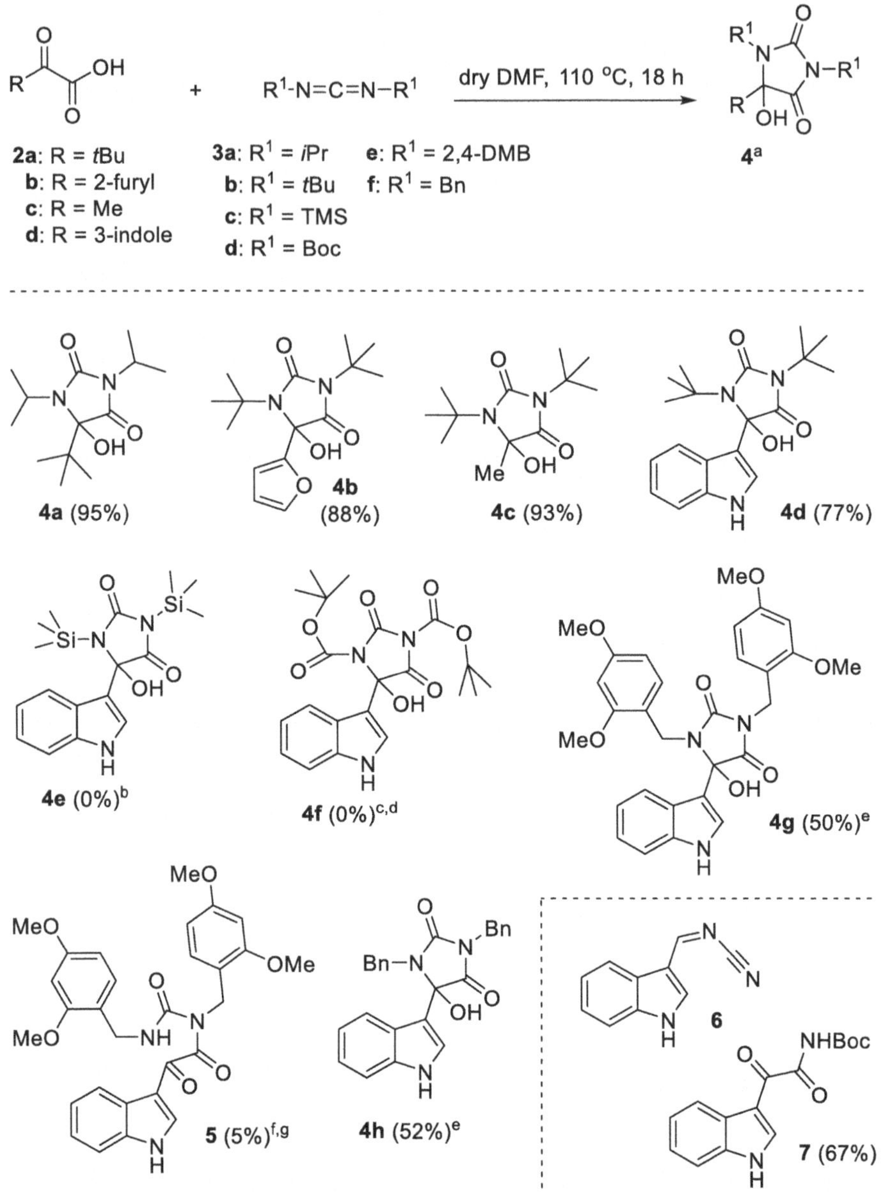

Fig. 3 One-pot synthesis of 1,3,5-trisubstituted 5'-hydroxyhydantoins. **a** Isolated yields are in parentheses. **b** Cyanimide **6** was the only product which could be isolated. **c** **2d** was added to in situ generated **3d** in anhydrous CH_2Cl_2, and stirred for 20 min at rt. **d** Keto amide **7** was formed exclusively. **e** Performed in dry DMF, 4 Å MS, 125 °C, 24 h. **f** Performed in dry DMF, 4 Å MS, 110 °C, 24 h. **g** **4g** was isolated along with **5** in 43% yield

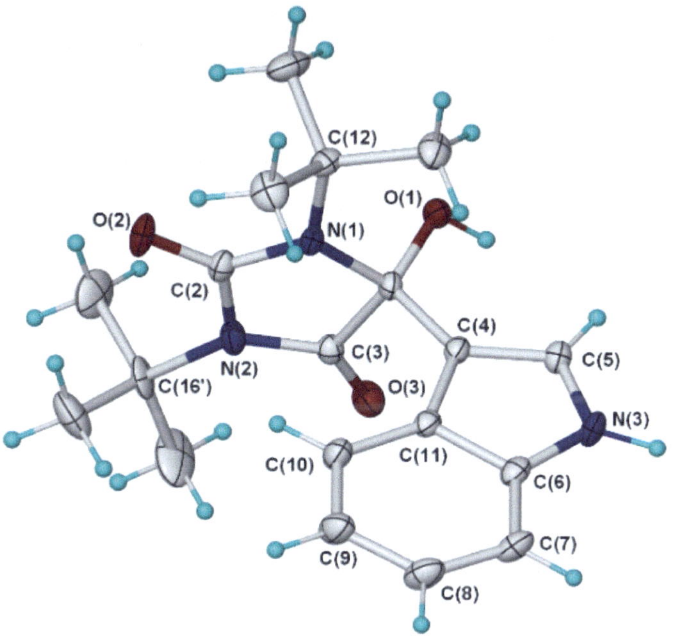

Fig. 4 X-ray structure of **4d**

systems [34–38] where a high degree of lability is desired [39]. Standard deprotection employing TFA-anisole mixture [34, 35, 38] facilitated the formation of the expected imidazoledione **9b** in a modest yield (entry 3), from which its structure was estimated from the LRMS data—the NMR spectra were not fully resolvable due to potential formation of its hydrated derivative. Raising the reaction temperature caused decomposition of the starting material (entry 4). Hydrogenolysis of **4g** failed to affect the deprotection (entries 5–6). Oxidative cleavage employing DDQ did not promote the desired deprotection of the DMB moiety (entry 7). However, replacing the oxidant to CAN was too corrosive to our hydantoin system leading to rapid degradation (entry 8). A protocol developed by Ozaki et al. [40] in the mild deprotection of PMB employing stannous chloride was also adopted and provided no reaction (entry 9). A system containing a stronger Lewis acid, $SnCl_4$, and cation scavenger thiophenol which has been successfully applied to the removal of PMB in carbohydrates [41], induced the formation of compound **10** in a high yield within a short reaction time (entry 10). Moving forward to the last substrate, **4h**, the same condition applied for **4g** as detailed in entry 6 was adopted (entry 11). Apparently, the starting material **4h** was partially transformed into another species with an m/z value of 394.1545, which upon further investigation, was relevant to a dehydrated scaffold **12** (calculated $m/z = 394.1550$). No trace of deprotected product was observed.

The proposed positively charged iminium **12**, in our experiment, was successfully collected through an ordinary column chromatographic purification as an inseparable

Table 1 Attempted deprotection of hydantoin **4d**, **4g&h**

Entry	4	Reagent(s)	Solvent	Temp (°C)	t (h)	Product (% yield)[a]
1	**4d**	TFA	$MeCN/H_2O$ 1/1	50	2	–[b]
2		HBr	AcOH	100	18	–[b]
3	**4g**	TFA	anisole	rt	24	**9** (38)[c]
4		TFA	anisole	90	15	-[b]
5		H_2 (1 atm) Pd/C, AcOH	MeOH	rt	48	NR
6		NH_4HCO_2, $Pd(OH)_2/C$	abs EtOH	95	20	NR
7		DDQ	DCM/H_2O 18/1	rt	2	NR
8		CAN	MeCN	rt	0.15	–[b]
9		$SnCl_2$, TMSCl	dry DCM/ anisole	rt	18	NR
10		$SnCl_4$, PhSH	DCM	-78	0.6	**10** (79)
11	**4h**	NH_4HCO_2, $Pd(OH)_2/C$	abs EtOH	95	18	**11&12** (ND)[d]

[a] Isolated yield. [b] Starting material decomposed. [c] The structure was estimated from its LRMS data. [d] In 1:0.42 ratio based on the ^{1}H NMR spectrum. NR = no reaction. ND = not determined

mixture with another compound previously thought as unreacted starting material **4h**. However, organic ions such as **13** should exist as a neutral ion pair [42] to be able to moving during the elution process. Hence, we hypothesized that instead of unreacted **4h**, anion **11** was most likely the species co-eluted with **12** during the chromatographic purification.

HRMS data obtained from the crude reaction detailed in entry 11 (Table 1) provided two values of m/z which were consistent with the M^+ of **12** and $[M + H]^+$ of starting material **4h** (for **4h**, $[M + H]^+$ calculated 412.1656, found 412.1643). Since the anion **11** is identical with **4h** after protonation, its actual presence is expected. Nevertheless, further evidence is required to support our idea about the presence of **11**–**12** ion pair.

Our unsuccessful efforts to obtain imidazolidinedione **8** was rather surprising as common *tert*-butyl-, more acid-labile 2,4-dimethoxybenzyl-, and benzyl groups in **4d**, **4g&h**, respectively, failed to be removed completely. The tendency to form the imino group conjugated with the carbonyl could be the driving force of this process. wever, it is difficult to explain the resistance to deprotection of the protecting group. The similar observation was experienced by Pyne et al. [43] in the incomplete deprotection of 1,3-diPMB hydantocidin precursor using ceric ammonium nitrate.

4 Conclusion

In summary, we have demonstrated a versatile and straightforward one-step construction of 1,3,5-trisubstituted 5'-hydroxyhydantoins from α-keto acids and carbodiimides in moderate to excellent yields through simple mixing and heating. In general, less bulky carbodiimides provided the hydantoin products in higher yield than the bulkier variants. Carbodiimides with strong electron donating and strong electron withdrawing substituents such as bis-TMS- (**3c**) and bis-Boc-carbodiimides (**3d**), respectively, failed to form the desired hydantoins. All attempts to deprotect hydantoins **4d**, **4g&h** were unsuccessful to achieve the desired compound **8**.

References

1. Cachet N, Genta-Jouve G, Regalado EL, Mokrini R, Amade P, Culioli G, Thomas OP (2009) Parazoanthines A–E, hydantoin alkaloids from the Mediterranean Sea anemone *Parazoanthus axinellae*. J Nat Prod 72(9):1612–1615. https://doi.org/10.1021/np900437y
2. Tasdemir D, Mallon R, Greenstein M, Feldberg LR, Kim SC, Collins K, Wojciechowicz D, Mangalindan GC, Concepción GP, Harper MK, Ireland CM (2002) Aldisine Alkaloids from the Philippine Sponge *Stylissa massa* Are Potent Inhibitors of Mitogen-Activated Protein Kinase Kinase-1 (MEK-1). J Med Chem 45(2):529–532. https://doi.org/10.1021/jm0102856
3. Davis RA, Aalbersberg W, Meo S, Rocha RMD, Ireland CM (2002) The isolation and synthesis of polyandrocarpamines A and B: two new 2-aminoimidazolone compounds from the Fijian ascidian, Polyandrocarpa sp. Tetrahedron 58(16):3263–3269. https://doi.org/10.1016/S0040-4020(02)00228-4
4. Nakajima N, Matsumoto M, Kirihara M, Hashimoto M, Katoh T, Terashima S (1996) Novel synthesis of (+)-hydantocidin based on the plausible biosynthetic pathway. Tetrahedron 52(4):1177–1194. https://doi.org/10.1016/0040-4020(95)00975-2
5. Ogawa J, Horinouchi N, Shimizu S (2012) Hydrolysis and formation of hydantoins. In: Drauz K, Groger H, May O (eds) Enzyme catalysis in organic synthesis, vol 1, 3rd ed. Wiley-VCH, Germany, pp 651–674
6. Olivieri R, Fascetti E, Angelini L, Degen L (1979) Enzymatic conversion of N carbamoyl-D-amino acids to D-amino acids. Enzyme Microb Technol 1(3):201–204. https://doi.org/10.1016/0141-0229(79)90030-9
7. Yamada H, Takahashi S, Kii Y, Kumagai H (1978) Distribution of hydantoin hydrolyzing activity in microorganisms: microbial transformation of hydantoins to amino acids (I). J Ferment Technol 56:484–491
8. Cecere F, Galli G, Morisi F (1975) Substrate and steric specificity of hydropyrimidine hydrase. FEBS Lett 57(2):192–194. https://doi.org/10.1016/0014-5793(75)80714-9
9. Olimpieri F, Bellucci MC, Marcelli T, Volonterio A (2012) Regioselective multicomponent sequential synthesis of hydantoins. Org Biomol Chem 10(48):9538–9555. https://doi.org/10.1039/C2OB26498F
10. Meusel M, Gütschow M (2004) Recent developments in hydantoin chemistry: a review. Org Prep Proced Int 36(5):391–443. https://doi.org/10.1080/00304940409356627
11. Byrtus H, Obniska J, Czopek A, Kamiński K (2011) Synthesis and anticonvulsant activity of new *N*-mannich bases derived from 5-cyclopropyl-5-phenyl-hydantoins. Arch Pharm 344(4):231–241. https://doi.org/10.1002/ardp.201000241
12. Thenmozhiyal JC, Wong PT-H, Chui W-K (2004) Anticonvulsant activity of phenylmethylene-hydantoins: a structure-activity relationship study. J Med Chem 47(6):1527–1535. https://doi.org/10.1021/jm030450c

13. Varia SA, Schuller S, Sloan KB, Stellax VJ (1984) Phenytoin prodrugs III: water-soluble prodrugs for oral and/or parenteral use. J Pharm Sci 73(8):1068–1073. https://doi.org/10.1002/jps.2600730812
14. Guella G, Mancini I, Zibrowius H, Pietra F (1988) Novel aplysinopsin-type alkaloids from scleractinian corals of the family dendrophylliidae of the mediterranean and the Philippines: configurational-assignment criteria, stereospecific synthesis, and photoisomerization. Helv Chim Acta 71(4):773–782. https://doi.org/10.1002/hlca.19880710412
15. Breton JJ, Chabot-Fletcher MC (1997) The natural product hymenialdisine inhibits interleukin-8 production in U937 cells by inhibition of nuclear factor-κB. J Pharmacol Exp Ther 282(1):459–466
16. Cafieri F, Fattorusso E, Mangoni A, Taglialatela-Scafati O (1996) Dispacamides, anti-histamine alkaloids from Caribbean Agelas sponges. Tetrahedron Lett 37(20):3587–3590. https://doi.org/10.1016/0040-4039(96)00629-6
17. Mancini I, Guella G, Pietra F, Debitus C, Duhet D (1994) Imidazolone and imidazolidinone artifacts of a pivotal imidazolthione, zyzzin, from the poecilosclerid sponge Zyzza massalis from the Coral Sea: the first thermochromic systems of marine origin. Helv Chim Acta 77(7):1886–1894. https://doi.org/10.1002/hlca.19940770719
18. Konnert L, Lamaty F, Martinez J, Colacino E (2017) Recent advances in the synthesis of hydantoins: the state of the art of a valuable scaffold. Chem Rev 117(23):13757–13809. https://doi.org/10.1021/acs.chemrev.7b00067
19. Olimpieri F, Bellucci MC, Volonterio A, Zanda M (2009) A mild, efficient approach for the synthesis of 1, 5-disubstituted hydantoins. Eur J Org Chem 2009(35):6179–6188. https://doi.org/10.1002/ejoc.200900868
20. van der Knaap M, Lageveen LT, Busscher HJ, Mars-Groenendijk R, Noort D, Otero JM, Llamas-Saiz AL, van Raaij MJ, van der Marel GA, Overkleeft HS, Overhand M (2011) Evaluation of readily accessible azoles as mimics of the aromatic ring of D-phenylalanine in the turn region of gramicidin S. ChemMedChem 6(5):840–847. https://doi.org/10.1002/cmdc.201000539
21. Trost BM, Fandrick DR (2005) DYKAT of vinyl aziridines: total synthesis of (+)-pseudodistomin D. Org Lett 7(5):823–826. https://doi.org/10.1021/ol047513l
22. Wang Z, Sueki S, Kanai M, Kuninobu Y (2016) Rhenium-catalyzed synthesis of 1, 3-diiminoisoindolines via insertion of carbodiimides into a C-H bond of aromatic and heteroaromatic imidates. Org Lett 18(10):2459–2462. https://doi.org/10.1021/acs.orglett.6b01012
23. Üngören ŞH, Sırça F (2017) Novel self-condensation of ammonium dithiocarbamates leading to symmetrical substituted thioureas. Phosphorus Sulfur Silicon Relat Elem 192(1):28–33. https://doi.org/10.1080/10426507.2016.1223075
24. Luo B, Cui Q, Luo H, Hu Y, Huang P, Wen S (2016) *N*-Benzyldithiocarbamate salts as sulfur sources to access tricyclic thioheterocycles mediated by Copper species. Adv Synth Catal 358(17):2733–2738. https://doi.org/10.1002/adsc.201600405
25. Kumar Y, Shaw M, Thakur R, Kumar A (2016) Copper (II)-mediated aerobic oxidation of benzylimidates: synthesis of primary α-ketoamides. J Org Chem 81(15):6617–6625. https://doi.org/10.1021/acs.joc.6b01262
26. Iwanowicz EJ, Poss MA, Lin J (1993) Preparation of N, N'-bis-tert-Butoxycarbonylthiourea. Synth Commun 23(10):1443–1445. https://doi.org/10.1080/00397919308011234
27. Munch H, Hansen JS, Pittelkow M, Christensen JB, Boas U (2008) A new efficient synthesis of isothiocyanates from amines using di-tert-butyl dicarbonate. Tetrahedron Lett 49(19):3117–3119. https://doi.org/10.1016/j.tetlet.2008.03.045
28. Zanda M, Volonterio A, Olimpieri F (2008) Three-component, one-pot sequential synthesis of 1, 3-disubstituted 5-arylhydantoins. Synlett 2008(19):3016–3020. https://doi.org/10.1055/s-0028-1087352
29. Porcheddu A, De Luca L, Giacomelli G (2009) A mild and inexpensive procedure for the synthesis of N, N′-di-Boc-protected guanidines. Synlett 2009(20):3368–3372. https://doi.org/10.1055/s-0029-1218365
30. Yong YF, Kowalski JA, Lipton MA (1997) Facile and efficient guanylation of amines using thioureas and Mukaiyama's reagent. J Org Chem 62(5):1540–1542. https://doi.org/10.1021/jo962196k

31. Kim KS, Qian L (1993) Improved method for the preparation of guanidines. Tetrahedron Lett 34(48):7677–7680. https://doi.org/10.1016/S0040-4039(00)61537-X
32. Zhao B, Du H, Shi Y (2008) A Cu (I)-catalyzed C−H α-amination of esters: direct synthesis of hydantoins. J Am Chem Soc 130(23):7220–7221. https://doi.org/10.1021/ja802242h
33. Simig G, Lempert K, Tamás J (1975) The reaction of N-cyanoamines with 1-(t-butyl)-3, 3-diphenylaziridinone: Czira, G. Tetrahedron 31(9):1195–1200. https://doi.org/10.1016/0040-4020(75)85056-3
34. Wilson LJ, Malaviya R, Yang C, Argentieri R, Wang B, Chen X, Murray WV, Cavender D (2009) Synthetic staurosporines via a ring closing metathesis strategy as potent JAK3 inhibitors and modulators of allergic responses. Bioorg Med Chem Lett 19(12):8–3333. https://doi.org/10.1016/j.bmcl.2009.04.039
35. Palmer BD, Thompson AM, Booth RJ, Dobrusin EM, Kraker AJ, Lee HH, Lunney EA, Mitchell LH, Ortwine DF, Smaill JB, Swan LM, Denny WA (2006) 4-Phenylpyrrolo [3, 4-c] carbazole-1, 3 (2 H, 6 H)-dione Inhibitors of the Checkpoint Kinase Wee1: structure—activity relationships for chromophore modification and phenyl ring substitution. J Med Chem 49(16):4896–4911. https://doi.org/10.1021/jm0512591
36. Shen L, Prouty C, Conway BR, Westover L, Xu JZ, Look RA, Chen X, Beavers MP, Roberts J, Murray WV, Demarest KT, Kuo G-H (2004) Synthesis and biological evaluation of novel macrocyclic bis-7-azaindolylmaleimides as potent and highly selective glycogen synthase kinase-3β (GSK-3β) inhibitors. Bioorg Med Chem 12(5):1239–1255. https://doi.org/10.1016/j.bmc.2003.09.047
37. Qian X, Zheng B, Burke B, Saindane MT, Kronenthal DR (2002) A stereoselective synthesis of BMS-262084, an azetidinone-based tryptase inhibitor. J Org Chem 67(11):3595–3600. https://doi.org/10.1021/jo010757o
38. Watson DJ, Dowdy ED, Li W-S, Wang J, Polniaszek R (2001) Electronic effects in the acid-promoted deprotection of N-2, 4-dimethoxybenzyl maleimides. Tetrahedron Lett 42(10):1827–1830. https://doi.org/10.1016/S0040-4039(01)00031-4
39. Chern C-Y, Huang Y-P, Kan WM (2003) Selective N-debenzylation of amides with p-TsOH. Tetrahedron Lett 44(5):1039–1041. https://doi.org/10.1016/S0040-4039(02)02738-7
40. Akiyama T, Shima H, Ozaki S (1992) Trimethylsilyl chloride-tin (II) chloride-anisole: a novel selective p-methoxybenzyl ether cleavage reagent. Synlett 1992(05):415–416. https://doi.org/10.1055/s-1992-21364
41. Yu W, Su M, Gao X, Yang Z, Jin Z (2000) A facile chemoselective deprotection of the p-methoxybenzyl group. Tetrahedron Lett 41(21):4015–4017. https://doi.org/10.1016/S0040-4039(00)00574-8
42. Ahuja S, Alsante KM (2003) Handbook of isolation and characterization of impurities in pharmaceuticals, vol 5. Academic Press
43. Pham TQ, Pyne SG, Skelton BW, White AH (2005) Regioselective and diastereoselective phosphine-catalysed [3+ 2] cycloadditions to 5-methylenehydantoins: reversal of regioselectivity using chiral N-2-butynoyl-(4S)-benzyloxazolidinone. J Org Chem 70(16):6369–6377. https://doi.org/10.1021/jo050827h

Protease Stabilities of *Lactobacillus fermentum* EN17-2 and *Lactobacillus plantarum* B110 Indigenous at Various Storage Temperatures and Times

T. Khusniati, M. Fajar, Y. S. Juariah, and Sulistiani

Abstract Protease as a protein hydrolyzer is an imported enzyme in Indonesia. To produce local enzyme, indigenous bacteria can be used as an alternative source of enzyme. This study aims to determine protease stability of indigenous *Lactobacillus fermentum* EN-17-2 and *Lactobacillus plantarum* B110 at time of storage. Storage times was carried out for 0, 7, 14, 21 and 28 days. The protease activity was detected using tyrosine method. Relative activity was stated stable with a value of $\geq$ 50%. Data were analyzed statistically using ANOVA. The results showed that the protease stability of *L. fermentum* EN17-2 at frozen temperature for 28 days with a relative activity of 60.06% was not significantly different from 0 day, on the contrary, at cold temperatures the protease stability showed a significant difference ($p < 0.05$). The protease stability of *L. plantarum* B110 for 28 days at cold temperature with a relative activity of 52.26%, and at freezing temperature with a relative activity of 50.51% was significantly different from 0 day ($p < 0.05$). Based on the stability of protease, the protease stability of *L. fermentum* EN17-2 was better than that of *L. plantarum* B110. It is recommended to use *L. fermentum* EN17-2 as a source of protease-producing lactic acid bacteria.

Keywords *Lactobacillus fermentum* EN17-2 · *Lactobacillus plantarum* B110 · Protease · Stability · Storage

---

T. Khusniati (✉) · Sulistiani
Research Center for Applied Microbiology, National Research and Innovation Agency, Cibinong, Bogor, West Java, Indonesia
e-mail: tatikkhusni@gmail.com

M. Fajar · Y. S. Juariah
Pharmacy Faculty, ISTN, Jakarta, Indonesia

M. A. Rahandi Lubis (ed.), *Proceedings of the 1st International Conference of Lignocellulose*, Green Energy and Technology,
https://doi.org/10.1007/978-981-19-1596-3_9

1 Introduction

Protease is a protein hydrolyzing enzyme which in its hydrolysis produces peptides and amino acids as simple compounds [1–3]. Protease is one of the enzymes that has the best potential to be developed, because it is considered quite broad, it can be used in various fields such as industry, both food and non-food products that are traded around the world [4]. The ability of proteases to accelerate reactions is influenced by several factors, including pH and temperature [5].

Economically, proteases from microorganisms are more profitable because microorganisms grow quickly, can grow on cheap substrates, are easily increased in yield through regulation of growth conditions and genetic engineering, and are able to produce enzymes with extreme conditions [6, 7]. The need for protease in Indonesia continues to increase but still depends on imported production. Almost all of the enzyme needs are met from imports.

Lactic acid bacteria (LAB) are safe microorganisms because they are non-toxic and do not produce toxins, so they are known as Generally Recognized as Safe (GRAS) microorganisms that are not a risk to health, even some types of bacteria are useful for health [8]. Various protease-producing LABs, including *Pediococcus pentosaceus* [9], *Lactococcus lactics* [10], *Lactobacillus plantarum* S31 [11], *Lactobacillus delbrueckii* subsp. lactis CRL 58 [12], and other lactic acid bacteria [13]. The character of Lactobacillus and the protease activities of various Lactobacillus species in hydrolyzing proteins in various media have been reported, although their protease stability at different temperatures and storage times have not been reported [14–16]. To reduce the import of enzymes in Indonesia which continues to increase every year, it is needed production of local enzyme from indigenous lactic acid bacteria as an alternative enzyme source.

Based on the above, protease-producing LAB needs to be examined so that its potential can be found, especially local bacteria, and experiments at different temperatures and storage times to detect the stability of proteases are important to do, to determine the best temperature and storage time of protease with optimum stability. To know the potential of two indigenous protease-producing bacteria whose best stability detected between treatments at various storage temperatures and times, and to determine the difference in protease stability of the two bacteria, this study focused on the protease stability of *Lactobacillus fermentum* EN17-2 and *Lactobacillus plantarum* B110 indigenous at various storage temperatures and times.

2 Materials and Methods

2.1 Sub-culture

An MRSB medium (Merck 1.10661.0500, Germany) was prepared, and the bacteria present on the solid medium were collected using a ossicle. Bacteria from solid media EN 17-11 were inoculated into liquid media using aseptic technique, then the liquid media filled with bacteria were vortexed (Sibata TTM-1, Japan), then stored in an incubator (Isuzu, Japan) for 24 h at temperature 35 °C [17].

2.2 Protease Production

A total of 50 ml of casein (Sigma 22090-500G, Japan) 2% was added to 200 ml of nutrient broth medium (Merck 1.0543.0500, Germany). After that, as much as 2% of each inoculum of *L. fermentum* EN 17-2 and *L. plantarum* B110 was put into protease production medium (nutrient broth) and then vortexed (Sibata TTM-1, Japan) until homogeneous. After homogenization, the solution was incubated (Bioshaker-BR 300 LF, Germany) at 37 °C for 24 h and the solution was then centrifuged (Kubota 6500 and Kubota 5910, Japan) for 15 min at 5 °C. The supernatant obtained was then measured for its protease activity [18].

2.3 Protease at Storage Temperatures and Times

Crude supernatant was put into pre-sterilized test tubes as much as 5 ml in each tube. Transfer of the crude supernatant into the tube was carried out in a laminar (CVB-1300M, Japan) aseptically. Protease on day 0 required 3 test tubes for 3 replications, as well as on days 7, 14, 21, and 28. The crude supernatant for room temperature and frozen was the same number of test tubes used, so the total used was 45 test tubes.. The crude supernatant at cold temperature was placed in the refrigerator, and the crude supernatant at freezing temperature was placed in the freezer.

2.4 Protease Activity Assay

The sample solution was prepared by means of 0.2 ml of the protease *L. fermentum* EN 17-2 into a test tube, then 0.4 ml of 2% casein and 0.4 ml of 0.05 M phosphate buffer pH 6.5 were added. The mixture was then incubated at 35 °C for 10 min. After that, 1.0 ml of 20% TCA solution was added and homogenized using vor-tex. The solution was then incubated at 35 °C for 10 min and centrifuged for 5 min.

The control solution was prepared by piping 0.2 ml of protease and added 0.4 ml of 0.05 M phosphate buffer pH 6.5 and 1.0 ml of TCA solution, then the solution was incubated at 35 °C for 10 min. The solution was then added with 0.4 ml of 2% (w/v) casein solution, homogenized with a vortex and again incubated at 35 °C for 10 min. The solution was then centrifuged for 5 min. The resulting supernatant was then measured for absorbance at 275 nm. The protease activity of *L. fermentum* EN 17-2 was calculated in units per milliliter (U/mL) of the enzyme extract. One protease unit was defined as the number of mL of protease required to produce 1 tyrosine per minute using casein as substrate [19].

$$\text{Protease Activity(Unit/mL)} = \frac{[\text{Tyrosin}]}{\text{Mr Tyrosin}} \times \frac{\text{V}}{\text{pxq}} \times \text{df} \quad (1)$$

Note: [Tyrosin] Tyrosin concentration; Mr Tyrosin = Mr Tyrosin; Volume total (V), Volume protease (p); Reaction time (q) Dilution factor (df).

2.5 *Protease Relative Activity*

The relative activity of proteases was measured by calculating the unit activity in a particular treatment divided by the highest unit activity among treatments with a relative activity value of more than or equal to 50% which was considered stable. Calculation using the formula:

$$\text{Protease Relative Activity} = \frac{\text{Treatment activity(U/mL)}}{\text{Highest activity(U/mL)}} \times 100\% \quad (2)$$

2.6 *Protease Statistical Tests at Various Temperatures and Storage Times*

The data were analyzed statistically using ANOVA (Analysis of Variance) with the SPSS method. The variables measured were protease activity (U/mL) against variations in temperature and storage time, and each treatment was carried out with 3 repetitions for each sample. The data obtained were analyzed using the one-way ANOVA (Analysis of Variance) method. The data were then analyzed by Duncan's test to determine whether there was a significant difference in the sample data [20].

3 Results and Discussion

3.1 Protease Stability of L. fermentum *EN 17-2 at Freezing and Cold Temperatures in Various Storage Times*

The activity of the protease unit of *L. fermentum* EN17-2 at freezing temperature at each storage time of 0, 7, 14, 21 and 28 days did not show any significant difference ($p < 0.05$). The activity of the protease unit *L. fermentum* EN 17–2 at freezing temperature at storage, 0 day amounted to 0.4470 U/ mL, 7 days was 0.2869 U/mL, 14 days was 0.2851 U/mL, 21 days was 0.2722 U/mL, and 28 days of 0.2685 U/mL. The relative activity of *L. fermentum* protease EN38-44 at freezing temperature at storage time of 7 to 28 days ranged from 64.18% to 60.06%.

The activity of the protease units of *L. fermentum* EN 17-2 at 28 days of storage (0.2851 U/mL) was not significantly different from the activity of the protease units of *L. fermentum* EN38-44 at storage time of 7 days (0.4856 U/mL) 14 days (0.3936 U/mL) and 21 days (0.3384 U/mL) ($p < 0.05$). The highest activity of *L. fermentum* EN17-2 protease unit occurred at cold temperatures with a storage time of 0 day, which was 0.5353 U/mL, on the other hand the lowest activity occurred at cold temperatures with a storage time of 28 days, which was 0.2851 U/mL. The activity of the protease unit *L. fermentum* EN17-2 at 0 day storage with a value of 0.5353 U/mL was not significantly different from the protease unit activity of *L. fermentum* EN17-2 at 7 days of storage with a value of 0.4856 U/mL, 14 days with a value of 0.3936 U/mL, and 21 days with a value of 0.3384 U/mL ($p < 0.05$).

The activity of the protease unit *L. fermentum* EN17-2 at 0 day storage with a value of 0.5353 U/ mL showed a significant difference with the protease unit activity of *L. fermentum* EN17-2 at 28 days of storage with a value of 0.2851 U/mL ($p < 0.05$). The relative activity of *L. fermentum* protease EN 17-2 at cold temperature at the storage time of 7 to 28 days ranged from 90.72% to 53.25%.

The unit activities and relative activities of *L. fermentum* protease EN 17-2 at freezing and cold temperatures at various storage times can be seen in Table 1.

In general, storage at freezing does not affect protease susceptibility. Most of the enzymes that survive freezing temperatures come from psychrophilic microorganisms. However, several enzymes that show high activity at freezing temperatures were obtained from mesophilic microorganisms, one of the mesophilic group bacteria was *L. fermentum* [21]). Bacteriocins from *L. fermentum* N 38-44 produce a maximum inhibition zone with storage at −20 °C [22]. Therefore, *L. fermentum* at this temperature was able to inhibit pathogenic bacteria so that it did not have a big effect ($p < 0.05$) on the activity of the protease unit at freezing temperature storage at various times of storage.

The decrease in unit activity and relative activity of the protease *L. fermentum* EN17-2 was influenced by temperature and storage time. During cold storage, lactic acid bacteria can still grow even though their growth was slow. The growth phase was more stable in the temperature range (15–20 °C) affecting the stationary phase of bacteria [23]. At lower temperatures (4–10 °C) the growth of Lactobacillus can

Table 1 Unit activity and relative activity of *L. fermentum* protease EN17-2 at freezing and cold temperatures at various storage times

Storage time (Days)	Protease activity (U/mL)	Protease relative activity (%)
Freezing temperature		
0	0.4470 ± 0.0994[a]	100
7	0.2869 ± 0.0166[a]	64.18
14	0.2851 ± 0.0828[a]	63.78
21	0.2722 ± 0.0749[a]	60.89
28	0.2685 ± 0.0138[a]	60.06
Cold temperature		
0	0.5353 ± 0.0276[b]	100
7	0.4856 ± 0.0335[ab]	90.72
14	0.3936 ± 0.0193[ab]	68.04
21	0.3384 ± 0.0386[ab]	63.22
28	0.2851 ± 0.0222[a]	53.25

Note Different letters in the same column indicated significant differences ($p < 0.05$)

be controlled through cooling because most microbes develop more slowly at lower environmental temperatures [24]. The decrease in relative activity of *L. fermentum* EN 38-44 protease to cold temperature at storage time of 7 to 28 days was still stable. The relative protease activity of *L. fermentum* EN17-2 with a value of ≥ 50% incubated for a certain time was expressed as protease stability.

3.2 *Unit Activity and Relative Activity of* L. plantarum *B110 Protease at Freezing and Cold Temperature at Various Storage Times*

The activity of the *L. plantarum* B110 protease unit at freezing temperature at 0 day of storage with a value of 0.4433 U/mL was not significantly different from the activity of the *L. plantarum* B110 protease unit at 7 days of storage with a value of 0.3679 U/mL, 14 days with a value of 0.3366 U/mL, and 21 days with a value of 0.3035 U/mL ($p < 0.05$). The activity of the protease unit *L. plantarum* B110 with a storage time of 28 days with a value of 0.2317 U/mL was not significantly different from the storage time of 7 days with a value of 0.3679 U/mL, 14 days with a value of 0.3366 U/mL, and 28 days with a value of 0.3035 U/mL ($p < 0.05$). The activity of the *L. plantarum* B110 protease unit at storage time of 0 day with a value of 0.4433 U/mL showed significantly different results when compared to the activity of the *L. plantarum* B110 protease unit at 28 days of storage with a value of 0.2317 U/mL) ($p < 0.05$). Relative activity of *L. plantarum* B110 protease at freezing temperature at the storage time of 7 to 28 days ranged from 52.26 to 82.99%.

The highest activity of *L. plantarum* B110 protease units at cold temperatures occurred at 0 day at 0.5389 U/mL. The activity of the *L. plantarum* B110 protease unit at 0 day with a value of (0.5389 U /mL was not significantly different from the 7th storage time with a value of 0.3973 U/mL ($p < 0.05$). The activity of the *L. plantarum* B110 protease unit at 7 days with a value of 0.3973 U/mL not significantly different from the activity of the protease unit *L. plantarum* B110 14 days with a value of 0.3623 U/mL, 21 days with a value of 0.3293 U/mL, 28 days with a value of 0.2722 U/mL ($p < 0.05$).

The activity of the protease unit *L. plantarum* B110 at cold temperature with a storage time of 0 day with a value of 0.5389 U/mL was significantly different from the activity of the protease unit *L. plantarum* B110 14 days with a value of 0.3623 U/mL, 21 days with a value of 0.3293 U/ml, 28 days with a value of 0.2722 U/mL ($p < 0.05$). Relative activity of *L. plantarum* B110 protease at cold temperature at storage time of 7 to 28 days ranged from 50.51% to 73.72%. This indicated that the protease activity of *L. plantarum* B110 at cold temperatures for up to 28 days was still stable.

The unit activity and relative activity of *L. plantarum* B110 protease at freezing and cold temperatures at various storage times can be seen in Table 2

The decrease in the activity of the protease unit *L. plantarum* B110 on the length of time of storage and the results that were significantly different could be due to the influence of the adaptation of the microorganisms concerned. The decrease in the activity of the protease unit *L. plantarum* B110 occurred gradually and stably. Low temperatures can have a detrimental effect on protein folding and denaturation. This protein folding and denaturation at freezing temperature is -20 °C [25]. The relative activity of *L. plantarum* B110 protease at freezing temperature to the storage time

Table 2 Unit activity and relative activity of *L. plantarum* B110 protease at freezing and cold temperatures at various storage times

Storage time (Days)	Protease activity (U/mL)	Protease relative activity (%)
Freezing temperature		
0	0.4433 ± 0.0993^{b}	100
7	0.3679 ± 0.0605^{ab}	82.99
14	0.3366 ± 0.0198^{ab}	75.93
21	0.3035 ± 0.0742^{ab}	68.46
28	0.2317 ± 0.0584^{a}	52.26
Cold temperature		
0	0.5389 ± 0.0689^{b}	100
7	0.3973 ± 0.0832^{ab}	73.72
14	0.3623 ± 0.0138^{a}	66.55
21	0.3293 ± 0.0975^{a}	61.10
28	0.2722 ± 0.0229^{a}	50.51

Note Different letters in the same column indicated significant differences ($p < 0.05$)

of the 7th to 28th day ranged from 52.26 to 82.99%. showed that the relative activity of *L. plantarum* B110 protease at freezing temperatures at various storage times was still stable because the relative activity value was $\geq$ 50%.

Temperature can affect microorganisms, decreasing protease activity slowly and not too significantly, and at cold temperatures, proteases tend to be stable [26]. The growth of lactic acid bacteria can also increase with increasing incubation time, temperature, humidity, light, pH and nutrients which will lead to optimum growth of lactic acid bacteria [24]. In addition, enzymes, which were proteins, may affect the temperature and duration of the incubation process so that they can cause the denaturation process to occur. When this happens, the reaction speed will decrease. In addition, the lactic acid produced by LAB will continue to increase which cause a decrease in pH. The low pH of the media was an important factor of the survival of lactic acid bacteria [27, 28].

Based on the unit activity and relative activity of protease, it showed that between 2 temperatures with a storage time of 7–28 days the unit activity and relative activity of protease from the two lactic acid bacteria used at freezing and cold temperatures had similar unit activity and protease activity. Unit activity and relative activity of *L. plantarum* B110 protease ranged from 0.2722–0.3973 U/mL and 50.51–73.72% at cold temperature, and 0.2317–0.3679 U/mL and 52.26–82.99% at freezing temperature, while the unit activity and relative activity of protease from *L. fermentum* EN17-2 ranged from 0.2851–0.4856 U/mL and 53.25–90.72% at cold temperatures, and 0.2685–0.2869 U/mL and 60.06–64.18% at freezing temperatures, respectively. This was because the storage of freezing and cold temperatures of bacteriocins from lactic acid bacteria can produce a maximum inhibition zone that can inhibit pathogenic bacteria [22].

The protease activity at cold and frozen temperatures of both *L fermentum* and *L. plantarum* during storage time of 7–28 days, respectively, showed no significant difference compared to the freezing temperatutr (Tables 1 and 2) ($p < 0.05$), while the protease activity of *L fermentum* at cold temperature at 0 day of storage showed a significant difference compared to the freezing temperature (Tables 1 and 2) ($p < 0.05$). This difference was probably due to the degradation of substrate, especially protein by protease at cold temperature less than at freezing temperature, so that protease activity at cold temperature was higher than at freezing temperature. The protease activity was influenced by the substrate components, especially the protein content, the higher the protein content, the more protease activity increased up to the activity reached its optimum. Several factors that influence protease activity included substrate with its nutrient components, and temperature [16, 29, 30].

4 Conclusion

The protease stability of *L. fermentum* EN17-2 with a relative activity of 60.06% at frozen temperature for 28 days was not significantly different from 0 day, on the contrary, the protease stability showed a significant difference at cold temperature

($p < 0.05$). The protease stabilities of *L. plantarum* B110 with a relative activity of 52.26% for 28 days at cold temperature, and with a relative activity of 50.51% at freezing temperature were significantly different from 0 day ($p < 0.05$). Based on the stability of protease, the protease stability of *L. fermentum* EN17-2 was better than that of *L. plantarum* B110. As a source of protease-producing lactic acid bacteria, to use *L. fermentum* EN17-2 was recommended.

Acknowledgements The author would like to thank the project of Priority DIPA, Research Center for Biology, Deputy of Life Sciences, Indonesian Institute of Sciences for providing research fund, and Mrs. Desy Septiani, S.Si. for her assistance in the Laboratory. We also thank the Japan Society for the Promotion of Science (JSPS) for partial funding of the publication fee.

References

1. Naidu K (2011) Characterization and purification of protease enzyme. J Appl Pharm Sci 1:107–112
2. Rezakhani N, Rad AM, Parivar K, Khayati M, Etemadzade S (2014) Immobilization of protease in biopolymers (mixture of alginate-chitosan). J Paramed Sci (JPS) 5:108–113
3. Tennalli G, Udapudi B, Naik P (2012) Isolation of proteolytic bacteria and characterization of their proteolytic activity. Int J Adv Eng Sci Technol (IJAEST) 2L185–195
4. Sawant R, Nagendran S (2014) Protease: an enzyme with multiple industrial applications. World J Pharm Pharm Sci 3:568–579
5. Afsharnezhad M, Shahangian S, Sariri R (2019) Protease: a novel milk-clotting cysteine protease from ficus jannis: purification and characterization. Int J Biol Macromol 121:173–182
6. Moran L, Horton H, Scrimgeour K, Perry M (2012) Principles of biochemistry, 5th edn. Prentice Hall, Pearson New Jersey, p 170
7. Singh R, Mittal A, Kumar M, Mehta PK (2016) Microbial proteases in commercial applications. J Pharm Chem Biol Sci 4:365–374
8. Thomas G, Neltner JD, Alger HM, James TO et al (2013) Conflicts of interest in approval of additives to food determined to be generally recognized as safe out of balance. Am Med Assoc 173:2032–3036
9. Suri WL, Syukur S, Jamsari (2013) Optimization of protease activity from lactic acid bacteria (LAB) *Pediococcus pentosaceus* isolated from soursop fermentation (*Annona muricata* L). Chemistry Department Univ Andls 2:18–25
10. Adi N, Guessas B (2016) Characterization of protease activity of *Lactococcus lactis* species Isolated from raw camel's milk. J Biol Sci 16:215–220
11. Budiarto B, Mustopa A, Idarmawan T (2016) Characterization of partially extracellular proteases from bekasam isolated *Lactobacillus plantarum* S31 and its application to hydrolyze skimmed-milk with antibacterial property. Int Food Res J 23:340–34
12. Villegas J, Brown E (2014) Characterization of the mature cell surface proteinase of Lactobacillus delbrueckii subsp. lactis CRL 58. Appl Microbiol Biotechnol 99:4277–4286
13. Sulthoniyah S, Hardoko TM, Nursyam H (2015) Characterization of extracellular protease lactic acid bacteria from shrimp paste. J Life Sci Biomed 5:01–05
14. Noohi N, Papizadeh M, Rohani M, Talebi M, Pourshafie MR (2021) Screening for probiotic characters in lactobacilli isolated from chickens revealed the intra-species diversity of *Lactobacillus brevis*. Anim Nutr 7:119–126
15. Oladejo TC, Olaniyi OO, Ayodeji AO, Akinyele BJ (2020) Protease produced by *Lactobacillus brevis* enhanced nutritional values of African yam beans and demonstrated improvement in the growth and blood indices of albino rats. Helliyon 6:e5123

16. Wang G-O, Jing-Pu, Yu X-K, Xia YJ, Ai L-Z (2020) Influence of freezing temperature before freeze drying on the viability of various *Lactobacillus plantarum* strains. J Dairy Sci 103:3066–3075
17. Public Health England (PHE) (2013) Inoculation of culture media for bacteriology. London, UK
18. Suberua Y, Akandea L, Samuela T, Lawalb A (2019) Optimization of protease production in indigenous Bacillus species isolated from soil samples in Lagos, Nigeria using response surface methodology. Biocatal Agric Biotechnol 18:1–13
19. Prasad MP, Rekha S, Arthirajan J (2014) Optimization studies of protease enzyme in in-vitro conditions from *Bacillus licheniformis*. Int Res J Biol Sci 3:34–39
20. Morado S, Jemei S, Hissel D, Gouriveau R, Zerhouni N (2015) ANOVA method applied to proton exchange membrane fuel cells aging forecasting using an echo state network. Math Comput Simul, 1–11
21. Santiago M, Sarmiento CAR, Zamora RA, Parra LP (2016) Discovery, molecular mechanisms, and industrial applications of cold-active enzymes. Front Microbiol 7:1–32
22. Adebayo FA, Afolabi OR, Akintokun AK (2014) Antimicrobial properties of purified bacteriocins produced from *Lactobacillus casei* and *Lactobacillus fermentum* against selected pathogenic microorganisms. Br J Med Med Res 4:3415–3431
23. Furhan J, Awasthi P, Sharma S (2019) Biochemical characterization and homology modeling of cold-active alkophilic protease from Northwestern Himalayas and its application indetergent industry. Biocatal Agrycurtural Biotechology 17:726–735
24. Mallesha SR, Selvakumar DJH (2010) Isolation and identification of lactic acid bacteria from raw and fermented products and their antibacterial activity. Rec Res Sci Technol 2:42–46
25. Wu R, Wu C, Liu D, Yang X et al (2017) Antioxidant and anti-freezing peptides from salmon collagen hydrolysate prepared by bacterial extracellular protease. Food Chem, 1–39
26. Mageswari A, Subramanian P, Chandrasekaran S et al (2017) Systematic functional analysis and application of a cold-active serine protease from a novel Chryseobacterium sp. Food Chem 217:18–27
27. Papadimitriou K, Alegria A, Bron PA et al (2016) Stress physiology of lactic acid bacteria. Microbiol Mol Biol Rev 80:837–890
28. Hayek S, Ibrahim S (2013) Current limitations and challenges with lactic acid bacteria: a review. Food Nutr Sci 4:73–87
29. Lario LD, Pillaca-Pulo OS, Sette LD, Converti A, Casati P, Spampinato C, Pessoa A (2020) Optimization of protease production and sequence analysis of the purified enzyme from the cold adapted yeast *Rhodotorula mucilaginosa* CBMAI 1528. Biotechnol Rep 28:e00546
30. Sukahara H, Nagayama K, Ikeda S, Hirota T, Nakamura Y (2021) D and L amino acid concentrations in culture broth of lactobacillus are highly dependent on the phylogenetic group of lactobacillus. Biochem Biophys Rep 27:101073

Formaldehyde Free Plywood Panel Bonded with Natural Rubber Latex Based Adhesives

Lisman Suryanegara, Muhammad Adly Rahandi Lubis, Yoana Tika Arpini, and Eko Setio Wibowo

Abstract A formaldehyde-free adhesive based on natural rubber latex (NRL) was prepared by mixing NRL (50% w/v) and poly(vinyl alcohol) (PVA, 10% w/v), which acted as adhesion promoter. Approximately 15% of polymeric 4,4- methylene diphenyl diisocyanate (pMDI) was used as cross-linker based on the adhesive's solids content and dextrin (20% w/v) was then added as filler. Basic and chemical properties of NRL-based adhesives were characterized using several techniques. The adhesion strength was examined in plywood using a universal testing machine. The non-volatile solids content, gelation time, and viscosity were suitable for making plywood. The results were comparable with neat pMDI conventional UF resins adhesives. Incorporation of pMDI and dextrin into NRL-PVA mixture altered the functional groups of the adhesives. Plywood bonded with NRL-based adhesives met the requirement of JAS No. 233:2003 with maximum moisture content of 14%. No alteration in density of plywood panels, indicating no influence of adhesive types on density of plywood. Plywood bonded with NRL-PVA/15% pMDI/20% dextrin had five times greater delamination compared to plywood bonded with NRL-PVA/15% pMDI owing to hydrolysis of dextrin. Dry tensile shear strength (TSS) of plywood bonded with NRL-based adhesives, neat pMDI and UF resins were higher than the requirement of JAS No. 233:2003 with minimum TSS value of 0.7 MPa. However, none of NRL-based adhesives could met the minimum requirement of wet TSS. This study showed that NRL-PVA/15% pMDI/20% dextrin could be used as formaldehyde free plywood adhesives for interior application and dry condition.

Keywords Dextrin · Formaldehyde free · Natural rubber latex · Plywood · pMDI

L. Suryanegara (✉) · M. A. R. Lubis · E. S. Wibowo
Research Center for Biomass and Bioproducts, National Research and Innovation Agency, Bogor, Indonesia
e-mail: l_suryanegara@yahoo.com

Y. T. Arpini
Department of Chemistry, IPB University, Bogor, Indonesia

M. A. Rahandi Lubis (ed.), *Proceedings of the 1st International Conference of Lignocellulose*, Green Energy and Technology,
https://doi.org/10.1007/978-981-19-1596-3_10

1 Introduction

Sustainability, performance and production cost of plywood panel is depending on wood adhesives and hot-pressing process. The most used wood adhesive to manufacture of plywood is urea-formaldehyde (UF) resins, calculating around 30–40 million metric tons are utilized globally in 2020 [1, 2]. UF resins adhesive is known as thermosetting resins which is inexpensive, has fast curing and a good adhesion strength [3]. However, it needs elevate temperature to form cross-linked thermosetting polymers and emits formaldehyde which is carcinogenic [4]. The sustainability in green manufacturing of plywood has become a concern since the 1990s, then the scientific publications with focus on topics such as life cycle assessment on adhesives production and hot-pressing process have been reported intensively [5, 6]. A way to achieve sustainability in green manufacturing of plywood is by using bio-based wood adhesives and cold-pressing process to manufacture the panel.

A bio-based wood adhesive, derived from natural rubber latex (NRL), is biopolymer harvested from rubber (*Hevea Brasiliensis* (Willd. ex A. Juss)) tree, consisting of repeated units of isoprene and has been used as adhesives due to its environmental friendliness and sustainability [7]. NRL has magnificent mechanical properties such as good static tensile strength, excellent elongation, elasticity, degradation stability and flexibility [8–10]. The properties of NRL-based adhesives can be improved by incorporating fillers, viscosity modifiers, tackifiers, adhesion promotors, and cross-linkers [8, 11, 12]. The lower cost and sustainability, which was around 1.4–2.0 USD/kg in 2021 [13], are the benefits of using NRL as an adhesive.

NRL have been widely used as an adhesive for non-polar materials, where the adhesion mechanism is built from strong mechanical strength of the materials. However, NRL does have sufficient adhesion for bonding high polarity and high surface energy materials such as wood [8–10]. Thus, the polar based adhesives are more suitable for wood based products. To solve this drawback, polymeric isocyanates can be added to functionalize the NRL as cold-setting adhesive. Polymeric isocyanates are high molecular weight resins derived from aliphatic and aromatic diisocyanates [14]. Polymeric isocyanates have been widely used as cross-linker for polymers bearing hydroxyl groups in two-component coatings, adhesives and sealants [15]. Water-based polymeric isocyanate adhesives are widely used in glued laminated timbers, laminated wood and plywood manufacturing. These adhesives are consisted of a water-based polymer bearing hydroxyl groups and polymeric isocyanates cross-linker [14, 16, 17]. Water-based polymeric isocyanate adhesives have shown excellent bond performance for decades.

However, incorporation of polymeric isocyanate results in highly viscosity adhesive which can reduce its storage time and limits the adhesive's application. Therefore, research on mixed of NRL with polymeric isocyanates and viscosity reducer is required for bonding wood. Incorporation of dextrin as adhesion promoter that provides hydroxyl groups into NRL-based adhesive system is necessary for cross-linking with polymeric isocyanates [18]. This study is aiming to formulate and evaluate the performance of NRL-based adhesive as cold-pressing adhesive for plywood.

Poly(vinyl alcohol) was specifically used as adhesion promoter to provide hydroxyl groups, polymeric 4,4-methylene diphenyl diisocyanate (pMDI) was selected as cross-linker, and dextrin was selected as filler and viscosity reducer for NRL-based adhesive.

2 Experimental Section

2.1 *Materials*

High ammonium natural rubber latex (60% of dry rubber content) was purchased from Garuda Indopacific Company (Bogor, West Java, Indonesia). Poly(vinyl alcohol) (PVA, ± 22.000 g mole^{-1}) was technical grade from PT. Brataco (Bogor, West Java, Indonesia) and polymeric 4-4 diphenylmethane diisocyanate (pMDI, ± 31% of NCO content) was purchased from Anugerah Raya Kencana Company (Tangerang, Banten, Indonesia). Technical grade of urea (99%) and formaldehyde (37%) was used to prepare UF resins according to the published works [19, 20]. Sengon (*Paraserianthes falcataria*) wood veneers, sizing 300 × 300 × 2 mm, was bought from local market (Cibinong, West Java, Indonesia) and was used to fabricate plywood.

2.2 *Formulation of Natural Rubber Latex (NRL) Adhesives*

In this study, neat pMDI was used as a control adhesive, which was labelled as A. Prior to the formulation of NRL-based adhesive, PVA solution was prepared by dissolving PVA beads with distilled water at 80 °C until reaching a concentration of 10% w/v [21, 22]. NRL then was mixed with PVA homogeneously at temperature of 27 ± 2 °C under stirring rate of 200 rpm with a ratio of 1:1 to promote adhesion. Approximately 15% of pMDI was added based on the NRL-PVA solids content before plywood manufacturing to enhance the adhesion, which was labelled as B. The mixing process was done at temperature of 27 ± 2 °C under stirring rate of 200 rpm for 20 min. Low viscosity of NRL-based adhesive was also prepared with addition of 20% dextrin, which was labelled as C. Convention plywood adhesive, namely UF resins, was also prepared according to published method and labelled as D. Illustration on the preparation of NRL-based adhesives is displayed in Fig. 1.

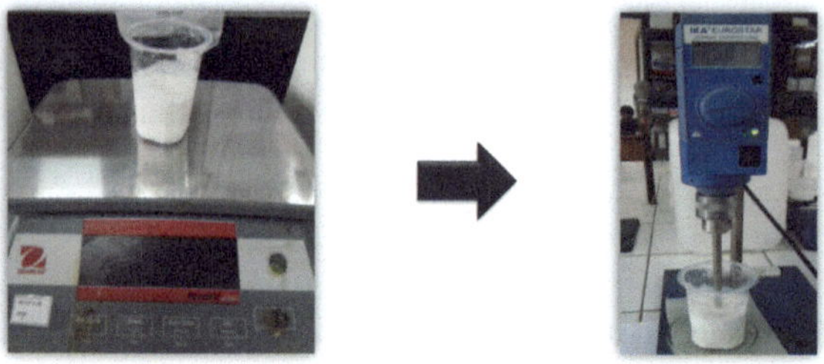

Fig. 1 Preparation of natural rubber latex adhesive

2.3 Investigation of NRL Adhesives Properties

Non-volatile solids content of NRL-based adhesives, neat pMDI and conventional UF resins was measured by drying the samples in an oven at 105 ± 2 °C for 3 h. The solids content of NRL-based adhesives was calculated by dividing the dry samples with wet samples. Average viscosity of NRL-based adhesives, neat pMDI and conventional UF resins were measured using a rotational rheometer (Rheolab QC, Anton Paar, Austria). Around 20 mL of NRL-based adhesives was poured in a measuring cup. The viscosity measurement was done using a cylinder concentric No. 27 under constant speed of 100 rpm and at 27 ± 2 °C. Gelation time of NRL-based adhesives and neat pMDI was determined using a gel time meter (Techne GT-6, Coleparmer, United States) at 27 ± 2 °C using spindle No. 27. In case of UF resins, the gelation time was measured at 103 ± 2 °C.

Functional groups NRL-based adhesives were investigated using UATR–FTIR spectrometer (Spectrum Two, Perkin Elmer, United States) in the range of 400–4000 cm^{-1} with a resolution of 2 cm^{-1}. An average of 16 scans were used for scanning the background and sample. The spectrum of NRL-based adhesive was then normalized using a min–max normalization.

2.4 Preparation and Evaluation of Plywood Bonded NRL Adhesives

Plywood samples were fabricated using Sengon wood veneers bonded NRL-based adhesives as depicted in Fig. 2. Three-layer plywood with a target size of 300 × 300 × 6 mm was fabricated by spreading the glue mixture on the veneer surface with a glue spread of 200 g/m^2. Plywood were cold pressed at 27 ± 2 °C under an average pressure of 5.0 MPa for 24 h. Subsequently, all samples were conditioned at 27 ± 2 °C for two weeks before cutting and testing. As a control, plywood bonded with pMDI with pMDI and urea-formaldehyde (UF) resins were also prepared according to the published works [19, 20].

Properties of plywood bonded with NRL-based adhesives, pMDI and UF resins were evaluated according to Japanese Agriculture Standard (JAS) No. 233:2003

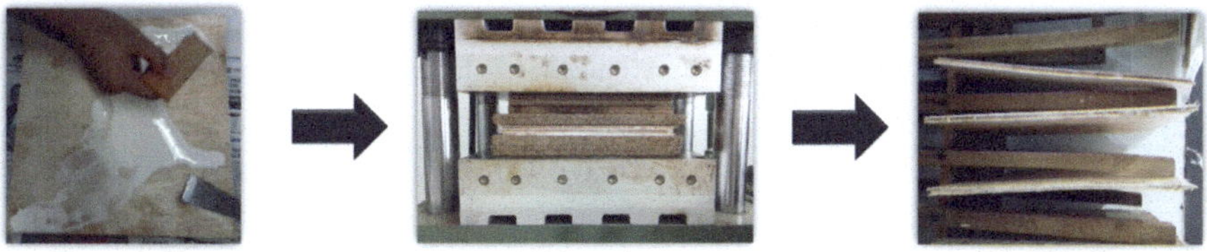

Fig. 2 Fabrication of plywood bonded with natural rubber latex adhesive

[23]. Plywood samples with a size of 75 × 75 × 6 mm were used for density, moisture content (MC) and delamination measurement. Briefly, the density of plywood was determined by comparing the mass and volume of plywood, while the MC was obtained by dividing the initial and final mass after drying at 103 ± 2 °C for 24 h. Meanwhile, delamination test was determined according to type I and type II methods. The percent delamination was obtained by dividing total length of delaminated lines and total length of glue lines. The tensile shear strength (TSS) measurement of plywood samples were determined according to a standard procedure [23]. The TSS values of ten specimens, measuring 80 × 25 × 6 mm, was measured using a universal testing machine (UTM 50kN, Shimadzu, Japan) at a crosshead speed of 2 mm/min. Dry TSS values was obtained by directly testing the samples after conditioning, while wet TSS was tested after immersing the samples in hot and cold water.

3 Results and Discussion

Basic properties of NRL-based adhesives, neat pMDI and UF resins are summarized in Table 1. The non-volatile solids content of NRL-based adhesives with 15% pMDI was around 41.5%. The solids content of NRL-based adhesives increased to 46.7% with addition of 20% dextrin. This value was lower than the neat pMDI and UF resins. Viscosity of NRL-based adhesives with 15% pMDI was around 907 mPa.s, which was in agreement with the published work, showing a range of viscosity of NRL-based adhesives around 900–1610 mPa.s [21]. Incorporation of 20% dextrin into NRL-PVA/15%pMDI decreased the viscosity to 230 mPa.s, indicating that the water soluble dextrin contained many water thus reduced the viscosity of NRL-based adhesives. The average viscosity value of those NRL-based adhesives were lower than the neat pMDI but higher that the conventional UF resins. The gelation time of both NRL-based adhesives with 15% pMDI and 20% dextrin was longer than the the neat pMDI and UF resins.

The change in viscosity of NRL based adhesives was monitored using a rotational rheometer at different times at room temperature as depicted in Fig. 3. Initial viscosity of NRL-based adhesives with 15% pMDI and 20% dextrin were lower than the neat pMDI, but higher than the conventional UF resins. The viscosity increased as a

Table 1 Basic properties of natural rubber latex based adhesives

Type of adhesive	Solids content (%)	Average viscosity (mPa·s)	Gelation time (min)
A*	41.5 ± 0.8	907.5 ± 48.4	102 ± 0.4
B	46.7 ± 1.0	330.0 ± 7.0	158 ± 1.5
C	98.6 ± 0.9	2272.5 ± 67.2	15.5 ± 0.5
D	60.5 ± 0.8	250.0 ± 7.0	4 ± 0.5

* A: NRL-PVA/15% pMDI, B: NRL-PVA/15% pMDI/20% dextrin, C: pMDI, D: UF

function of time for NRL-based adhesives with 15% pMDI, mixed with 20% dextrin, and neat pMDI. But no alteration was detected in viscosity of UF resins, showing no reaction occurred in UF resins system at room temperature. This indicated that addition of pMDI into NRL-PVA built up higher viscosity by forming cross-linked structure. The –NCO groups of pMDI is known with its high reactivity to –OH groups of NRL-PVA and moisture [21]. The reaction of –NCO and –OH groups resulted in urethane linkages [14].

The change in functional groups of NRL-based adhesives was recorded using FTIR analysis (Fig. 4). The neat pMDI has specific functional groups of –NCO at wavenumber of 2272 cm^{-1}. In addition, pMDI showed typical peak of C=O of –NCO, $-CH_2$ of pMDI and C=C of aromatic pMDI at wavenumber of 1728 cm^{-1}, 2954 cm^{-1} and 1520 cm^{-1}, respectively. These peaks were also found in the published works [14, 24, 25]. Incorporation of NRL-PVA and 15% pMDI in NRL-based adhesive altered the functional groups as depicted by the black line of FTIR spectrum. The peaks of $-CH_3$ and $-CH_2$ of cis-1,4-polyisoprene were detected at wavenumber of 2961 cm^{-1} and 2852 cm^{-1}, respectively. This results were in agreement with the published works [26, 27]. The free –NCO and urethane (R-NH-C=O-R) groups were also identified in NRL-based adhesives mixed with 15% pMDI at wavenumber of 2272, 3326, and 1660 cm^{-1}. Addition of 20% dextrin into NRL-based adhesives mixed with 15% pMDI slightly changed the functional groups as depicted by the green line of FTIR

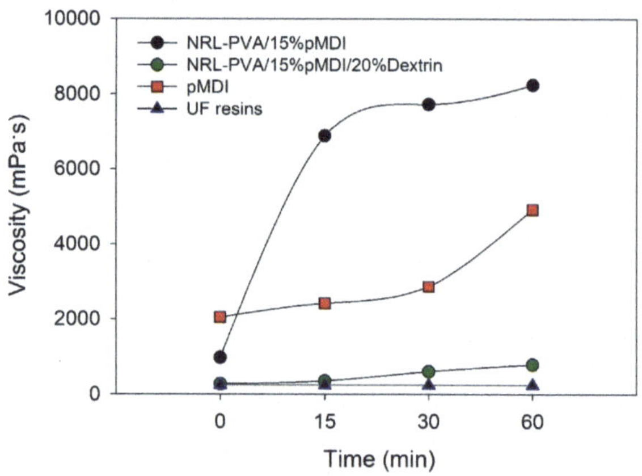

Fig. 3 Viscosity of natural rubber latex based adhesive as a function of time

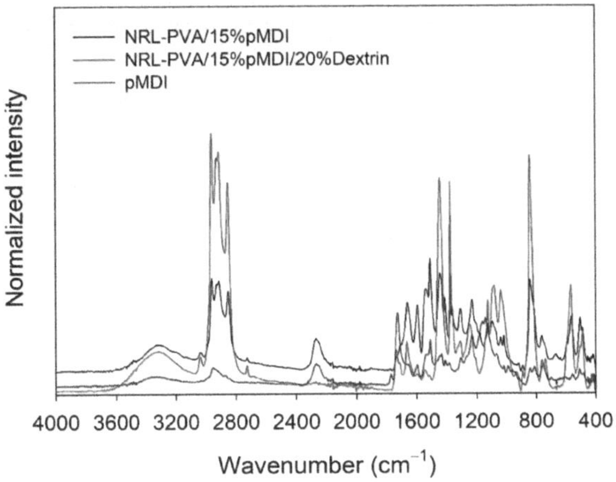

Fig. 4 FTIR spectra of natural rubber latex based adhesive

spectrum. The free NCO groups at 2272 cm^{-1} did not detected in the spectrum, showing that the –NCO has been completely reacted with –OH of NRL-PVA and dextrin. As a result, two new peaks appeared at 1737 and 1663 cm^{-1} attributed to urethane (R-NH-C=O-R) groups. This results were in agreement with the published works, demonstrating the reaction of –NCO of pMDI with –OH of NRL and wood [14, 16].

Physical properties of plywood bonded with NRL-based adhesives, neat pMDI and conventional UF resins are presented in Table 2. Moisture content (MC) of plywood bonded with NRL-based adhesives was greater than those bonded with neat pMDI and UF resins. Plywood bonded with NRL-PVA/15% pMDI has around 10.3% of MC, while addition of 20% into NRL-PVA/15% pMDI increased slightly the MC of plywood to 10.8%. By contrast, plywood panels bonded with neat pMDI and UF resins had MC of around 7.1% and 8.3%, respectively. The results showed that all plywood panels met the requirement of JAS with maximum MC value of 14% [23]. No alteration in density of plywood panels, with value of around 0.45 g/m^3, showing no influence of adhesive types on density of plywood. Significant differences were found in delamination of plywood. Both NRL-based adhesives mixed with 15% pMDI and 20% dextrin had greater delamination value compared to neat pMDI and UF resins bonded panels. Plywood bonded with NRL-PVA/15% pMDI/20% dextrin had five times greater delamination compared to plywood bonded with NRL-PVA/15% pMDI. This could be due to that dextrin is soluble to water, therefore some of the adhesive parts containing dextrin was hydrolyzed during delamination test, resulting in higher delamination value [18, 28]. By contrast, no delamination occurred in plywood bonded with neat pMDI, and small delamination for UF bonded panel.

Mechanical properties of plywood by means of tensile shear strength (TSS) was evaluated by comparing dry and wet TSS of plywood (Fig. 5). In general, dry TSS values of plywood bonded with NRL-based adhesives, neat pMDI and UF resins were higher than the requirement of JAS with minimum value of 0.7 MPa [23]. The highest dry TSS value was obtained in plywood bonded with NRL-PVA/15% pMDI/20%

Table 2 Physical properties of plywood bonded natural rubber latex based adhesive

Type of adhesive	Moisture content (%)	Density (g/m^3)	Delamination	
			Type I (%)	Type II (%)
A*	10.3 ± 0.02	0.45 ± 0.04	10.63 ± 0.07	2.50 ± 0.03
B	10.8 ± 0.05	0.45 ± 0.03	54.97 ± 0.11	7.53 ± 0.07
C	7.1 ± 0.02	0.45 ± 0.04	0.00 ± 0.00	0.0 ± 0.00
D	8.3 ± 0.02	0.45 ± 0.04	2.50 ± 0.05	0.47 ± 0.03

* A: NRL-PVA/15% pMDI, B: NRL-PVA/15% pMDI/20% dextrin, C: pMDI, D: UF

dextrin of around 1.47 MPa, better than that of bonded with neat pMDI and UF resins. The lowest dry TSS was around 0.85 MPa for plywood bonded with NRL-PVA/15% pMDI. This results indicated that the addition of dextrin remarkably increased dry TSS values of plywood. This could be due to the formation of cross-linked structure of NRL-PVA/15% pMDI/20% dextrin was greater that that of NRL-based adhesives without addition of dextrin. As explained by FTIR result (Fig. 4), the free NCO groups at 2272 cm^{-1} did not detected in the spectrum, showing that the –NCO has been completely reacted with –OH of NRL-PVA and dextrin. As can be seen in Fig. 5, the wet TSS of plywood bonded with NRL-based adhesives remarkably decreased, indicating that hydrolysis could deterior the bonding of NRL-based adhesives. Only plywood bonded with neat pMDI could met the requirement of JAS with minimum value of 0.7 MPa [23]. pMDI is a kind of thermosetting polymers that has a high water resistance properties owing to the presence of –NCO groups that can form a cross-linking with different functional groups [14, 16, 17].

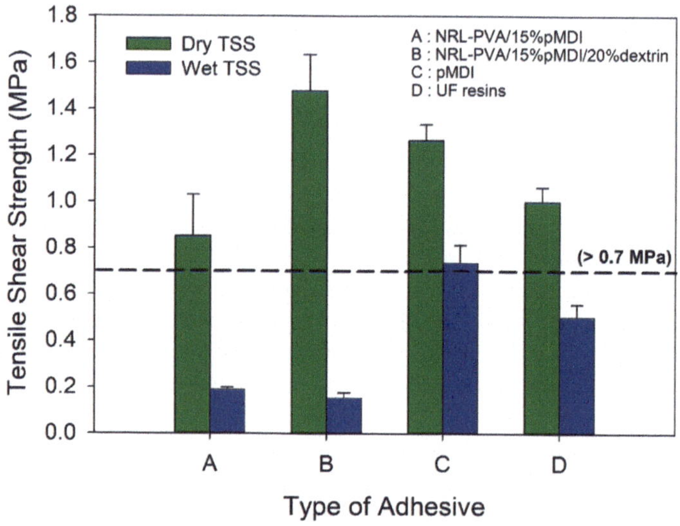

Fig. 5 Tensile shear strength (TSS) of plywood bonded with natural rubber latex

4 Conclusion

This study showed that a formaldehyde free adhesive for plywood could be prepared from natural rubber latex (NRL) mixed with poly(vinyl alcohol), polymeric isocyanate (pMDI) and dextrin. Basic properties of NRL-based adhesives such as non-volatile solids content, gelation time, and viscosity were suitable for making plywood. The results were comparable with neat pMDI conventional UF resins adhesives. Incorporation of pMDI and dextrin into NRL-PVA mixture altered the functional groups of the adhesives. The free –NCO and urethane (R-NH-C=O-R) groups were identified in NRL-based adhesives mixed with 15% pMDI. Further, addition of 20% dextrin into NRL-PVA/15% pMDI resulted in complete reaction of –NCO of pMDI with –OH of NRL-PVA and dextrin. Plywood bonded with NRL-based adhesives met the requirement of JAS with maximum moisture content of 14%. No alteration in density of plywood panels, indicationg no influence of adhesive types on density of plywood. Plywood bonded with NRL-PVA/15% pMDI/20% dextrin had five times greater delamination compared to plywood bonded with NRL-PVA/15% pMDI owing to hydrolysis of dextrin. By contrast, no delamination occurred in plywood bonded with neat pMDI, and small delamination for UF bonded panel. Dry tensile shear strength (TSS) of plywood bonded with NRL-based adhesives, neat pMDI and UF resins were higher than the requirement of JAS with minimum value of 0.7 MPa. However, none of NRL-based adhesives could met the minimum requirement of wet TSS. This study suggested that NRL-PVA/15% pMDI/20% dextrin could be used as formaldehyde free plywood adhesives for interior application and dry condition.

Acknowledgements This work was supported by integrated laboratory of bio-products (iLaB), Research Center for Biomaterials, National Research and Innovation Agency. This study was also supported the Japan-ASEAN Science and Technology Innovation Platform (JASTIP) for FY 2020-2021 with the title "Development of High Performance of Bio-Based Adhesives for Durable Wood-Based Composites". Authors are also grateful for the publication support from Japan Society for the Promotion of Science (JSPS) and JSPS Alumnae Association Indonesia (JAAI).

References

1. FAO (2019) Forest products 2019. Rome
2. IHS Markit (2020) Amino resins—chemical economics handbook, p 124
3. Lubis MAR, Park B-D (2020) Influence of initial molar ratios on the performance of low molar ratio urea-formaldehyde resin adhesives. J Korean Wood Sci Technol 48:1–18
4. Lubis MAR, Park B-D (2021) Enhancing the performance of low molar ratio urea–formaldehyde resin adhesives via in-situ modification with intercalated nanoclay. J Adhes 97:1271–1290
5. Liu M, Wang Y, Wu Y, He Z, Wan H (2018) "Greener" adhesives composed of urea-formaldehyde resin and cottonseed meal for wood-based composites. J Clean Prod 187:361–371
6. Jia L, Chu J, Ma L, Qi X, Kumar A (2019) Life cycle assessment of plywood manufacturing process in China. Int J Environ Res Public Health. https://doi.org/10.3390/ijerph16112037

7. Shybi AA, Varghese S, Thomas S (2021) Natural rubber latex-based adhesives: role of nanofillers. J Adhes Sci Technol 35:406–418
8. Radabutra S, Khemthong P, Saengsuwan S, Sangya S (2020) Preparation and characterization of natural rubber bio-based wood adhesive: effect of total solid content, viscosity, and storage time. Polym Bull 77:2737–2747
9. Hermiati E, Fatriasari W, Risanto L, Darmawan T, Sari FP, Jayadi S, Laksana RPB, Akbar F (2019) Optimization of application of natural rubber based API adhesive for the production of laminated wood. IOP Conf Ser Earth Environ Sci 374:012007
10. Agrebi F, Ghorbel N, Bresson S, Abbas O, Kallel A (2019) Study of nanocomposites based on cellulose nanoparticles and natural rubber latex by ATR/FTIR spectroscopy: the impact of reinforcement. Polym Compos 40:2076–2087
11. Öter M, Karaağaç B (2020) Epoxidised natural rubber as adhesion promoter in natural rubber based compounds. J Rubber Res 23:333–341
12. Zhang Y, Cunningham MF, Dubé MA (2020) Modification of adhesive and latex properties for starch nanoparticle-based pressure sensitive adhesives. Macromol React Eng. https://doi.org/10.1002/mren.201900023
13. Statista (2020) Rubber price monthly 2021|Statista
14. Lubis MAR, Park BD, Lee SM (2020) Microencapsulation of polymeric isocyanate for the modification of urea-formaldehyde resins. Int J Adhes Adhes 100:102599
15. Parodi F (2012) Isocyanate-derived Polymers. In: Comprehensive polymer science, pp 387–412
16. Lubis MAR, Suryanegara L, Pramesti MA (2020) Synthesis of polymeric isocyanate microcapsules via interfacial polymerization and their characterisation using spectroscopy and microscopy techniques. IOP Conf Ser Mater Sci Eng 935:012053
17. Nuryawan A, Alamsyah EM (2018) A review of isocyanate wood adhesive: a case study in Indonesia. In: Applied adhesive bonding in science and technology. InTech, p 13
18. Alamsyah EM, Sutrisno NA, Widyorini R (2020) Identifying best parameters of particleboard bonded with dextrin-based adhesives. Open Agric 5:345–351
19. Wibowo ES, Lubis MAR, Park BD (2021) In-situ modification of low molar ratio urea–formaldehyde resins with cellulose nanofibrils for plywood. J Adhes Sci Technol 35:1–14
20. Yadav SM, Lubis MAR, Park B-D (2021) Modification of nanoclay with different methods and its application in urea-formaldehyde bonded plywood panels. Wood Mater Sci Eng 17:1–10
21. Lubis MAR, Sari FP, Laksana RPB, Fatriasari W, Hermiati E (2021) Ambient curable natural rubber latex adhesive cross-linked with polymeric isocyanate for bonding wood. Polym Bull. https://doi.org/10.1007/s00289-021-03845-0
22. Lubis MAR, Falah F, Harini D, Sudarmanto KA, Tjahyono B, Fatriasari W, Subiyanto B, Suryanegara L, Iswanto AH (2021) Enhancing the performance of natural rubber latex with polymeric isocyanate as cold-pressing and formaldehyde free adhesive for plywood. J Adhes 1:1–16
23. Japanese Agricultural Standard (JAS) for Plywood (2003)
24. Lubis MAR, Park B-D, Lee S-M (2017) Modification of urea-formaldehyde resin adhesives with blocked isocyanates using sodium bisulfite. Int J Adhes Adhes 73:118–124
25. Lubis MAR, Park B-D, Lee S-M (2019) Performance of hybrid adhesives of blocked-pMDI/melamine-urea-formaldehyde resins for the surface lamination on plywood. J Korean Wood Sci Technol 47:200–209
26. Ninjan R, Thongnuanchan B, Lopattananon N, Thitithammawong A, Nakason C (2019) Ambient curable latex films and adhesives based on natural rubber bearing Acetoacetoxy functionality. Polym Adv Technol 30:598–607
27. Rolere S, Liengprayoon S, Vaysse L, Sainte-Beuve J, Bonfils F (2015) Investigating natural rubber composition with Fourier Transform Infrared (FT-IR) spectroscopy: a rapid and non-destructive method to determine both protein and lipid contents simultaneously. Polym Test 43:83–93
28. Daud S, Afiq MM, Azura AR (2018) Studies on the influenced of dextrin filler on the mechanical properties of carboxylated nitrile butadiene rubber (Xnbr) films. J Phys Conf Ser. https://doi.org/10.1088/1742-6596/1082/1/012013

Synthesis of Polyacrylamide-Grafted Cassava Starch Hydrogel by Microwave-Assisted Method

Maudy Pratiwi Novia Matovanni, Sperisa Distantina, and Mujtahid Kaavessina

Abstract The grafting of natural and synthetic polymers to form hydrogels can be done using a microwave-assisted process. It is faster, more dependable, accurate, and is able to produce product with a higher quality than the conventional approach. In this study, microwave-assisted method was applied to make polyacrylamide-grafted cassava starch (CSt-g-PAM) hydrogels using free radical potassium peroxodisulfate (KPS) as the initiator. The influence of acrylamide as the monomer and KPS as the initiator on the characteristics of the resulting hydrogel was investigated. Chemical structure of the hydrogel was studied using FTIR spectroscopy. The FTIR spectra of the CSt-g-PAM showed additional peaks indicating the presence of C=O, N–H, and C–N stretching vibrations, which confirmed the successful grafting of the polyacrylamide chains onto the backbone of the cassava starch. The absorption ability of CSt-g-PAM was evaluated using the swelling test in distilled and saline water. In distilled water, the increasing concentration of the acrylamide monomer in the reaction system improved the swelling ratio of cassava starch from 2.97 to 15.04 g/g. The improved concentration of the initiator KPS also improved the swelling ratio of the resulting polyacrylamide-grafted cassava starch from 5.17 to 10.42 g/g. The results of swelling test for CSt-g-PAM performed in the saline water provided significantly lower value than that of performed in distilled water, with the highest swelling ratio obtained of 10.57 g/g.

Keywords Biopolymer · Microwave-assisted · Monomer · Initiator · Swelling ratio

M. P. N. Matovanni · S. Distantina (✉) · M. Kaavessina
Department of Chemical Engineering, Universitas Sebelas Maret, Surakarta, Indonesia
e-mail: sperisa_distantina@staff.uns.ac.id

M. A. Rahandi Lubis (ed.), *Proceedings of the 1st International Conference of Lignocellulose*, Green Energy and Technology,
https://doi.org/10.1007/978-981-19-1596-3_11

1 Introduction

Hydrogel is a macromolecule that can swell and absorb solvents in water or biological liquids. The polymeric structure of hydrogel consists of hydroxyl (OH), carboxylic acid (COOH), amide ($CONH_2$) and sulfonic acid (SO_3H) groups, which makes it soft, hydrophilic, porous and elastic [1]. It is well known that the presence of hydrophilic groups and the flexibility of the polymer chains increase the swelling capacity of hydrogels. Hydrogels can be synthesized from natural or synthetic polymers or a combination of them. The advantages of hydrogels made from natural materials are inexpensive, readily available, biocompatible and biodegradable (environmentally friendly), but they have low mechanical properties, cannot stand high temperatures, and have a complex structure [2]. In the other hand, synthetic polymers provide greater flexibility in controlling the chemical structure of the polymers obtained, increasing the strength and absorption capacity of the hydrogels. However the use of synthetic polymers has less impact on the environment because they provide toxicity and carcinogenicity coming from the residual monomers of the hydrogels [3, 4]. Therefore, hydrogel synthesis from mixture of natural and synthetic polymers can be one of the choices to obtain eco-friendly and inexpensive superabsorbent polymers with excellent mechanical strength.

Hydrogel synthesis from a mixture of natural and synthetic polymers can be achieved by combining copolymerization grafting, physical crosslinking, and chemical crosslinking [5]. Existing redox grafting techniques, microwave irradiation, gamma-ray irradiation, and electron beams are often used to achieved grafted polymers. Microwave irradiation of any of these methods has the greatest potential for the synthesis of graft polymers. This is because the microwave energy can be used to selectively heat materials almost instantly. In addition, the use of microwaves increases the formation of copolymers, resulting in higher polymer products with very short time required to polymer formation compared to the conventional heating methods. To control the percentage of grafting, microwave power and irradiation exposure time are electronically controlled [6].

Previous studies have reported the synthesis of hydrogel from cassava starch-poly(acrylate acid)-poly(vinyl alcohol) by conventional polymerization methods using potassium peroxodisulfate (KPS) as initiator [7], cassava starch-polyacrylamide by conventional polymerization methods using ceric ammonium nitrate (CAN) as initiator and *N,N'*-methylene bis-acrylamide (MBA) as crosslinker [2], acrylate acid-dimethyl aminopropyl metacrylamide (DMAPMA)-cassava starch hydrogel by conventional methods using KPS as initiator and MBA as crosslinker [8]. The ratio of cassava starch to monomer and monomer concentration can affect its absorption to water or saltwater. Previous studies showed that the synthesis of polyacrylamide-grafted cassava starch hydrogel was successfully carried out using potassium persulfate (KPS) as a free radical initiator. In this study, grafting was performed by utilizing energy from the domestic microwave that has, to the best of our knowledge, never been reported previously.

In this study reported here, we investigated the effect of acrylamide and KPS amounts to the properties of the hydrogel obtained. The chemical structure of the hydrogel was studied using FTIR spectra. The absorption ability of the grafted starch was evaluated using the swelling test in distilled and saline water.

2 Materials and Method

2.1 Materials

Cassava starch was obtained from a local market in Surakarta, Indonesia. Acrylamide, sodium chloride, and potassium peroxodilsulfate (KPS) were purchased from E. Merck, Germany. Acetone was purchased from Saba Kimia, Surakarta, Indonesia. All the chemicals and reagents were used without additional treatment. All mixtures were formulated using distilled water.

2.2 Synthesis of Polyacrylamide-Grafted Cassava Starch

In 50 mL of distilled water, 1 g of cassava starch was added. The mixture was stirred with the addition of a certain amount of acrylamide and potassium peroxodilsulfate (KPS). The mixture was well stirred before being placed to the reaction vessel (1000 mL beaker). The reaction vessel was put on the turntable of a domestic microwave oven (Krisbow 23 L) at 250 W for the microwave-assisted grafting procedure. Microwave irradiation of the reaction mixture was paused before boiling of the reaction solvent (about 65 °C), and the reaction mixture was cooled by soaking the reaction vessel in cold water. The cycle of microwave irradiation followed by cooling was continued until the viscosity structure was obtained. After the microwave irradiation process, the reaction mixture was cooled and left for 24 h to complete the grafting reaction. The leftover viscous material in the reaction vessel was soaked in excess acetone until it precipitated. The polyacrylamide-grafted cassava starch precipitate was dried in the oven at 50 °C for 24 h. Equation (1) was used to calculate the percentage of grafting of the polyacrylamide-grafted cassava starch obtained.

$$\%\text{grafting} = \frac{\text{mass of graft copolymer} - \text{mass of cassava starch}}{\text{mass of cassava starch}} \times 100\% \qquad (1)$$

The details of acrylamide and KPS amounts for the synthesis of polyacrylamide-grafted cassava starch is shown in Table 1.

Table 1 Various initial compositions and the resulting of polyacrylamide-grafted cassava starch (CSt-g-PAM)

Grades	Mass of starch (g)	Mass of acrylamide (g)	Mass of KPS (g)	Irradiation time to gel formed (s)	%grafting
CSt-g-PAM 1	1	2	0.3	210	284.96
CSt-g-PAM 2	1	5	0.3	60	634.16
CSt-g-PAM 3	1	10	0.3	30	1598.72
CSt-g-PAM 4	1	5	0.1	150	348.90
CSt-g-PAM 5	1	5	0.2	90	564.41

2.3 Characterization

FT-IR Spectroscopy

Fourier-transform infrared (FT-IR) spectra of cassava starch, CSt-g-PAM 1, CSt-g-PAM 2, and CSt-g-PAM 3 were analyzed on a Frontier FT-IR/FIR spectrophotometer (Shimadzu, IRSpirit). The polyacrylamide-grafted cassava starch obtained from various conditions were analyzed directly without forming the pellets with KBr.

Swelling Measurements

Dried polyacrylamide-grafted cassava starch was weighted and recorded as W_1 (gram). The swelling test was determined by soaking W_1 in 100 mL of distilled or saline water (NaCl solution) at 4000 ppm at 25 °C until the swelling equilibrium was established. Swollen polyacrylamide-grafted cassava starch was measured as W_2 (gram). The swelling ratio was calculated using Eq. (2):

$$S = \frac{W_2 - W_1}{W_1} \quad (2)$$

3 Results and Discussion

3.1 Synthesis of Polyacrylamide-Grafted Cassava Starch Using Microwave-Assisted Method

Microwave-assisted method was used to produce polyacrylamide-grafted cassava starch. Various polyacrylamide-grafted cassava starch have been produced by changing the amount of acrylamide and potassium peroxodisulfate (KPS). This method was based on free radical mechanism using microwave irradiation and potassium peroxodisulfate as the initiator to create free radicals on the starch backbone.

Initiation

S- OH + M $\xrightarrow{\text{Microwave irradiation}}$ S-OH* + M*

Propagation

S-OH* + M → S-OHM*

S-OHM* + M → S-OHMM*

S-OHMM$_{n-1}$ + M → S-OHM*$_n$

Termination

S-OHM*n + S-OHM*n → grafted copolymer

Homopolymer Formation (Side reaction)

M* + M → MM*

M*$_{n-1}$ + M → M*$_n$

M$_n$ + S-OHM → S-OH* + M$_n$H

S : Starch

M: Monomer (Acrylamide)

* : free radical

n : sum of acrylamide monomer

Fig. 1 The plausible mechanism of the synthesis of polyacrylamide-grafted cassava starch using microwave-assisted method [11]

Mechanism of polyacrylamide-grafted cassava starch using microwave-assisted method is shown in Fig. 1.

The microwave irradiation—cooling cycle of the reaction mixture was periodically repeated until viscous mass was achieved. The details of polyacrylamide-grafted cassava starch synthesis are shown in Table 1. The optimum result is defined by the highest value of the grafting percentage. From Table 1, the reaction mixture with microwave power maintained at 250 W demonstrated the highest grafting percentage at the amount of acrylamide and KPS of 10 and 0.3 g, respectively.

The initiation phase occurred when a certain amount of initiator was added and the reaction mixture was irradiated on a turntable of the domestic microwave oven. The formation of complexes between the hydroxyl groups of the polysaccharides and oxidants occurred due to the presence of KPS which generates free radicals. Microwave irradiation increased the spin frequency of the acrylamide molecules, which resulted in the chemical bonds within the molecules to stretch. The electron cloud of the pi bond, where the C–C double bond would be extended, split into two local clouds (i.e. free radical sites of the constituent carbon atoms). When free radical was formed on the cassava starch (i.e. the backbone polymers), monomers were reacted by chains propagation until the free radical was inactivated through the chain termination step. Microwave irradiation-cooling cycle was performed to minimize the competitive reaction of homopolymer formation and also to prevent the formation of water vapor, which might contain acrylamide. Termination step occurred when excess acetone was added until the precipitate of polyacrylamide-grafted cassava starch was formed. Here, unlike thermal energy and high energy irradiation, microwaves selectively excited polar bonds only [9, 10]. Polyacrylamide-grafted cassava starch was synthesized by the interaction between the free radical generated in the backbone of cassava starch (by KPS) and the free radical generated in acrylamide (by microwave irradiation) through free radical reaction mechanism.

Effect of Monomer Concentration

The grafting percentage increased as the monomer concentration increased. The irradiation time required for the gel to formed decreased as the monomer concentration increased. The grafting percentage increased from 284.96 to 1598.72% as the content of acrylamide monomer increased from 2 to 10 g. The irradiation time required for the gel formation decreased from 210 to 30 s as the content of the acrylamide monomer increased from 2 to 10 g. The increase of grafting percentage and the decrease of irradiation time for the gel formation might be presumably due to the presence of more monomer molecules in the reaction mixture for the chain propagation. The highest grafting percentage in the synthesis of polyacrylamide-grafted cassava starch in this study was achieved when the ratio of cassava starch to acrylamide was 1:10. Furthermore, at this ratio, the shortest irradiation time was also achieved.

Effect of Initiator Concentration

The amount of the initiator KPS was adjusted from 0.1 to 0.3 g for the grafting reaction. As the initiator increased, the grafting percentage increased as well. The grafting percentage increased from 348.90 to 634.16% as the KPS content increased from 0.1 to 0.3 g. The required irradiation time for the gel formation decreased from 150 to 60 s as the KPS content increased from 0.1 to 0.3 g. This observation indicated that as the initiator concentration increased, the required time for the gel to generation decreased. The improved availability of the free radical generated in the backbone of the cassava starch by the initiator might explain the increase in grafting percentage and hence lower the required irradiation time for gel formation. The highest value of grafting percentage was 634.16% and the shortest irradiation time for the formation of gel-like mass was achieved when the amount of initiator was set at 0.3 g.

3.2 FTIR Spectroscopy

Figure 2 presents the FTIR spectra of the cassava starch (CSt) and the polyacrylamide-grafted cassava starch (Cst-g-PAM) obtained in various conditions. The FTIR spectrum of CSt showed absorption peaks at 3289.02 cm^{-1} representing to the O–H strain vibration, 2945.29 cm^{-1} representing to the C–H strain vibration, and 1155.29 cm^{-1}, 1082.55 cm^{-1}, 1012.67 cm^{-1} indicating the strain vibrations of the C–O–C bond.

All the grafted products displayed an identical profile on the FTIR spectra. Hence, the discussion was made for sample CSt-g-PAM 1 only as the representative for the whole grafted samples. In sample CSt-g-PAM 1, the strain vibration of both O–H group of the starch and N–H groups of the PAM overlapped, and resulted in

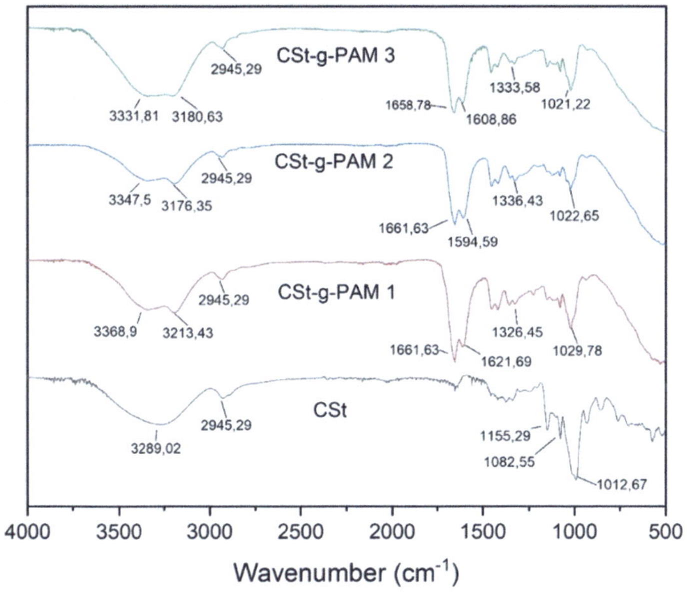

Fig. 2 FTIR spectra of cassava starch before and after grafting with acrylamide monomer under microwave irradiation

a peak at 3368.9 cm^{-1} and a shouldered peak at 3213.43 cm^{-1}. A small peak at 2945.29 cm^{-1} related to the C–H strain vibration. The absorption band at 1029.78 cm^{-1} corresponded to the C–O–C strain vibration. The presence of a sharp peak at 1661.63 cm^{-1} is related to the stretching of C=O bond and the peak at 1621.69 cm^{-1} related to the N–H bond (i.e. stretching of amide I and amide II). Furthermore, there was another additional peak in the grafted product, at 1326.45 cm^{-1}, signifying the stretching of C–N bond.

The presence of the C=O, N–H, and C–N stretching vibrations in the $-CONH_2$, detailed above indicated the success of the grafting process. The FTIR data displayed in Fig. 1 showed that the graft copolymer was produced by interactions between the free radical generated in the backbone of cassava starch and acrylamide through the free radical reaction mechanism. The relevant literature data reported that the stretching of the C=O bond corresponded to the peak at 1791.87 cm^{-1}, and the same vibration for N–H and C–N bonds corresponded to the peaks centered at 1504.48 at 1334.74 cm^{-1}, respectively [9]. These peaks appeared in the FTIR spectra of CSt-g-PAM 1, CSt-g-PAM 2, and CSt-g-PAM 3. Thus, we concluded that the polyacrylamide chain was successfully grafted onto the backbone of the cassava starch.

3.3 Swelling Behavior

Figure 3 presents the equilibrated swelling ratio of cassava starch (CSt) and polyacrylamide-grafted cassava starch (Cst-g-PAM) in distilled and saline water influenced by monomer concentration. The amount of monomer in the reaction system clearly affected the equilibrated swelling ratio of the samples (after 4 days) both in distilled and saline water. The swelling equilibrium was achieved when the swelling ratio reached a constant value. The swelling ratio of the grafted samples in distilled and saline increased from 2.97 to 15.04 g/g and 2.59 to 10.57 g/g, respectively, as the amount of monomer applied increased from 0 to 10 g as compared to the untreated one. This observation could be explained by improved hydrophilic properties of polyacrylamide-grafted cassava starch with increased monomer concentrations. When monomer concentration increased, the hydrophilic nature of polyacrylamide-grafted cassava starch significantly improved. In both distilled water and saline solution, the swelling ratio of polyacrylamide-grafted cassava starch was higher than that of the original cassava starch. In conclusion, the swelling ratio was improved with the increased the grafting percentage [12].

The swelling test in distilled and saline water (NaCl solution) was conducted to determine the ability of the polyacrylamide-grafted cassava starch material to absorb liquids. The swelling test performed in the saline water with the concentration of 4000 ppm showed a relatively lower swelling ratio as compared to that performed in distilled water. The maximum swelling ratio in distilled water reached 15.04 g/g, while that performed in saline water reached 10.57g/g only. This observation could be due to the presence of NaCl in water, which caused the osmotic pressure for water penetration into the hydrogel structure to decrease. As the consequence, the ability of the copolymers to absorb water or swelling decreased [8].

The similar observation of the decreased swelling ratio of hydrogel in saline water solution is found in the literature, such as in hydrogel of cassava starch/acrylic acid/dimethylaminopropyl methacrylamide [8], acrylic acid/acrylamide/

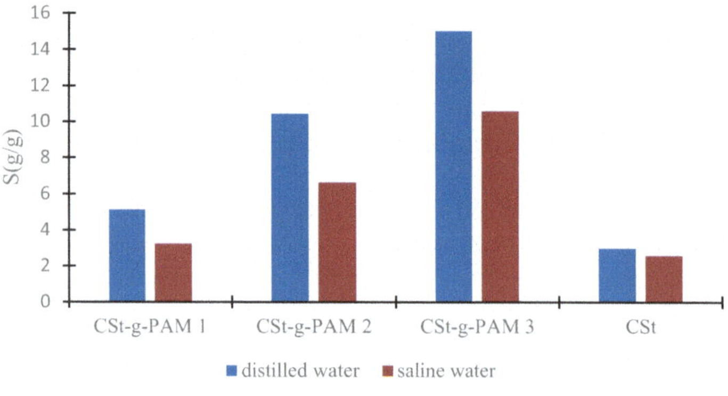

Fig. 3 Equilibrated swelling ratio influenced by monomer concentration

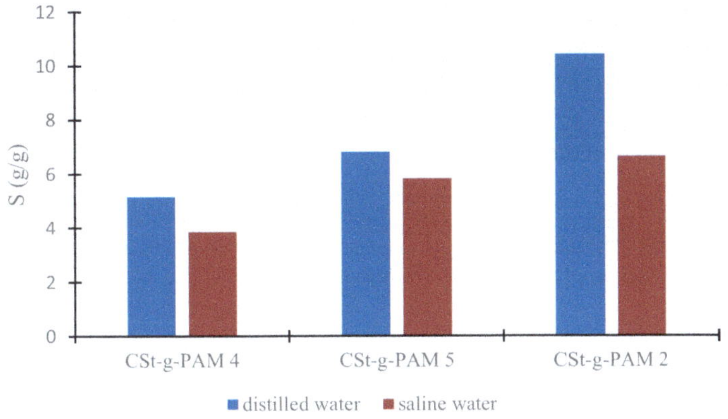

Fig. 4 Equilibrated swelling ratio influenced by initiator concentration

bentonite [13], CMC/acrylamide [14], acrylamide-acrylic acid–starch [15], and acrylamide/*N,N*-dimethyl acrylamide/2-acrylamido-2-methylpropane sulfonic sodium salt/*N*-vinylpyrrolidone (NVP) [16], which further support our hypothesis.

Figure 4 presents the equilibrated swelling ratio of polyacrylamide grafted cassava starch (Cst-g-PAM) in distilled and saline water influenced by initiator concentration. The amount of initiator also affected the swelling ratio in distilled and saline water. The swelling ratio increased as the initiator concentration increased. The swelling ratio performed in distilled and saline water increased from 5.17 to 10.42 g/g and 3.87 to 6.66 g/g, respectively, with the increased amount of initiator from 0.1 to 0.3 g. When the concentration of the initiator is low in the reaction mixture, the amount of radicals produced is minimum. Consequently, the polymerization process was incomplete and this caused the copolymer tissue to form inefficiently [17]. However, the maximum swelling ratio was obtained when the amount initiator used was 0.3 g. when more initiators are added, more radicals are formed and this leads to more heterogeneous characteristic of the copolymer structure. In turn, more porous space are obtained which affects the swelling behaviors.

4 Conclusion

The synthesis of graft copolymer of polyacrylamide-grafted cassava starch (CSt-g-PAM) has been performed utilizing the microwave-assisted method and potassium peroxodisulfate (KPS) as the initiator. FTIR spectra confirmed the success of the polyacrylamide chain grafting onto the backbone of the cassava starch. The effects of acrylamide monomer and KPS content to the water absorption of polyacrylamide-grafted cassava starch were evaluated in both distilled and in saline water. In distilled water, cassava starch has a swelling ratio of 2.97 g/g. The increased amount of monomer (from 0 to 10 g) improved the swelling ratio of the polyacrylamide-grafted cassava starch to 15.04 g/g. The initiator concentration also improved the swelling ratio of polyacrylamide-grafted cassava starch from 5.17 to 10.42 g/g as increased amount of initiator from 0.1 g to 0.3 g. The result of swelling test of polyacrylamide-grafted cassava starch performed in the saline water provided relatively lower value than that performed in distilled water, with the highest swelling ratio in saline water is 10.57 g/g.

Acknowledgements The authors sincerely thank to The Directorate General of Higher Education, Indonesia, for financial support of this work through Penelitian Fundamental 2021 PNBP Universitas Sebelas Maret.

References

1. El-hoshoudy AN, Mohammedy MM, Ramzi M, Desouky SM, Attia AM (2019) Experimental, modeling and simulation investigations of a novel surfmer-co-poly acrylates crosslinked hydrogels for water shut-off and improved oil recovery. J Mol Liq 277:142–156. https://doi.org/10.1016/j.molliq.2018.12.073
2. Junlapong K, Maijan P, Chaibundit C, Chantarak S (2020) Effective adsorption of methylene blue by biodegradable superabsorbent cassava starch-based hydrogel. Int J Biol Macromol 158:258–264. https://doi.org/10.1016/j.ijbiomac.2020.04.247
3. Pu W, Shen C, Wei B, Yang Y, Li Y (2018) A comprehensive review of polysaccharide biopolymers for enhanced oil recovery (EOR) from flask to field. J Ind Eng Chem 61:1–11. https://doi.org/10.1016/j.jiec.2017.12.034
4. Abbas S, Sanders AW, Donovan JC (2013) Applicability of hydroxyethylcellulose polymers for chemical EOR. In: Society of petroleum engineers—SPE enhanced oil recovery conference, pp 981–989. Society of Petroleum Engineers, Kuala Lumpur. https://doi.org/10.2118/165311-ms
5. Xiao X, Yu L, Xie F, Bao X, Liu H, Ji Z, Chen L (2017) One-step method to prepare starch-based superabsorbent polymer for slow release of fertilizer. Chem Eng J 309:607–616. https://doi.org/10.1016/j.cej.2016.10.101
6. Pal S, Sen G, Ghosh S, Singh RP (2012) High performance polymeric flocculants based on modified polysaccharides—microwave assisted synthesis. Carbohyd Polym 87(1):336–342. https://doi.org/10.1016/j.carbpol.2011.07.052
7. Arayaphan J, Maijan P, Boonsuk P, Chantarak S (2021) Synthesis of photodegradable cassava starch-based double network hydrogel with high mechanical stability for effective removal of methylene blue. Int J Biol Macromol 168:875–886. https://doi.org/10.1016/j.ijbiomac.2020.11.166
8. Wang Z, Shi H, Wang F, Wang A, He Q, Cuan S (2021) Synthesis of cassava starch-g-acrylic acid/dimethylaminopropyl methacrylamide: a new hydrogel for brine solution. Carbohyd Polym 266:118109. https://doi.org/10.1016/j.carbpol.2021.118109
9. Mishra S, Mukul A, Sen G, Jha U (2011) Microwave assisted synthesis of polyacrylamide grafted starch (St-g-PAM) and its applicability as flocculant for water treatment. Int J Biol Macromol 48(1):106–111. https://doi.org/10.1016/j.ijbiomac.2010.10.004
10. Mishra S, Usha Rani G, Sen G (2012) Microwave initiated synthesis and application of polyacrylic acid grafted carboxymethyl cellulose. Carbohyd Polym 87(3):2255–2262. https://doi.org/10.1016/j.carbpol.2011.10.057
11. Singh AV, Nath LK, Guha M (2011) Microwave assisted synthesis and characterization of Phaseolus aconitifolius starch-g-acrylamide. Carbohyd Polym 86(2):872–876. https://doi.org/10.1016/j.carbpol.2011.05.029

12. Abd El-Ghany NA, Abdel Aziz MS, Abdel-Aziz MM, Mahmoud Z (2019) Antimicrobial and swelling behaviors of novel biodegradable corn starch grafted/poly(4-acrylamidobenzoic acid)copolymers. Int J Biol Macromol 134:912–920. https://doi.org/10.1016/j.ijbiomac.2019.05.078
13. Abidin AZ, Noezar I (2011) Ridhawati: synthesis and characterization of superabsorbent polymer composites based on acrylic acid, acrylamide and bentonite. Indonesian J Mater Sci 12(2):114–119
14. Sultana S, Rahaman S, Hasnine SM (2018) Effect of salinity on swelling behaviors of super-water absorbent hydrogel prepared from carboxymethyl cellulose/acrylamide blends by gamma radiation. Am J Appl Ind Chem 2(2):20–26. https://doi.org/10.11648/j.ajaic.20180202.13
15. Erizal, Perkasa DP, Abbas B, Sudirman, Sulistioso GS (2014) Fast swelling superabsorbent hydrogels starch based prepared by gamma radiation techniques. Indonesian J Chem 14(3):246–252. https://doi.org/10.22146/ijc.21235
16. Farasat A, Sefti MV, Sadeghnejad S, Saghafi HR (2017) Effects of reservoir temperature and water salinity on the swelling ratio performance of enhanced preformed particle gels. Korean J Chem Eng 34(5):1509–1516. https://doi.org/10.1007/s11814-017-0017-1
17. Abidin AZ, Puspasari T, Graha HPR (2014) Utilization of cassava starch in copolymerisation of superabsorbent polymer composite (SAPC). J Eng Technol Sci 46(3):286–298. https://doi.org/10.5614/j.eng.technol.sci.2014.46.3.4

Physicomechanical Characteristics of Citric Acid Crosslinking Coated on Old Corrugated Paper

Wida Banar Kusumaningrum, Theresa Stefany, Anggraini Wulansari, Fazhar Akbar, Yudhi Dwi Kurniawan, and Nanang Masruchin

Abstract Paper-based food packaging is increasingly used nowadays related to its biodegradability and recyclability. However, this kind of paper packaging is still coated using non-biodegradable plastics such as polyethylene and polypropylene for maintaining water and moisture resistance. These research aims are to study the effect of citric acid crosslinking on the paper surface and to investigate the effect of sodium dihydrogen phosphate (SDP) as a food grade catalyst on physic and mechanical properties of coated papers. Old corrugated paper layers were disaggregation using a laboratory waring blender in 3 min, then mold in the round paper molding. Papers were coated with a citric acid solution in 5, 10, and 15% (w/w) and then activated at 150, 170, and 190 °C for 3, 5, and 7 min. In addition, 5 g/m^2 of SDP was reacted in 10% of citric acid solution. Papers that were processed at 190 °C provide the best water absorption and moisture resistance up to 40.67 and 50% respectively than untreated paper. In the activation time effect, 7 min in 10% of concentration result in water absorption reduction and moisture resistance up to 57 and 75% than untreated paper. Catalyst addition does not affect the physical properties, yet increases the mechanical properties that could reach 8.13 MPa for tensile strength and 1.11 GPa for modulus of elasticity.

Keywords Crosslinking · Citric acid · Catalyst · Old corrugated paper

W. B. Kusumaningrum (✉) · F. Akbar · Y. D. Kurniawan · N. Masruchin
Research Center for Biomass and Bioproduct, National Research and Innovation Agency, Bogor, West Java, Indonesia
e-mail: wida002@brin.go.id

T. Stefany · A. Wulansari
Agricultural Technology, Universitas Brawijaya, Malang, East Java, Indonesia

M. A. Rahandi Lubis (ed.), *Proceedings of the 1st International Conference of Lignocellulose*, Green Energy and Technology,
https://doi.org/10.1007/978-981-19-1596-3_12

1 Introduction

The food and beverage industry has become widely developed in the last decade in line with the changing of people's lifestyles. As a result, the demand for food packaging following by this trend. Unfortunately, commercial food packaging which is commonly used is from fossil-based plastics such as polyethylene, polypropylene, polyester, polyaniline, etc. which are not biodegradable and accumulate in the environment [1–3]. Paper-based packaging is potentially eco-friendly for food packaging because of its biomass-derived materials which are biodegradable and recyclable. In another hand, the application of paper-based food packaging is limited because of its low mechanical properties, less water and vapor resistance. To overcome those problems, commercial paper-based food packaging commonly coated with fossil-based plastics such as polyethylene and polypropylene that could be another serious environmental problem in the case of microplastics. Some works have been done to solve this problem by biofilm coating in the papers such as chitosan [4–6], modified organic oil [7], varnish and bio wax [6–8], etc. However, those method does not meet the technical application yet. Therefore, it needs to find another method to overcome those limitations especially for water and vapor resistance of paper-based food packaging.

Sustainability and circular economy outlook, recycle paper such as corrugated board, liner board, kraft paper, etc. could be as a versatile cellulose source for paper-based food packaging. Somehow, extrusion in the re-pulping process enables the fibers to become shorter then decrease the mechanical properties of paper. The tensile modulus of recycling paper increases up to twofold than untreated ones by blending with plasticized polyvinyl alcohol [3]. Hydrophobic recycled paper has been produced by deep coated with polymethyl methacrylate (PMMA) and cyclic olefin copolymer (COC) [9]. It should be noted, that eco-technology should be performed as consideration for sustainable materials. One of the most potential methods is polycarboxylic acid (PCA) crosslinking which could maintain mechanical and water barrier properties. Butanetetracarboxylic (BTCA) and citric acid (CA) are known to have good ability for water resistance in paper [10, 11]. Even BTCA is more reactive and results in better wet breaking length than CA, it does not suitable for food packaging due to its toxicity. The proposed mechanism of crosslinking process between cellulose on paper with citric acid by the thermochemical reaction is represented in Fig. 1. The key factor of this process is the formation of citric acid anhydride then esterified with cellulose and promoting further cross-linked [12]. Both the formation of citric acid anhydride and esterification process need some thermal initiation, hence activation temperature and time be the crucial factors to initiate the crosslinking process. Some studies report for different activation condition, depend on the matrix, optimum curing condition for biofilm are about 110–190 °C and 5–15 min [13–15], cotton fabrix are about 140–210 °C and 3 min [16, 17], pulp fiber and paper are about 160–215 °C and 5–10 min [10, 11].

The activation energy for citric acid crosslinking is relatively high [18], therefore needs some catalyst to minimize the activation condition. The most common

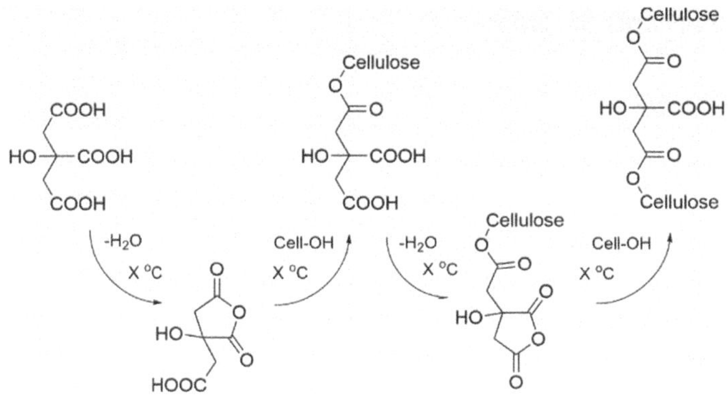

Fig. 1 Crosslinking mechanism of citric acid and cellulose

catalyst used for citric acid crosslinking is sodium hypophosphite (SHP) [10, 11, 13, 16]. However, this catalyst is categorized as toxic materials, so it does not considerably use for food-packaging applications. As the key factor for crosslinking is citric acid anhydride formation, other reactive phosphate groups such as sodium dihydrogen phosphate (SDP) could be applied as a catalyst. The mechanism of citric acid anhydride formation by SDP is shown in Fig. 2.

As the best known of authors, the work of coated paper with citric acid crosslinking into old corrugated paper with SDP catalyst has not been done yet. Widsten et al. [11] was coated the commercial old corrugated board with citric acid crosslinking by SHP catalyst. The SDP has been used for citric acid crosslinking catalyst on water-soluble matrix hydroxypropylmethylcellulose (HPMC) [14]. This research aim is to study the effect of activation temperature and time on physicomechanical properties of old corrugated paper as well as the effect of catalyst addition.

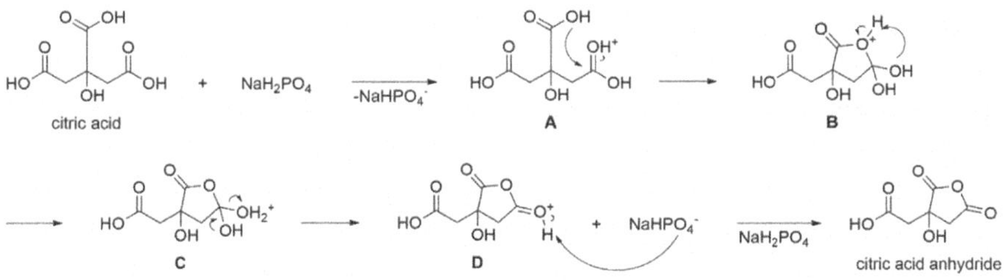

Fig. 2 Mechanism of sodium dihydrogen phosphate catalization

2 Material and Methods

2.1 Materials

The anhydrous citric acid as crosslinking agent was purchased from Merck, Indonesia. Sodium dihydrogen phosphate dihydrate as catalyst and sodium hydroxide were purchased from Merck, Indonesia. Chemicals were used as received without further purification. Corrugated and inner layers of corrugated box were used for paper source. Commercial paper for food packaging with polyethylene film coating was used for this experiment.

2.2 Re-pulping Process of Old Corrugated Box

Papers were cut into small pieces about 1 × 1 cm then immersed in distilled water for 24 h. Laboratory waring blender was used for disaggregation process in 3 min on 22.000 rpm. Pulp was molded in round paper molding with 22.5 cm in diameter and 150 gsm in gramatur. Molded papers were concitioned at room temperature for 3 days then oven-dried at 80 °C for 6 h.

2.3 Preparation of Coated Paper Crosslinking Citric Acid

Citric acid solution (CA) in determining concentrations 5, 10, and 15% (w/w) was gradually sprayed on the paper surface, with a sprayed distance of about ± 15 cm. The sprayed paper was activated in the set oven at 150, 170, and 190 °C for 3, 5, and 7 min. As a comparison for the effect of the catalyst, sodium dihydrogen phosphate (SDP) as a crosslinking catalyst in 5 g/m^2 was added into 10% (w/w) citric acid solution then activated in determining condition. Activated papers were conditioned at room temperature for 5 days then oven-dried at 60 °C for 6 h.

2.4 Charterization of Coated Papers

Moisture content. The moisture content of coated papers was analyzed according to SNI 638:2014. Samples were cut into 30 × 30 mm and testing was conditioned at 105 °C for 24 h. Moisture content was calculated as weight difference before and after drying conditions.

Water absorption. Water absorption was tested according to the Cobb method of SNI 0499:2008. Samples were prepared in 30 × 30 mm then placed on the perforated metal plate. About 5 mL of distilled water was poured into the papers then weighed after 60 s of absorption. Water absorption was calculated as weight difference before and after water contact at conditioning time as represented in Eq. 1. a is the paperweight before contact (g), b is the paperweight after contact (g), c is a testing area (cm^2), and F is the conversion factor for area unit.

$$Cobb_x = \frac{(a - b)}{c} xF \tag{1}$$

Water vapor transmission rate (WVTR). Water vapor transmission rate is a testing method to analyze the ability of papers to restrain water vapor. WVTR testing was conducted according to the dish method of ASTM E 95-96 1995. Some of the aquadest was placed into the petri dish then covered by perforated aluminum foils in 10% of the total area to place the testing papers. Samples were cut in 50 × 50 mm then tested at 37 °C and RH 81%. Samples were weighed every 1 h for 7 h. WVTR value was calculated with Eq. 2. Flux is the slope of increased weight in a petri dish ($g/h/m^2$).

$$\text{WVTR} = \frac{\text{Flux}}{\text{surface area}} \tag{2}$$

Mechanical properties. Universal testing machine (UTM), Shimadzu, Japan with load cell 5 kN was used for mechanical properties testing. The testing method for mechanical properties was referred to SNI 1924-2:2016 for a single paper with a constant elongation rate in 20 mm/s. Samples were cut into 100 × 15 mm with spawn about 50 mm. Testing was conducted for six replications of control and coated papers. Mechanical properties were covered for tensile strength and modulus of elasticity.

Chemical component. Analysis of functional groups was conducted using FTIR-UATR, Spectrum Two, Perkin Elmer, USA. Paper was placed into the reflector then scanned under wavelength 400–4000 cm^{-1} with scanning replication for 16 scans. Prior to analysis, samples were immersed in NaOH 0.1 M for 15 min then rinsed in aquadest for 5 min as conducted by Quellmalz et al. [12]. Subsequently, samples were oven-dried at 100 °C for 10 min then conditioned at room temperature for 3 days before testing.

3 Result and Discussion

3.1 *Effect of Activation Temperature*

The physical properties of a coated paper by crosslinked CA include moisture content, water absorption, and water vapor transmission rate. Activation conditions to analyze the effect of activation temperature were determined in 5 min of activation time and CA concentration in 5, 10, and 15%. Activation temperatures were varied at 150, 170, and 190 °C. Moisture content that corresponds to the bounded water in the microporous paper is represented in Fig. 3a. In the case of the same CA concentration, activation temperature enhancement decrease the moisture content. Contrary, moisture content increase as rising CA concentration in case of the same activation temperature. That is possibly due to excessive carboxyl groups of CA that have not crosslinked yet in 5 min of activation time so could easily bound with the hydroxyl groups of water vapor. The uncrosslinked CA in the biopolymer matrix may increase the free hydroxyl and carboxyl groups [19]. The paper that was activated at 190 °C in 5% of CA concentration has the lowest moisture content in 2.41% among other coated papers and is slightly better than untreated paper in 2.6%. Furthermore, the moisture content of coating papers is still under the commercial paper that coating with polyethylene film in 2.1%.

Water absorption (WA) is the ability of treated paper to absorb some of the contacted waters on the paper surface. The water absorption of treated papers is decreasing as the temperature increase at the same CA concentration as shown in Fig. 3b. WVTR that corresponds to the ability of treated paper to restrain the water vapor penetration also tends to decrease as temperature increase up to 190 °C. This result seems to be similar to the moisture content trend including the effect of CA concentration. Treated paper processed at 190 °C has the lowest WA in 423.07 g/m^2 for 5% of CA concentration while WVTR in 0.02106 g/h/m^2 for 15% of CA concentration. This result remarkably shows for the WA and WVTR improvement in about 40.67 and 50% respectively than untreated paper. The higher activated temperature of treated linerboard produces more crosslinked CA that has more ability to limit the water absorption [11]. The surface area of crosslinked cellulose with CA drops up to more than 60% than the untreated one that suggests water vapor restriction due to less microporous availability [15]. Ester group formation improves as elevating temperature up to 200 °C that indicates for successfully crosslinking process [10] and further contributes to decreasing the free hydrophilic groups so may inhibit water absorption [19]. Some other works provide a similar result on crosslinking improvement as rising temperature [13, 16].

Tensile strength (TS) and modulus of elasticity (MOE) of treated papers increase as temperature increases as represented in Fig. 4. The highest TS and MOE value obtained by treated paper processed at 190 °C and 15% CA concentration in 7.34 MPa and 1.35 GPa. Overall, the treated papers have lower mechanical properties than the untreated ones. It probably caused by the higher moisture content of treated papers that attributed to the plasticising effect of water [19]. The reduction of stress–strain to

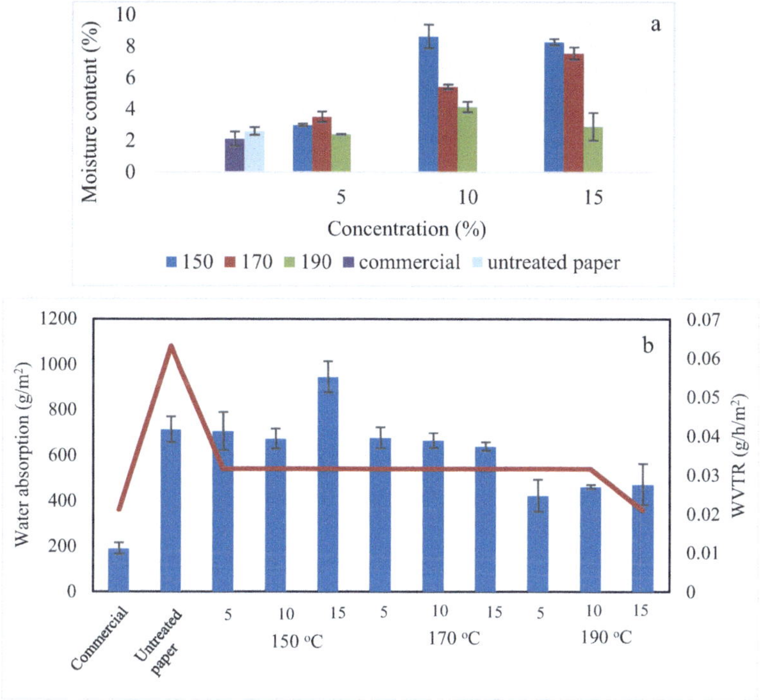

Fig. 3 Physical properties of coated papers as a function of activation temperature and citric acid concentration

failure of crosslinked dry filter paper indicates brittle behavior [12]. CA crosslinking on the paper surface is unable to improve mechanical properties. The mechanical properties enhancement as increasing CA concentration might promote the esterification reaction between cellulose and citric acid. Compressive strength improvement with an increase of CA dose is affected by esterification rather than citric acid anhydride intermediates formation to propose crosslinking process [11, 20]. Based on the data, the activation temperature at 190 °C produces the best physicomechanical properties of treated papers.

3.2 *Effect of Activation Time*

According to the activation temperature optimization, the best temperature is at 190 °C which then was applied to study the effect of activation time. Activation time in 3, 5, and 7 min with 5, 10, 15% of CA concentration were conducted as the same previous procedure. Longer activation times produce lower moisture content, WA, and WVTR as represented in Fig. 5a, b. A similar result has occurred for physical properties that the best value is in 7 min of activation time and 10% of CA

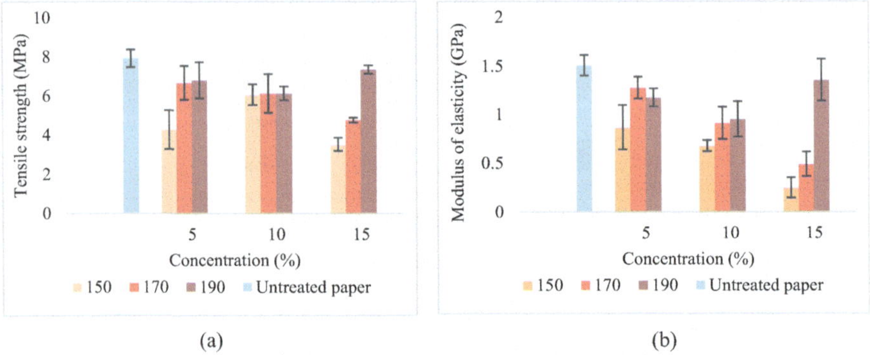

Fig. 4 Mechanical properties of coated papers as a function of activation temperature and citric acid concentration

concentration with 2.46%, 302.35 g/m^2, and 0.0158 g/h/m^2 for moisture content, WA, and WVTR respectively. WA drops up to 57.6%, while WVTR is up to 75% than those untreated paper that attributed to better crosslinked process on the paper surface. Furthermore, the WVTR of treated paper in the maximum result is lower than commercial paper in 0.0211 g/h/m^2. It could be ensured due to the substitutions of hydroxyl groups to hydrophobic ester groups of crosslinked paper [19, 21, 22].

The TS and MOE of treated papers improve as rising of activation time especially for 5% and 10%, while relatively drop by 15% as show in Fig. 6. Unfortunately, this result remains under the untreated papers. The value of TS and MOE reach 6.62 MPa and 1.21 GPa that occurred in 10% of CA in 7 and 3 min respectively. The optimum compressive strength of crosslinked paper perform when cured in 400 s and remains constant above those times that further degrade into a chromophoric compound as elevating time and temperature [11]. Awadhiya et al. [13] report that curing time above the optimum may decrease the mechanical properties of bioplastic film due to charing of hydrocarbons. Overall, treated paper in 7 min of activation time with 10% of CA concentration provide better physicomechanical properties than others.

3.3 *Effect of Catalyst*

In comparison to catalyst effectivity, SDP was added to 10% of the CA solution. Physical properties including moisture content, WA, and WVTR tend to have a similar result. The addition of SDP affect MC, WA, and WVTR improvement as rising time up to 7 min and temperature 170 °C. Meanwhile, at a temperature above that (190 °C) the physical properties are prone to have higher water and moisture-holding for all activation time as seen in Fig. 7. These phenomena could be affected by decomposition of SDP into sodium acid pyrophosphate (SAPP) at a temperature above 170 °C. The effectivity of temperature and time perform for the effectivity

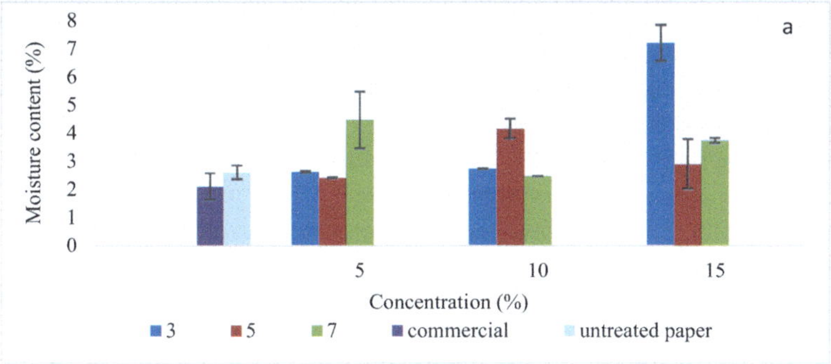

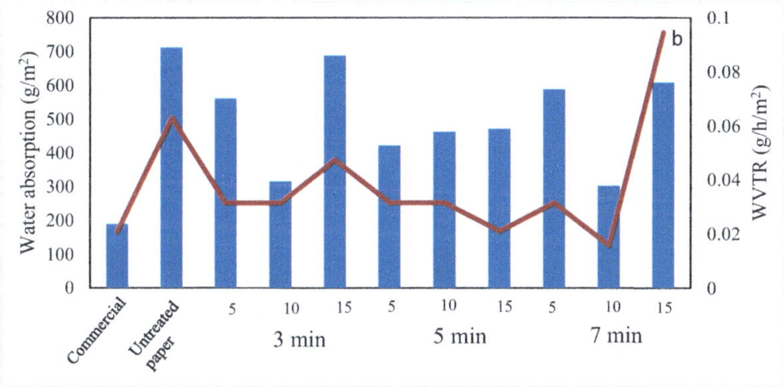

Fig. 5 Physical properties of coated papers as a function of activation time and citric acid concentration

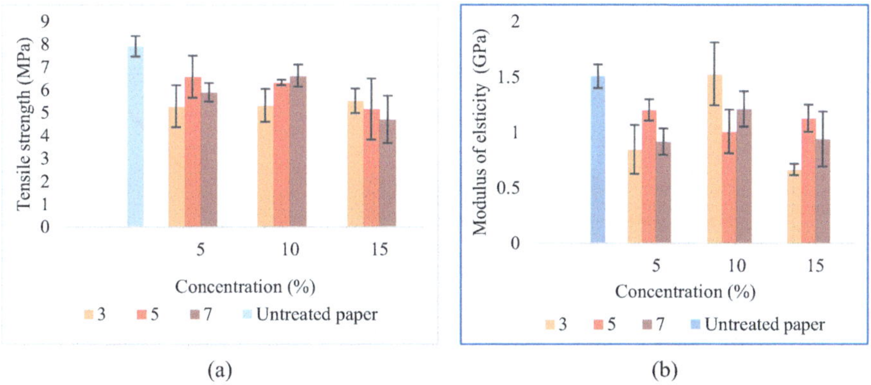

Fig. 6 Mechanical properties of coated papers as a function of activation time and citric acid concentration

of catalyst, for instance in the same temperature (150 °C) the WA of treated papers without catalyst in 7 min results in a slight difference of about 1–9% compared to papers with catalyst in 3–5 min. That result seems to indicate that catalyst has no significant effect on CA crosslinking process. Coma et al. [14] report that WVTR performance improves about 34% with the addition of SDP up to 50% by weight based on CA content, meanwhile this study use 3.5% by weight.

In contrary, TS value increase as the temperature level up into 190 °C in all activation time. This result improves even slightly higher in 8.13 MPa than untreated paper. Overall, the addition of SDP improves the TS in all temperature and time compared to treated paper without catalyst as seen in Fig. 8a. In another hand, MOE starts to drop after temperature 190 °C as represented in Fig. 8b. This could

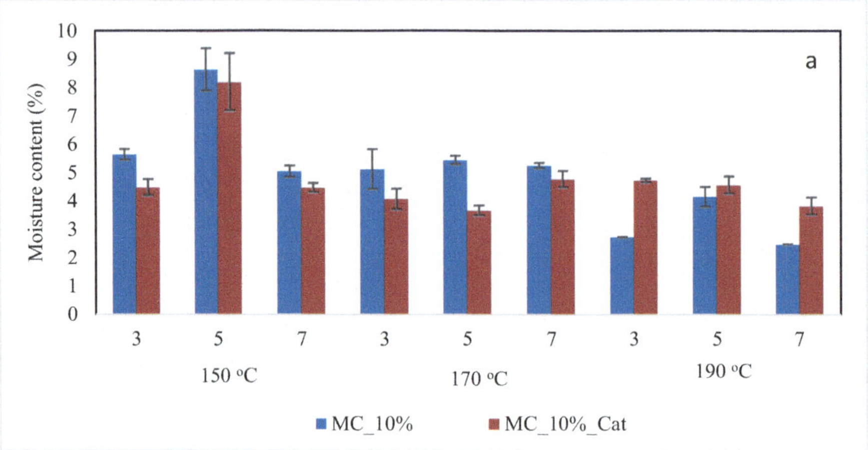

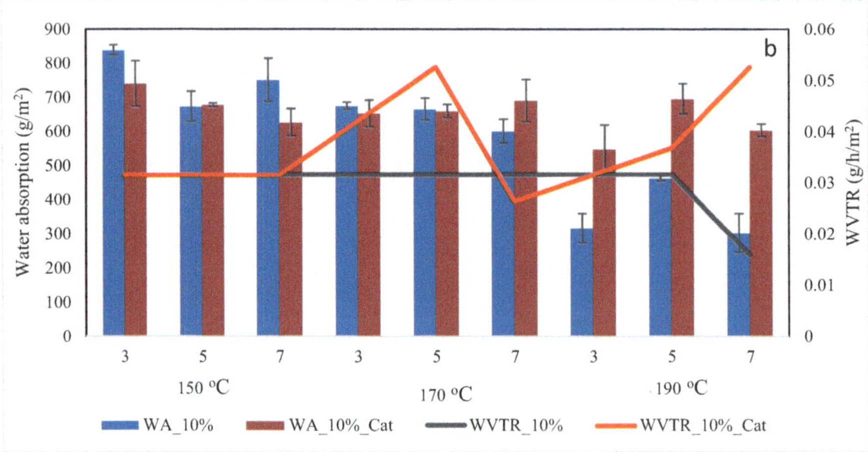

Fig. 7 Physical properties of coated papers with catalyst as a function of activation temperature and time

be inferred that treated paper with catalyst is quite strong but easy to break down. Widtsen et al. [11] report of compressive strength enhancement of about 60% by addition of 5 g/m^2 of SHP catalyst into 12% of CA dose. In addition, Awadhiya et al. [13] found that higher CA concentration and longer curing time with the addition of SHP provide a better degree of crosslinking that correlate to mechanical properties enhancement than at lower CA concentration. Moreover, it needs further study to obtain the effective process for CA crosslinking into the paper surfaces.

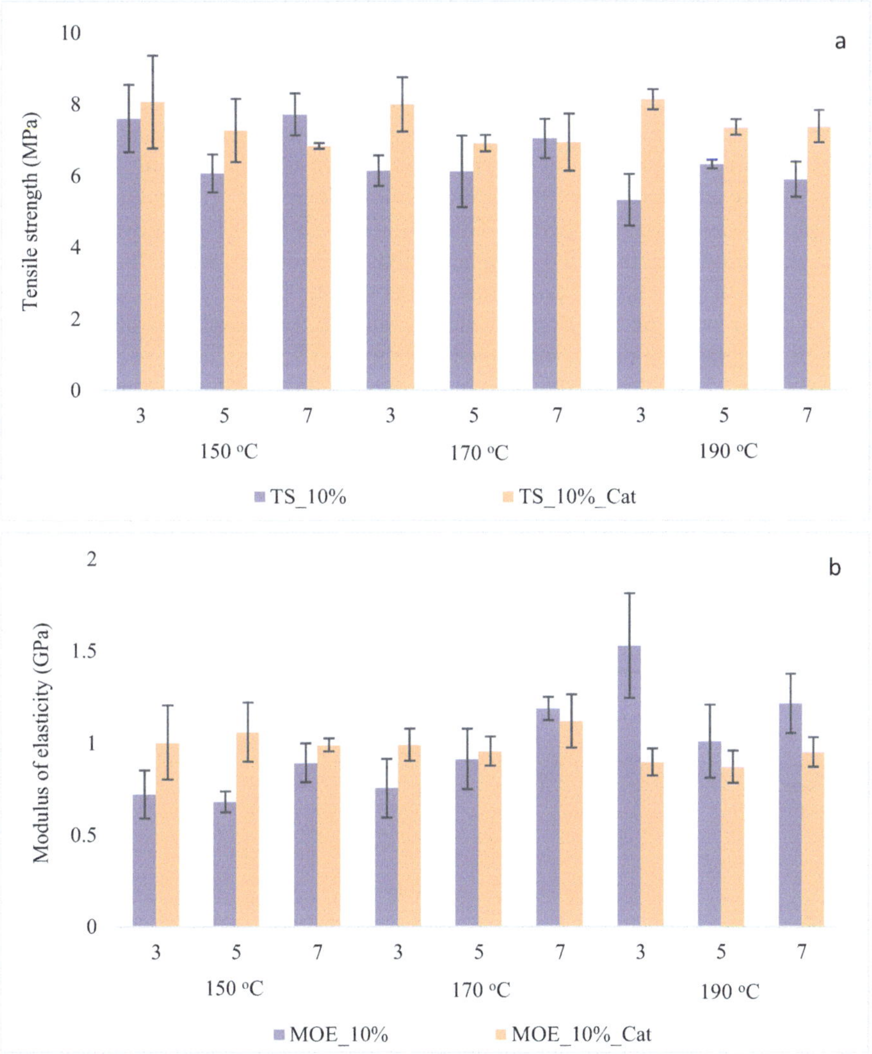

Fig. 8 Mechanical properties of coated papers with catalyst as a function of activation time and time

3.4 Chemical Component of Coated Papers

The free carboxylic groups and crosslinker carboxyl content may identify at the same spectra in 1730 cm^{-1}, therefore it needs saponification treatment by washing the papers with NaOH 0.1 M to eliminate unreacted polycarboxylic acid. Those free carboxylic acid, after treatment, perform into carboxylate which strongly appears in 1589 cm^{-1} [10, 12, 15, 23]. The spectra of carboxylate groups are clearly indentified after washing for all CA concentrations and with catalyst, as seen in Fig. 9, that unidentified in untreated paper and without treatment. Strong spectra in 1730 cm^{-1} indicate free carboxylic acid of citric acid. The intensity of that band decrease in 5% of CA concentration while quite broad in 1580 cm^{-1} which corresponds to more unreacted citric acid is converted. A technical approach for indicating the degree of crosslinking could perform by intensity ratio between 1730 and 1589 cm^{-1} band [15]. The ratio for 5% of CA is 0.41, 10% of CA is 2.3, and 15% of CA is 1.3. This result emphasizes that a higher ratio degree of crosslinking provides better physicomechanical properties. Treated paper with 15% of CA proves for broad carboxylate band that correlated to higher unreacted citric acid which further affects for lower physicomechanical properties. Moreover, the intensity of the carboxylate band in paper with the addition of catalyst tends to be lower than others which might be correlated to more crosslinked carboxylic acid. The new spectra in 1678 and 1638 cm^{-1} have clearly appeared in the paper with a catalyst that may identify as unsaturated polycarboxylic acid that simultanously comes with citric anhydride formation at high temperature [16].

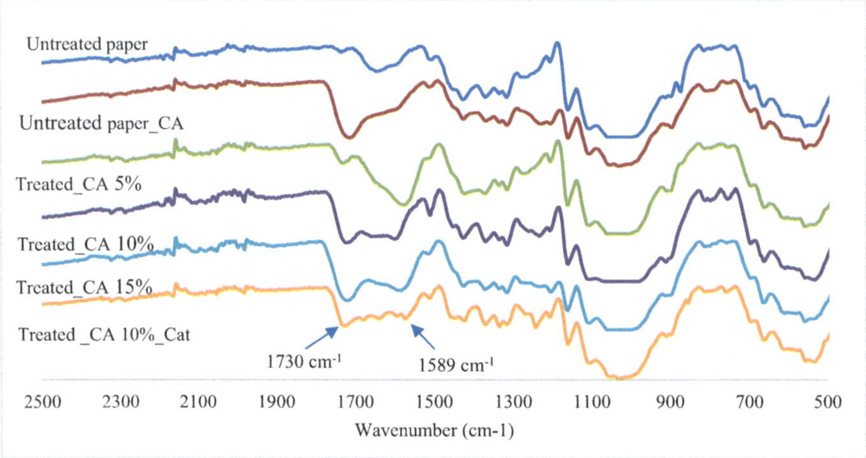

Fig. 9 FTIR Spectrograph of coated papers without and with catalyst

4 Conclusion

This research successfully produces coated paper with citric acid crosslinking that having good resistance in water and water vapor. Activation temperature and time remarkably affect for physical and mechanical properties of the papers. Activation temperature at 190 °C in 7 min and 10% of CA concentration provide the best physicomechanical properties with water absorption and water vapor transmission rate reduction in 57% and in 75% than untreated paper. Sodium dihydrogen phosphate as a catalyst has no significant effect on the effectivity of crosslinking process for limiting the water penetration but result in better mechanical properties which could reach 8.13 MPa than untreated paper in 7.9 MPa. Further study is considerably needed to improve the mechanical properties and the effectivity of crosslinking process with and without catalyst.

Acknowledgements This research funding is from Processed Food Packaging National Priority Program, Deputy for Scientific Services, LIPI. The authors would like to thanks for the support from JSPS Alumni Association of Indonesia (JAAI) and the facilities of Integrated Laboratory of Bioproduct, LIPI so this research could well undergo.

References

1. Zhu R, Liu X, Song P, Wang M, Xu F, Jiang Y, Zhang X (2018) An approach for reinforcement of paper with high strength and barrier properties via coating regenerated cellulose. Carbohydr Polym 200:100–105. https://doi.org/10.1016/j.carbpol.2018.07.069
2. Xiong J, Sheng C, Wang Q, Guo W (2019) Toughened and water-resistant starch/TiO_2 bio-nanocomposites as an environment-friendly food packaging material. Mater Res Express 6. https://doi.org/10.1088/2053-1591/ab058b
3. Del Curto B, Barelli N, Profaizer M, Farè S, Tanzi MC, Cigada A, Ognibene G, Recca G, Cicala G (2016) Poly-paper: a sustainable material for packaging, based on recycled paper and recyclable with paper. J Appl Biomater Funct Mater 14:e490–e495. https://doi.org/10.5301/jabfm.5000335
4. Bordenave N, Grelier S, Coma V (2010) Hydrophobization and antimicrobial activity of chitosan and paper-based packaging material. Biomacromol 11:88–96. https://doi.org/10.1021/bm9009528
5. Tanpichai S, Witayakran S, Wootthikanokkhan J, Srimarut Y, Woraprayote W, Malila Y (2020) Mechanical and antibacterial properties of the chitosan coated cellulose paper for packaging applications: effects of molecular weight types and concentrations of chitosan. Int J Biol Macromol 155:1510–1519. https://doi.org/10.1016/j.ijbiomac.2019.11.128
6. Khwaldia K, Arab-Tehrany E, Desobry S (2010) Biopolymer coatings on paper packaging materials. Compr Rev Food Sci Food Saf 9:82–91. https://doi.org/10.1111/j.1541-4337.2009.00095.x
7. Tambe C, Graiver D, Narayan R (2016) Moisture resistance coating of packaging paper from biobased silylated soybean oil. Prog Org Coatings 101:270–278. https://doi.org/10.1016/j.porgcoat.2016.08.016
8. Zhang D, Yu G, Long Z, Xiao H, Qian L (2016) Bio-wax latex-modified paper as antimicrobial and water-vapor-resistant packaging material. J Wood Chem Technol 36:182–191. https://doi.org/10.1080/02773813.2015.1104545

9. Paul UC, Fragouli D, Bayer IS, Mele E, Conchione C, Cingolani R, Moret S, Athanassiou A (2017) Mineral oil barrier sequential polymer treatment for recycled paper products in food packaging. Mater Res Express 4. https://doi.org/10.1088/2053-1591/4/1/015501
10. Awada H, Montplaisir D, Daneault C (2014) Cross-linking of papers based on thermomechanical pulp fibers by polycarboxylic acids: in fluence on the wet breaking length. Ind Eng Chem Res 53:4312–4317
11. Widsten P, Dooley N, Parr R, Capricho J, Suckling I (2014) Citric acid crosslinking of paper products for improved high-humidity performance. Carbohydr Polym 101:998–1004
12. Quellmalz A, Mihranyan A (2015) Citric acid cross-linked nanocellulose-based paper for size-exclusion nano filtration. ACS Biomater Sci Eng 1:271–276. https://doi.org/10.1021/ab500161x
13. Awadhiya A, Kumar D, Verma V (2016) Crosslinking of agarose bioplastic using citric acid. Carbohydr Polym 151:60–67. https://doi.org/10.1016/j.carbpol.2016.05.040
14. Coma V, Sebti I, Pardon P, Pichavant FH, Deschamps A (2003) Film properties from crosslinking of cellulosic derivatives with a polyfunctional carboxylic acid. Carbohydr Polym 51:265–271
15. Cuadro P, De Belt T, Kontturi KS, Reza M, Kontturi E, Vuorinen T, Hughes M (2015) Reactive & functional polymers cross-linking of cellulose and poly (ethylene glycol) with citric acid. 90:21–24
16. Cai Z, Ji B, Yang K, Zhu Q (2019) Investigation on reaction sequence and group site of citric acid with cellulose characterized by FTIR in combination with two-dimensional. Polymers (Basel) 11:1–13
17. Yang CQ, Wang D (2000) Evaluating ester crosslinking of cotton fabric by a polycarboxylic acid. Text Res J 615–620
18. Xu H, Shen L, Xu L, Yang Y (2015) Low-temperature crosslinking of proteins using non-toxic citric acid in neutral aqueous medium: mechanism and kinetic study. Ind Crop Prod 74:234–240. https://doi.org/10.1016/j.indcrop.2015.05.010
19. Wu H, Lei Y, Lu J, Zhu R, Xiao D, Jiao C, Xia R, Zhang Z, Shen G, Liu Y, Li S, Li M (2019) Food Hydrocolloids Effect of citric acid induced crosslinking on the structure and properties of potato starch/chitosan composite films. Food Hydrocoll 97:105208. https://doi.org/10.1016/j.foodhyd.2019.105208
20. Yan L-Y, Li S-Y, Song X-L (2017) Preparation of paper strengthening agent by esterification of cellulosic fines. Bioresources 12:469–477
21. Das A, Uppaluri R, Das C (2019) Feasibility of poly-vinyl alcohol/starch/glycerol/citric acid composite films for wound dressing applications. Int J Biol Macromol 131:998–1007. https://doi.org/10.1016/j.ijbiomac.2019.03.160
22. Wu Z, Wu J, Peng T, Li Y, Lin D, Xing B, Li C, Yang Y, Yang L, Zhang L, Ma R, Wu W, Lv X, Dai J, Han G (2017) Preparation and application of starch/polyvinyl alcohol/citric acid ternary blend antimicrobial functional food packaging films. Polymers (Basel) 9:1–19. https://doi.org/10.3390/polym9030102
23. Morris NM, Catalano EA, Andrews BAK (1995) FT-IR Determination of degree of esterification in polycarboxylic acid cross-link finishing of cotton. Cellulose 2:31–39

Cellulose Waste in Electrochemical Treatment of Bacterial Cellulose

Nugraeni Fahrunisa, Yeyen Nurhamiyah, Rachmadhani, Goldha Maula Hildayani, Syifa Annisa Rachmawati, and Sandi Sufiandi

Abstract Bacterial cellulose (BC) has advantage such high purity. The alkali treatment and electrolysis process for bacterial cellulose leaving waste water. The objective of this research to investigate the precipitated solution as water waste of electrolysis. Through after treatment for water waste can be discovered the effect and product waste of electrolysis from bacterial cellulose. The after treatment was done by centrifuge and oven to get water waste become powder ready for characterization. From Fourier Transmittance Infra-Red result shows the characteristic of cellulose. The microcrystalline has peak at 3417 cm^{-1} as –OH stretching and 976 as β-Glycosidic bond. From X-Ray Diffraction result shows the powder has impurity and amorphous phase from bacterial cellulose. In XRD result showing the powder contain impurity exhibit peaks 2θ at 27.5°, 31.85° and 45.67° known as sodium chloride. This result showing electrolysis using sodium chloride can be two-step purification for bacterial cellulose. The precipitated solution contains amorphous phase of BC, the leftover of sodium chloride and other impurity.

Keywords Water waste · Bacterial cellulose · Electrolysis · Alkali treatment · Sodium chloride

---

N. Fahrunisa · Y. Nurhamiyah · Rachmadhani · G. M. Hildayani · S. A. Rachmawati
Research Centre for Biomass and Bioproducts, National Research and Innovation Agency, Cibinong, Bogor, West Java, Indonesia

N. Fahrunisa
Department of Metallurgical and Materials Engineering, Faculty of Engineering, University of Indonesia, Depok, West Java, Indonesia

G. M. Hildayani
Department of Chemistry, Faculty of Mathematics and Science, University of Indonesia, Depok, West Java, Indonesia

S. Sufiandi (✉)
Research Center for Smart Mechatronics, National Research and Innovation Agency, Bandung, West Java, Indonesia
e-mail: sand003@brin.go.id

M. A. Rahandi Lubis (ed.), *Proceedings of the 1st International Conference of Lignocellulose*, Green Energy and Technology,
https://doi.org/10.1007/978-981-19-1596-3_13

1 Introduction

Cellulose was synthesized natural polymer by plant, fungi, algae and bacteria. On the plant, cellulose known as the main component wall cell plant discovered in wood, seed shell, etc. [1]. Bacterial cellulose (BC) is a type of polysaccharides which produces from microorganism [2]. As bio-compatible material, bacterial cellulose commonly used in wide application such drug delivery, bio-medic, water treatment and nanofiller [3]. Several bacteria were reported to successfully produce BC such as Bacterium xylinodes [4], Gluconacetobacter xylinus [5], and Acetobacter xylinum [6]. Based on the variety of bacteria, cellulose can be produced with different morphology, structure and properties [7, 8]. BC was used in bio-medic because of high purity cellulose without any toxic compound such as pectin, lignin and hemi-cellulose [9].

In order to improves properties of bacterial cellulose, some researcher found alkali treatment help to improves mechanical properties with proper concentration [10–14]. In addition, alkali treatment can be done for sterilized and purification of BC. The side note for alkali treatment in high concentration can swelled and leading cellulose fibrils to be brittle. Looking on Yiying Yue etc. result, they successfully used NaOH and $NaClO_2$ treatment to produce cellulose I and cellulose II and cellulose I/II hybrid [15]. The effective concentration came with 20 wt% NaOH for 10h and $NaClO_2$ for 2h. On the XRD results exhibit the convertion from cellulose I–II with decreased of percentage cellulose I and crystallinity index from 100 to 5% and 58.2 to 68.8%, respectively. Another research came from Gea etc. used 2.5 wt% NaOH and 2.5 wt% NaOCl as two-step purification for BC [16]. The purification for bacterial cellulose is aimed to removing leftover organic material such protein and nucleic acids [17]. The result on SEM morphology confirm after two-step purification using NaOCl the nanofibrils of BC can be clearly seen. On the FTIR result, after two-step treatment the hydrogen bond OH stretching shift to higher peak number occur and increases the intensity of O–H stretch.

In the last report we did alkali treatment used sodium chloride with result can improve and promote inter-molecular hydrogen bonding of BC with 3.5% NaCl [18]. According to the page Prof. Kirthi Tennakone posted, the NaOCl solution can be produced from electrolysis process using sodium chloride [19]. Previously many researchers successfully did purification used alkali solution such NaOH and NaOCl. In this work, we continue our previous work with addition of NaOH treatment and used electrolysis process with sodium chloride as two-step purification. The after treatment process of electrolysis leaving precipitate solution as water waste. The aimed of this paper to characterized the water waste produced by electrolysis process.

2 Material and Methods

2.1 Material

Coconut water and sucrose were obtained from the market, Indonesia. Acetobacter xylinum was purchased from e-Commerce, Biotechno store, Indonesia. Sodium chloride and sodium hydroxide was purchased from Sigma-Aldrich, Singapore.

2.2 Methods

Culture Method

The culture method is explained in our previous work [18]. Briefly, Coconut water was used as medium sterilized and boiling at 100 °C. Adding sucrose, acetic acid and CH_4N_2O as nutrient at boiling medium and stirred until homogeneous. Then, the medium was cooled until room temperature. A cultured container was sterilized by a 96% alcohol and wiped by tissue papers. Then, A. xylinum solution was poured into a culture container and covered by a sheet of newspapers and sealed by rubber bands. The container then left in room temperature for 7 days which cultured bacterial cellulose was harvested afterwards.

Electrochemical Treatment

Bacterial cellulose was cut with the dimension of 8 cm × 8 cm. Then, the BC was submerged in 1 M sodium hydroxide for 45 min while heated at 80 °C. After washed with water to neutralize, the material was submerged in sodium chloride solution 3.5 wt. % for 45 min at 90 °C in a beaker. The electrochemical process was added simultaneously with the voltage of 1.3 V. The apparatus used for electrochemical was power supply, zink plate as electrode, hot plate and beaker glass. The waste was considered as the material precipitated in the bottom of the beaker. The waste was separated using centrifuge with rate 6000 rpm at 25 °C in 10 min. The last produce after separating was paste and dried in oven until became powder. The powder sample was prepared using mixer milling (Fig. 1).

2.3 Characterization

Fourier Transform Infrared (FTIR) spectroscopy was employed on Perkin Elmer Spectrum Two (Perkin Elmer Spectrum Two, PerkinElmer, Waltham, MA USA) with attenuated reflectance mode in the wave region of 4000–400 cm^{-1}, with a resolution of 1 cm^{-1}. X-Ray Diffraction (XRD) was measured on (model) at 1.54

Fig. 1 Electro-chemical apparatus

radiation wavelength at a voltage of 40 kV and filament emission of 30 mA. The samples were scanned in a 2 between 10 and 80° with the speed of 2° min^{-1}

3 Result and Discussion

The cellulose waste was characterized using Fourier Transmission Infra-Red and recorded as Fig. 2. The spectrum shows characteristic for cellulose. In cellulose band at 976 cm^{-1} indicates the β-glycosidic bond, 1065 cm^{-1} as ring vibration and C–OH bending, 1635 cm^{-1} as (–O–) tensile vibration neighboring hydrogen atoms, and the broad peak at 3417 cm^{-1} shows OH stretching, bands of H bonds [20, 21]. There is no sign peak at 1430 cm^{-1} shifted to 1391 cm^{-1} this peak known as the "crystallinity band", this indicating decrease in intensity [22]. This result shows the electrolysis loosen amorphous phase from cellulose. The amorphous phase with weak bond came from cellulose after alkali treatment.

The FTIR result shows that during electrolysis process of bacterial cellulose, some part of it was break down and precipitated in the bottom of the beaker. This could be because the electric voltage shakes up some of the cellulose bond that make some of it to break down.

The powder X-Ray diffraction spectra exhibit peaks 2θ at 27.5°, 31.85° and 45.67° known as sodium chloride, which can be assign to the 110, 200 and 220, respectively [23]. The crystallinity of bacterial cellulose and cellulose waste was 78 and 62%. This result indicated that the powder impurities was confirmed (Fig. 3).

The impurities can be seen from many intensity peaks of cellulose waste compared to bacterial cellulose. The hump in the cellulose graph described as amorphous phase loosen by bacterial cellulose.

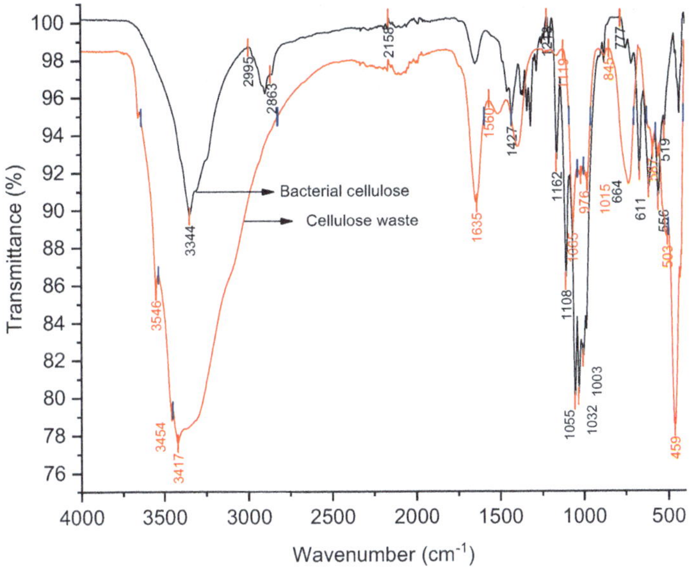

Fig. 2 The Infrared spectrum of bacterial cellulose and cellulose waste powder

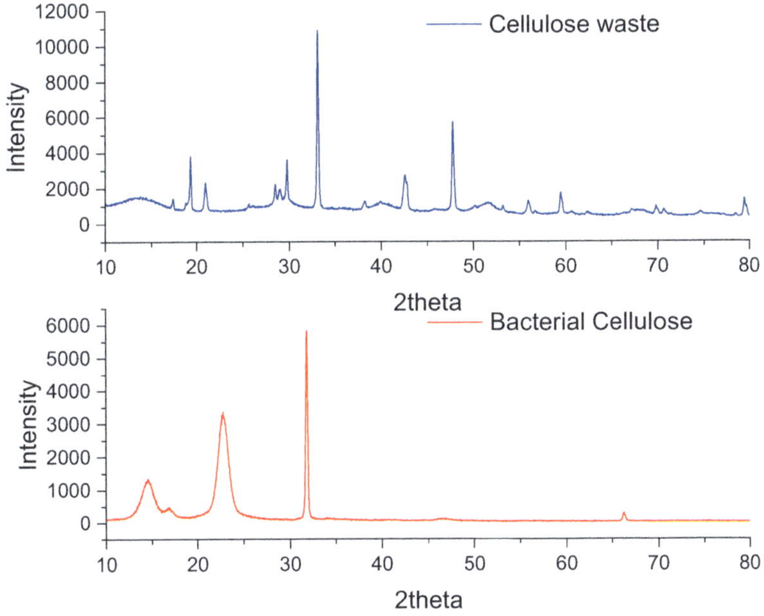

Fig. 3 XRD graph of cellulose waste and bacterial cellulose

4 Conclusion

In this work the two-step purification was successfully done by using alkali solution 1% NaOH and electrolysis 3.5% NaCl. After electrolysis treatment, precipitated solution was characterized using FTIR and XRD. The result from FTIR shows the water waste contain cellulose, comparing the graph with bacterial cellulose. In XRD result showing the powder contain impurity exhibit peaks 2θ at 27.5°, 31.85° and 45.67° known as sodium chloride.

Acknowledgements This research was funded by the Ministry of Research and Technology/ National Research and Innovation Agency (BRIN) through Mandatory RISPRO Funding themed National Research Priority (PRN) 2020 and Nugraeni Fahrunisa was supported by Indonesian Institute of Sciences as Research Assistant for Fiscal Year 2021. All authors contributed equally to this work

References

1. Niyazbekova ZT, Nagmetova GZ, Kurmanbayev AA (2018) An overview of bacterial cellulose applications. Biotechnol Theory Pract 1–15
2. Esa F, Tasirin SM, Rahman NA (2014) Overview of bacterial cellulose production and application. Agric Agric Sci Procedia 2:113–119
3. Huang Y, Zhu C, Yang J, Nie Y, Chen C, Sun D (2014) Recent advances in bacterial cellulose. Cellulose 21(1):1–30
4. Breed RS, Murray EGD, Smith NR (1957) Bergey's manual of determinative bacteriology. In: Bergey's manual of systematic bacteriology, 7th edn
5. Vazquez A, Foresti ML, Cerrutti P, Galvagno M (2013) Bacterial cellulose from simple and low cost production media by Gluconacetobacter xylinus. J Polym Environ 21(2):545–554
6. Krieg NR, Holt JG (1984) Bergey's manual of systematic bacteriology. Yi Hsien Publishing Co
7. Mazotto AM, de Ramos SJ, de Brito LAA, Rocha NU, de Souza SA (2021) How can microbiology help to improve sustainability in the fashion industry? Environ Technol Innov 1(23):101760
8. Wang J, Tavakoli J, Tang Y (2019) Bacterial cellulose production, properties and applications with different culture methods—a review. Carbohydr Polym 219(April):63–76
9. Keshk SM (2014) Bacterial cellulose production and its industrial applications. J Bioprocess Biotech 4(2)
10. McKenna BA, Mikkelsen D, Wehr JB, Gidley MJ, Menzies NW (2009) Mechanical and structural properties of native and alkali-treated bacterial cellulose produced by Gluconacetobacter xylinus strain ATCC 53524. Cellulose 16(6):1047–1055
11. Sederavičiūtė F, Bekampienė P, Domskienė J (2019) Effect of pretreatment procedure on properties of Kombucha fermented bacterial cellulose membrane. Polym Test 78
12. Liu Y, Hu H (2008) X-ray diffraction study of bamboo fibers treated with NaOH. Fibers Polym 9(6):735–739
13. George J, Ramana KV, Sabapathy SN, Bawa AS (2005) Physico-mechanical properties of chemically treated bacterial (Acetobacter xylinum) cellulose membrane. World J Microbiol Biotechnol 21(8–9):1323–1327
14. Široký J, Blackburn RS, Bechtold T, Taylor J, White P (2010) Attenuated total reflectance Fourier-transform Infrared spectroscopy analysis of crystallinity changes in lyocell following continuous treatment with sodium hydroxide. Cellulose 17(1):103–115

15. Yue Y, Han J, Han G, Zhang Q, French AD, Wu Q (2015) Characterization of cellulose I/II hybrid fibers isolated from energycane bagasse during the delignification process: morphology, crystallinity and percentage estimation. Carbohydr Polym 133:438–447
16. Gea S, Reynolds CT, Roohpour N, Wirjosentono B, Soykeabkaew N, Bilotti E et al (2011) Investigation into the structural, morphological, mechanical and thermal behaviour of bacterial cellulose after a two-step purification process. Bioresour Technol 102(19):9105–9110
17. Klemm D, Schumann D, Kramer F, Heßler N, Hornung M, Schmauder H-P et al (2006) Nanocelluloses as innovative polymers in research and application. Adv Polym Sci 205(1):49–96
18. Fahrunisa N, Nugroho A, Hadiansyah D, Hidayat I, Syahrial AZ, Sufiandi S (2021) Mechanical properties improvement of bacterial cellulose film by thermo-hydro-mechanical and electrochemical treatment. In: AIP conference proceedings. AIP Publishing LLC, p 40013
19. Tennakone K (2020) How to make sodium hypochlorite solution. National Institute of Fundamental Studies (NIFS), Sri Lanka. Available from: https://www.nifs.ac.lk/blog/how-make-sodium-hypochlorite-solution
20. Toğrul H, Arslan N (2003) Flow properties of sugar beet pulp cellulose and intrinsic viscosity–molecular weight relationship. Carbohydr Polym 54(1):63–71
21. Azubuike CP, Odulaja JO, Okhamafe AO (2012) Physicotechnical, spectroscopic and thermogravimetric properties of powdered cellulose and microcrystalline cellulose derived from groundnut shells. J Excipients Food Chem 3(3):106–115
22. Ciolacu D, Ciolacu F, Popa VI (2011) Amorphous cellulose—structure and characterization. Cellul Chem Technol 45(1–2):13–21
23. McMurdie HF, Morris MC, Evans EH, Paretzkin B, Wong-Ng W, Zhang Y et al (1987) Standard X-ray diffraction powder patterns from the JCPDS research associateship. Powder Diffr 2(1):41–52

The Effect of Harvesting Time and Finishing Treatment on Properties of Degummed Ramie Fibers

Asri Peni Wulandari, Budi Irawan, Nanang Masruchin, Maya Ismayanti, Rr. Srie Gustiani, Abdul Rohmat, Alifa Ramadhanti, and Marcus Sumadi

Abstract This paper investigated the influence of different harvesting times of stem ramie at 50 and 60 days on the degummed ramie properties. The decorticated fibers degummed with conventional chemical-physical degumming along with finishing treatment and advanced cleaning agents. The analysis of physical and mechanical properties such as tenacity, breaking-elongation, fineness, moisture content, and moisture regain of degummed fibers were measured. The results showed the 60 days-old fibers were more mature than the 50 days-old so the fibers component is more complicated. The cleaning agent and finishing agents worked well to improve fibers quality. The microstructure analysis showed the degradation of a non-cellulosic component on degummed fibers, the XRD analysis showed the increment of crystallinity after degumming. The results indicate that ramie fibers can be harvested at the age of 50 days which reducing the harvesting time, it has fiber tensile 29.5 ± 1.82 g/tex, elongation 15.55 ± 2.9%, fibers fineness 0.69 ± 0.08 tex, moisture content 8.46 ± 0.08%, and fiber humidity 9.24 ± 0.09%. The advance agent can serve as an effective finishing treatment without damaging the properties of the degummed fibers.

Keywords Degumming · Fibers · Harvesting time · Ramie · Finishing treatment

A. P. Wulandari (✉) · B. Irawan
Department of Biology, Faculty of Mathematics and Natural Sciences, Padjadjaran University, Jatinangor, Indonesia
e-mail: asri.peni@unpad.ac.id

N. Masruchin · M. Ismayanti
Research Center for Biomass and Bioproducts, Cibinong, Indonesia

Rr. S. Gustiani
Balai Besar Tekstil, Bandung, Indonesia

A. Rohmat · A. Ramadhanti · M. Sumadi
CV. Haramai Permai Abadi, Bandung, Indonesia

M. A. Rahandi Lubis (ed.), *Proceedings of the 1st International Conference of Lignocellulose*, Green Energy and Technology,
https://doi.org/10.1007/978-981-19-1596-3_14

1 Introduction

Ramie (*Boehmeria nivea*) is one of the ancient natural fibers categorized as bast fibers with excellent quality compare to hemp, jute, or even cotton [1]. Ramie has better tensile strength compared to any other natural fibers [2]. The cellulose on ramie fibers bundles up with the gum contained with pectin 4–5%, lignin 0.8–1.5%, and hemicellulose 14–16% that make the ramie fibers becomes so stiff and coarse [3]. Ramie fibers build from a cellulose long chain that can be an alternative for textiles to substitute linen, cotton, or other fibers [4].

Based on the research by [5] showed that the harvesting time has a significant effect ($p < 0.05$) on the protein content, crude fibers, and lignin forage content of the ramie plant. The Later harvesting time of ramie could decrease the protein content and increase the raw fibers contents and lignin content. Based on this study, every different clone of ramie has a different harvesting time. The Lembang, Indochina, and Ramindo clones need to be harvested on their 6–8 weeks based on their morphological characteristic [5]

Gum content on ramie fibers becomes one of the problems faced to process ramie fibers. Therefore, degumming technology is used in ramie fibers processing to degrade the gum content in ramie fibers. There are a lot of ways to degummed fibers include ramie fibers degumming. (1). Biological degumming that utilize biological agents such as bacteria, fungus, or enzymes (mostly xylano-pectinolytic enzymes [6]) produced from those organisms [7]. (2) Chemical degumming applies the use of a strong base to degrade gum around its ramie cellulose fibers [8]. Another degumming method is physical degumming that obtains in high pressure and happened in certain atmosphere pressure. A lot of degumming methods combine more than 1 type of degumming way to process fibers. The principle of degumming methods is to degrade non-cellulose content and open the micro-fibril structure of fibers [9]. This paper investigated the possibility of the use of the harvesting time of 50 days in the ramie cultivation system and examine the effect of treatment with a further cleansing process on the degummed fibers.

2 Methods

2.1 Materials

The decorticated ramie (harvesting time 50 days-old and 60-days old) was supplied by Central of Ramie, CV RABERSA Farm, Wonosobo, Central Java, Indonesia. The main chemicals used in this study which composed for degumming-scouring-bleaching (DGSB) was supplied CV Haramai Permai Abadi, Bandung, West Java, and the finishing agent as filasoft was purchased from PT Lautan Warna Sari, Bandung, Indonesia.

2.2 *Methods*

2.2.1 Degumming

Pretreatment of the decorticated fibers was soak in the water for around 15 h. The weight of ramie sample used for each degumming process was around 5 g/L and the dosages of chemicals were the weight percentage on raw ramie fibers (w/w) Alkaline scouring was performed using a solution comprising NaOH (10 g/l) and surfactant (2 g/l). This method was adapted from jute fibers processing with minor modification [10], with a slight modification. Bleaching agent of H_2O_2 also include in the degumming solution. Chemical scouring is done at higher temperature high pressure (HTHP) chamber at 100 °C at 4 atm for 2 h.

2.2.2 Finishing Treatment

The degummed-fibers from previous step soaking the fibers in mixed solutions contain (5 g/l) for 24 h with wet finishing agent filasoft that composition are wax dispersion, emulsifier, and component other additives.

2.2.3 Fibers Characterization

Degummed ramie fibers were tested at Balai Besar Tekstil, Bandung, Indonesia and Indonesian Institute of Sciences (LIPI). The tensile strength, fiber fineness, and fiber elongation were tested by MESDAN Tensolab 5000, Italy machine based on SNI 08-1112. The characterization of ramie fibers microstructure were carried out using Scanning Electron Microscope (SEM) using FE-SEM Thermo Scientific Quattro S high vacuum 4 kV. The fiber was tested in Fourier-transform infrared spectroscopy (FTIR)-ATR (Spectrum two, Perkin Elmer, USA) by the Universal Attenuated Total Reflectance (UATR) method. The ramie fibers samples were scanned from 4000 to 400 cm^{-1} at 23 °C $\pm$ 2 °C with an accumulative of 16 scans. The XRD analysis were tested using Shimadzu XRD-7000 MAXima-X, angle 10–35° scan speed 2°/s 40 kV 30 mA, Japan.

3 Result and Discussion

3.1 *Mechanical Characterization of Ramie Fibers*

The experiments carried out that the ramie harvesting time can determine the quality of the fibers produced. An early harvesting time will produce brittle fibers and low yields while the later harvesting times will result in stiff fibers even though the fiber yield is higher [11]. Based on mechanical quality, some parameters might be influenced by the harvesting time of ramie fibers. The tenacity of both raw fibers and degummed fibers was smaller than the 60-days old fibers. It could happen because the harvest age is too fast and resulting in fragile and immature fibers.

The research has shown that mechanical characteristic of the finishing treatment and harvesting time has differences characteristics on different parameters. The tenacity of the fibers treated using the degumming technique and finishing agent has the highest tenacity. The tenacity of the chemically degummed fibers is 26.55 ± 3.51 gram/tex and 29.5 ± 1.82 gram/tex, which is higher than the control fibers which is 27.17 ± 3.23 and 24.67 ± 2.32 gram/tex. The results indicate that the chemical degumming treatment was able to increase the tenacity of the decorticated ramie fibers. The fiber tenacity showed both control and degummed fibers are having better results than the degumming experiment conducted by Biswas et al. [12] (Table 1).

The mechanical properties of the fibers such as the elongation or stretch-ability of the fibers tested. Elongation is the ability of the fibers to stretch compared to their original length. Fibers elongation is closely related to fibers' strength. Elongation is influenced by the binding and elasticity of the fibers while strong fibers will have a high elongation. When compared between raw fibers and fibers processed by chemical degumming, it shows that the stretch of treated fibers is greater than that of china grass fibers. These results indicate that chemical degumming treatment does not damage the strength of the fibers but strengthens and improves the elongation of the fibers. Both china grass fibers and degummed fibers have a higher fibers elongation percentage than cotton which is only around 7–8% [13].

The fineness of the fibers is increased by degumming treatment and the addition of a finishing agent. The purpose of using the agent is to increase the fineness of the fibers and open up the micro fibrils of the fibers. The data observed from the effect of degumming treatment on the fiber fineness shows that degumming affected the fineness of the fibers. The fibers which have the fineness of 3.5 ± 0.22 tex and 2.5 ± 0.23 tex becomes smoother after degumming treatment, the fibers fineness be- comes around of 0.69 ± 0.08 and 0.89 ± 0.18 tex. The fineness of this ramie is certainly smoother than that of Sansiviera and Agave fibers [14]. The water content of ramie fibers can show how much the ability of ramie fibers to absorb water. This

Table 1 Characteristic of Ramie fibers

	Tenacity (gram/tex)	Elongation (%)	Fineness (tex)	Moisture content (%)	Moisture regain (%)
Raw fibers 60 days	27.17 ± 3.23	11.9 ± 2.05	3.5 ± 0.22	9.7 ± 0.007	10.78 ± 0.02
Degummed-finished fibers 60 days	26.55 ± 3.51	12.56 ± 1.70	0.89 ± 0.18	7.8 ± 0.31	8.5 ± 0.30
Raw rami fibers 50 days	24.67 ± 2.32	15.48 ± 1.96	2.5 ± 0.23	9.62 ± 0.40	10.65 ± 0.50
Degummed-finished fibers 50 days	29.5 ± 1.82	15.55 ± 2.90	0.69 ± 0.08	8.46 ± 0.08	9.24 ± 0.09

characteristic is one of the crucial things to determine the quality of the fibers. Fibers moisture can be seen by assessing the form of moisture content and moisture regain. Moisture regain is a parameter that determine a type of fibers can absorb water in the environment. The greater the moisture regain possessed by fibers, the greater the ability of the fibers to absorb water and can be called hygroscopic [8].

3.2 Thermal Characteristics of Ramie Fibers

The thermal stability of ramies fibers was tested with TGA and DTG analysis. TGA-DTG showed 3 steps of degradation. All four samples of ramie fibers have different thermal qualities. On Fig. 1. The first degradation occurs at 100 °C for all samples which is the evaporation of the liquid present in the fibers, the decomposition happened around 10%. The second degradation of fiber occur at 350–400 °C with mass loss around 12–13.5%. Thermal decomposition that occurs in cellulose generally occurs at temperatures of 325–400 and 250–350 °C [15]. The third decomposition happened at > 700 °C that might be some residual degradation. In the results of the thermal decomposition observations, all the tested fibers decomposed in the range of 300–400 °C The difference is the percentage of the weight loss of each sample. Degummed fibers from 50 and 60-day-old decomposed at 12.5%/m, which was lower than control fibers that 50 and 60 day-old, which around 17.5–20%/m^2. This shows that the decomposition of the degummed fibers is smaller than that of the control fibers.

3.3 Morphology of Ramie Fibers

Morphological observations of ramie fibers showed mixed results. This morphological observation was carried out to compare the untreated fibers and degummed fibers treated with WFA. Based on SEM observations, The 50 and 60 days-old fibers showed gum on the surface of the cellulose fibers, the chemical degumming process can remove most of the non-cellulose content from the surface of the ramie fibers. The decay of the non-cellulose content can be observed that the surface of the degummed fibers looks smoother when compared to the control fibers. In addition, the chemical degumming process also makes the cellulose micro fibrils exposed and visible in the image. However, it was also seen that the gum content was not completely dissolved in the degumming process. The results of these observations indicate that chemical degumming is able to degrade the gum content in ramie fibers. In both age variations, it was proven that the degumming treatment was able to dissolve the gum content in the fibers (Fig. 2).

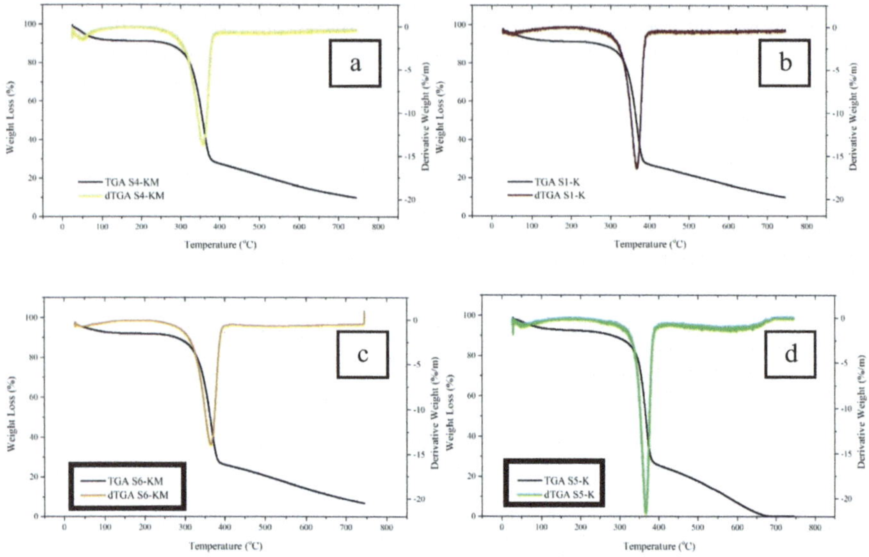

Fig. 1 Thermal analysis of Ramie fibers. **a** Degumming-finished fibers 60 days; **b** raw fibers 60 days; **c** degumming-finished fibers 50 days; **d** raw fibers 50 days

3.4 Ramie Fibers XRD Analysis

The x-ray diffraction spectrum (XRD) (in Fig. 3) shown the XRD analysis of 5-days old and 6-days-old fiber, along with finishing agent treated fibers. The samples showed two peaks observed at 2θ angles of 15.1° and 16.8° indicate high cellulose content. The mechanical properties of cellulose fibers are highly related to their crystallinity. The fibers tenacity, initial modulus, and hardness increase along with the flexibility and softness. Most plant-derived cellulose is highly crystalline while all the other components are amorphous [16]. The wave peaks at harvest age 50 and 60 days after experiencing the end showed a higher increase than the untreated fibers. That is approved that more gummies have been removed that covered the surface of the fibers.

3.5 FT-IR Analysis

The result of FTIR can show changes in the chemical composition of ramie fibers harvesting time at 50 and 60 days, including the effect of the finishing treatment. Figure 4 could identify the presence of an oxidation reaction that can transform the macromolecular structure of cellulose in the range of values between 3000 and 3600 cm^{-1} associated with –OH stretching vibrations, especially the peak at 3339 cm^{-1} which indicates the presence of a hydroxyl group in the polysaccharide

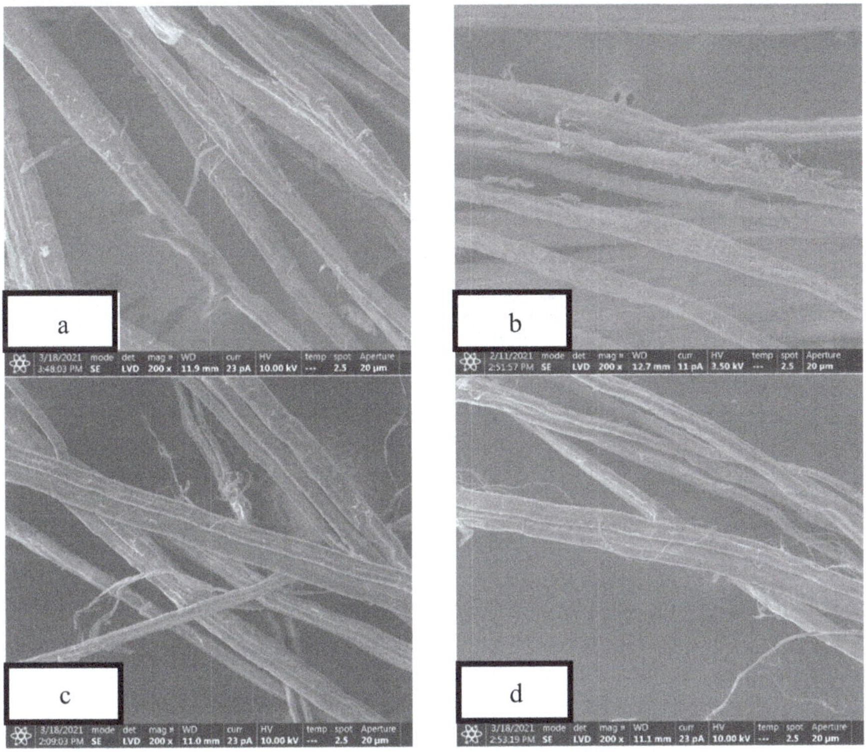

Fig. 2 SEM of Ramie fibers, **a** degumming-finished fibers 60 days; **b** raw fibers 60 days; **c** degumming-finished fibers 50 days; **d** raw fibers 50 days

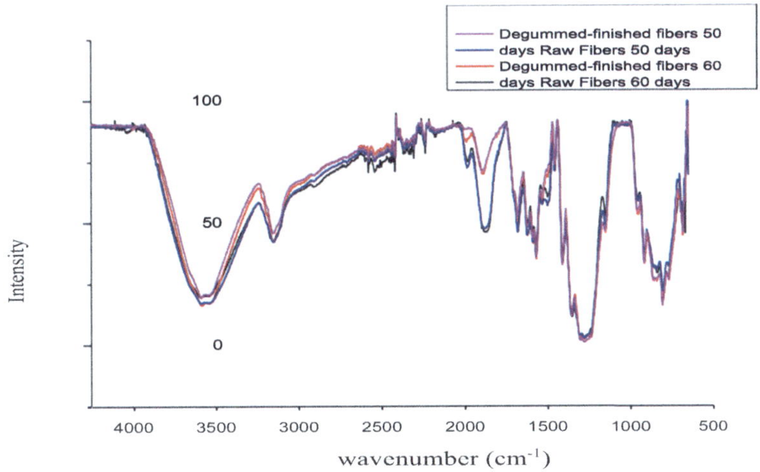

Fig. 3 X-ray spectra of ramie fibers

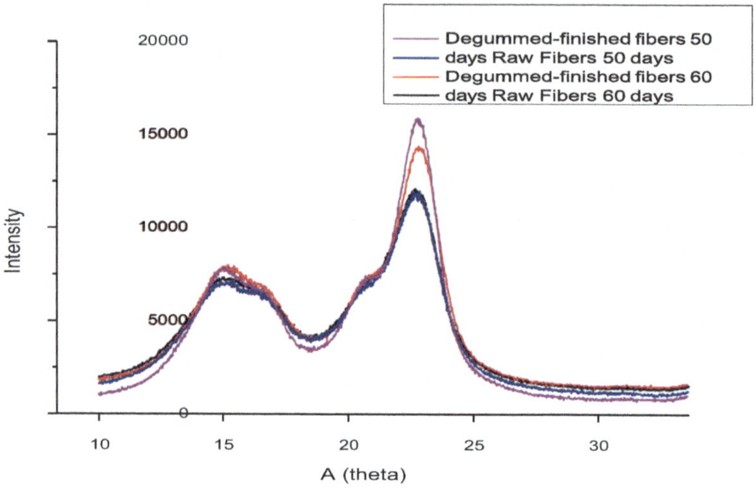

Fig. 4 Comparisons of FT-IR spectra of raw ramie and degummed fibers: 50 days and 60 days

[17], hydroxyl groups with large amounts of cellulose and lignin in fibers [18]. In addition, the fibers also show reactive intensities at 2286 and 2290 cm^{-1}. Those peaks are indicated by the presence of C–H groups that can be found in polysaccharides [19]. After the reaction with the degumming method and finishing-agent treatment, the relative intensity at 3600–3000 cm^{-1} changed in the fibers treatment at 50 days and 60 days, indicating that the content of hydroxyl groups increased significantly after treatment. The hydrogen structure in the cellulose-lose bonds is likely to cause an increase in the mechanical properties of the degummed fibers.

After the degumming process, the FTIR results showed the presence of C–O groups. The peak at 1765–1715 cm^{-1} indicates the presence of an ester or acetyl group from non-cellulose compounds from pectin, uronic, or hemicellulose or the presence of p-coumarate groups, functional groups of lignin or hemicellulose [20]. The decrease of CO group intensity in the infrared spectrum can be an indication of non-cellulose compounds degradation in the degumming process.

The degummed fibers processes with oxidizing agent show a peak around 1650 cm^{-1} as the characteristic of pectate. This peak decreased on treated fibers infrared spectrum indicates that the pectate content was significantly reduced. It means that pectin decrease in the final cleansing treatment. The infrared spectrum analysis showed the gum were successfully removed from ramie by chemical degumming technology and finishing treatment process.

4 Conclusion

This paper provides information on the novel steps in the degumming process with a finishing treatment agent. This additional cleaning effect causes the non-cellulose composition to effectively remove from ramie fibers confirmed by FTIR, SEM, and XDR analysis. The FTIR analysis shows the decrement of pectin content. It confirmed by the SEM analysis as the micro fibrils of fibers showed after the degumming process and the non-cellulose component removed from the surface of the fibers. The XRD analysis showed that the degumming process could increase the crystallinity of ramie fibers. The best results from this study were Degummed hemp fibers aged 50 days had ductility 29.5 ± 1.82 g/tex, elongation 15.55 ± 2.9%, fibers fineness 0.69 ± 0.08 tex, moisture content 8.46 ± 0.08%, and return humidity 9.24 ± 0.09%. The 50 days harvesting time method can shorten the harvest time and will be a benefit for the ramie plant's productivity. The advance agent can serve as an effective finishing treatment without damaging the property of the degummed fibers.

Acknowledgements The authors are grateful to the Ministry of Finance of the Republic of Indonesia, with PRN-LPDP funding for the 2020–2021 Fiscal Year. We also thank the Japan Society for the Promotion of Science (JSPS) for partial funding of the publication fee.

References

1. Sathishkumar TP, Navaneethakrishnan P, Shankar S, Rajasekar R, Rajini N (2013) Characterization of natural fiber and composites—a review. J Reinf Plast Compos 32(19):1457–1476
2. Ramesh M (2018) Hemp, Jute, Banana, Kenaf, Ramie, Sisal Fibers. Elsevier Ltd.
3. Cheng L et al (2020) Ramie-degumming methodologies: a short review. J Eng Fiber Fabr 15
4. Goda K, Sreekala MS, Gomes A, Kaji T, Ohgi J (2006) Improvement of plant based natural fibers for toughening green composites—effect of load application during mercerization of ramie fibers. Compos J 37:2213–2220
5. Maideliza T, Mayerni R, Rezki D (2017) Comparative study of length and growth rate of ramie (Boehmeria nivea L. Gaud.) bast fiber of Indonesian clones. Int J Adv Sci Eng Inf Technol 7(6):2273–2278
6. Singh A, Kaur A, Kumar A, Mahajan R (2016) Xylano-pectinolytic synergism from a bacterial isolate for ecofriendly processing of jute fabric. J Clean Prod 10414:6–11
7. Gong J, Zhang Q, Lou J, Zhang T, Li H, Li Z (2018) Investigation of the degradation of biorecalcitrance in Apocynum venetum fiber biodegumming. J Nat Fibers 00(00):1–12
8. Fan XS, Liu ZW, Liu ZT, Lu J (2010) A novel chemical degumming process for ramie bast fiber. Text Res J 80(19):2046–2051
9. Jiang W, Han G, Zhang Y, Liu S, Zhou C (2017) Monitoring chemical changes on the surface of Kenaf fiber during degumming process using infrared microspectroscopy. Sci Rep 7:1–8
10. Ibrahim NA, Amr A, Eid BM (2010) Innovative multi-functional treatments of ligno-cellulosic jute fabric. Carbohydr Polym 82(4):1198–1204
11. Lee CH, Khalina A, Lee SH, Liu M (2020) A comprehensive review on bast fibre retting process for optimal performance in fibre-reinforced polymer composites. Adv Mater Sci Eng
12. Biswas D, Chakrabarti SK, De S, Paral R, Biswas D (2016) Eco-friendly degumming technology for Ramie fiber eco-friendly degumming technology for Ramie fiber. 0478

13. Arul Marcel Moshi A, Ravindran D, Sundara Bharathi SR, Suganthan V, Kennady Shaju Singh G (2019) Characterization of new natural cellulosic fibers—a comprehensive review characterization of new natural cellulosic fibers—a comprehensive review
14. Widyasanti A, Napitupulu L, Thoriq A (2020) Physical and mechanical properties of natural fiber from Sansevieria trifasciata and Agave sisalana. Earth Environ Sci 1(1)
15. Chen M (2016) Chapter 12—Thermal analysis. Tsinghua University Press Limited
16. Bhattacharya SD, Shah SR (2002) Degumming of ramie: the role of the individual constituents of the gummy material coloration technology
17. Poletto M, Pistor V, Zeni M, Zattera AJ (2011) Crystalline properties and decomposition kinetics of cellulose fibers in wood pulp obtained by two pulping processes. Polym Degrad Stab 96(4):679–685
18. Choi HY, Lee JS (2012) Effects of surface treatment of ramie fibers in a ramie/poly (lactic acid) composite. 13(2):217–223
19. Wulandari AP, Septarini D, Zainuddin A (2017) Study on structure and morphology (Boehmeria nivea) in the irregular and regular parts of the fiber after biodegumming. IOP Conf Ser Earth Environ Sci 65(1):1–7
20. Zheng L, Du Y, Zhang J (2018) Degrading enzymes degumming of Ramie® bers by alkalophilic bacteria and their polysaccharide-degrading enzymes, 8524

Effect of Ramie Particle Content to the Physical and Mechanical Properties of Rigid Polyurethane Foam Composite

Rihab Stafian Nisa, Sasa Sofyan Munawar, Dede Hermawan, Deni Purnomo, Bambang Subiyanto, Ismadi, Ahmad Syahrir, Fazhar Akbar, and Deni Poniman

Abstract The ramie particle was used as a filler for the manufacture of rigid polyurethane foam composites (RPUFC). The purpose of this research is to investigate the effect of ramie particle contents to physical and mechanical properties of RPUFC. The free rise foaming was observed to determine the reactivity of rigid polyurethane foam (RPUF) with the addition of ramie particles. The RPUFC with five different volume contents of ramie particle (0, 2.5, 5, 7.5 and 10 wt%), with a board density of 50 kg/m^3 were manufactured. The polyols, ramie particles, and isocyanate were mixed, then poured and formed in a closed mold. The moisture content (MC), water absorption (WA) thickness swelling (TS), and internal bonding strength (IB) properties were determined according to JIS A 5908-2003. The flexural strength (FS) properties was determined according to ASTM D790. The results showed that the addition of ramie particles in the RPUF has accelerated the free rise foaming. The physical properties (MC, WA, TS) were increased with increasing ramie particles in the RPUFC. The IB value was increased by increasing the ramie particles in the RPUFC. The modulus of rupture (MOR) and modulus of young (MOE) values of the RPUFC between the addition and without the addition of ramie particle were not significantly different. In fact, with the addition of 2.5% ramie particles, the MOR value of RPUFC was higher than RPUFC without using ramie particles. Thereby, the addition of ramie particles in the RPUFC means reducing the use of polyurethane, hence the product is more environmentally friendly and can reduce production costs.

R. S. Nisa · D. Hermawan
Faculty of Forestry and Environment, IPB University, Bogor, Indonesia

S. S. Munawar (✉) · D. Purnomo · B. Subiyanto · Ismadi · A. Syahrir · F. Akbar
Research Center for Biomass and Bioproducts, National Research and Innovation Agency, Bogor, Indonesia
e-mail: sasasofyanm@biomaterial.lipi.go.id

D. Poniman
Faculty of Engineering, Subang University, Subang, Indonesia

M. A. Rahandi Lubis (ed.), *Proceedings of the 1st International Conference of Lignocellulose*, Green Energy and Technology,
https://doi.org/10.1007/978-981-19-1596-3_15

Keywords Rigid polyurethane foam composite · Ramie particle content · Polyol · Isocyanate

1 Introduction

The population in Indonesia reaches 270.20 million people in 2020 [1]. This case conduce a high demand for house building components, including panels for walls, floors, and roofs. Along with the increasing demand for housing, new innovations in the construction world have emerged to achieve cost and time effectiveness, as well as good quality. The high demand for building components is the reason for the development of prefabricated components. Pre-fabricated components are structural components that are manufactured to factory standards that located far from the building, and then transported to the location for assembly. These components are mass produced to construct buildings in a short time [2]. Pre-fabricated building components are widely used because the construction time is faster [3]. Pre-fabrication is widely used in the pharmaceutical industry, clean rooms, modern factories, food processing industries and storage warehouses, cold rooms and cold storage, mining industries, base camps, project mess and site offices and so on [4].

Pre-fabricated materials that are currently being developed are sandwich panels. Sandwich panels can be made with a core material of polyurethane foam (PUF) and a coating of plywood, zincalume, glass fiber reinforced cement (GRC), and others. This sandwich panel product is designed to get a lightweight structure but has high rigidity and strength, making it suitable to withstand bending, impact loads, dampening vibrations and sound. Sandwich panels can be used or applied to building walls, tables, doors, ceilings and wooden floors [5].

One of the core materials for sandwich panel products that is commonly developed is RPUF. RPUF has low thermal conductivity and good mechanical properties, hence RPUF can be used as an insulating material in pipelines, automotive equipment, refrigeration equipment, and building materials [6]. RPUF products are lightweight, comfortable and durable. RPUF in industry is usually modified by using fillers to improve properties such as density, dimensional stability, and retraction of the mold [7]. On the other hand, the fillers are useful for modifying mechanical properties and reducing production costs [6].

One of the RPUFC filler come from lignocellulosic materials, such as wood fibers, natural fibers, and others. The advantage of using natural fibers on manufacture of RPUFC is due to the presence of hydroxyl groups on the surface of lignocellulosic fibers that are able to react with isocyanate groups. This interaction triggers the formation of an excellent interfacial bond between the fiber and the polyurethane [8]. The use of treated natural fibers can improve the thermal stability, energy absorption, physical and mechanical properties of PUF [9]. Previous study [10] also investigated the effect of straw fiber content on the characteristics of the RPUF and resulted in increased sound absorption properties, decreased thermal conductivity, increased insulating ability, and decreased compressive strength.

Another lignocellulosic material that can be used for composites is ramie (*Boehmeria nivea* (L.) Gaud) bast fiber [11]. Ramie stem that have been separated/ decorticated from the bast fiber are usually used for fertilizer and animal feed [12]. Ramie particles made from decorticated ramie stem have been used as reinforcement in polyester resin composites. The results showed that the tensile strength of a composite increased up to 33.8% by adding 45% of ramie particles [13]. Particleboard was also developed by using ramie particles and showed good mechanical properties based on standard [14]. However, there has been no research on the manufacture of rigid polyurethane foam composites using ramie particles as filler. Therefore, investigating the effect of ramie particle to the physical and mechanical properties of RPUFC was conducted in this study.

2 Materials and Methods

2.1 Preparation of Materials

The ramie chips from decorticated ramie stem purchased from PT. Ramindo (Wonosobo Regency) was used. The chips were dried under sunlight and then oven-dried at 70 °C for 12 h. The dried chips were ground to particles with a ring flaker (Pallman Maschinenfabrik, Zweibrucken, Germany). The particles were screened with a vibrating sieve, and the particles that remained between the aperture sizes of 4 and 14 mesh were used as the filler materials. The particles were oven-dried at 60 °C for 24 h to reach a moisture content less than 7%.

Isocyanate (ARK R32040/P) and Polyol (ARK R32040/P) purchased from PT. Anugerah Raya Kencana (Tangerang, Indonesia) were used. The appearance of isocyanate was dark brown liquid, with specific gravity @ (25 °C) of 1.22–1.26 g/ml and viscosity @ (25 °C) about 150–250 mPas. Whereas, the appearance of polyol was Light Yellowish Liquid, with specific gravity @ (25 °C) of 1.04–1.06 g/ml and viscosity @ (25 °C) about 110–210 mPas. The materials used was shown in Fig. 1.

(a) (b) (c)

Fig. 1 The materials used for RPUFC: OPT particles (**a**), isocyanate (**b**), and polyol (**c**)

Table 1 Ramie particle content on the RPUFC

Type of composite	D50R0	D50R25	D50R50	D50R75	D50R100
Particle content	0%	2.5%	5%	7.5%	10%

2.2 RPUF Reactivity Test

Rigid polyurethane is a mixture of isocyanates and polyols before being used for various applications. With the addition of lignocellulosic material, there will be an effect on the polyurethane foaming process. Therefore, the free rise foaming test was conducted to determine the effect of adding ramie particles to the polyurethane foaming process. Parameters observed was tack free time. Tack free time is the time it takes polyurethane to stop reacting and the surface is not sticky [15].

The polyol solution and ramie particles were mixed by stirrer in a 1000 ml plastic container until polyol evenly distributed over the surface of ramie particles. Then, isocyanate was added to the mixture of polyol and ramie particles and stirred using a stirrer at a speed of about 2000 rpm for 10 s. The ratio of polyol and isocyanate was 1:1.1 (w/w %). The ratio ramie particle to polyurethane are shown in Table 1. After that, the foam was expanded freely. Cream time and tack free time were measured by using a stopwatch and recorded [16].

2.3 Manufacture of RPUFC

The manufacture of RPUFC began by mixing polyols with ramie particles using a stirrer in a plastic container until polyol evenly distributed over the surface of ramie particles. Then, isocyanate was added to the mixture of polyol and ramie particles and stirred using a stirrer at a speed of about 2000 rpm for 10 s. The ratio of polyol and isocyanate was 1:1.1 (w/w %) according to the technical instructions of the polyurethane manufacturer. The mixture of polyol, ramie particles and isocyanate was poured and formed into a closed mold with a size of 40 × 40 × 5 cm for 45 min (Fig. 2). RPUFC were made with a target density of 50 kg/m^3 and 4 differences in ramie particle content (2.5, 5, 7.5 and 10% of the composite weight). RPUFC without using ramie particles was made as a control. RPUFC were conditioned two week prior testing. The types of RPUFC shown in Table 1. Each type of RPUFC made 3 replications.

2.4 RPUFC Testing

The RPUFC were tested for their physical and mechanical properties. The MC, WA, TS and IB tests were carried out using samples with the sizes of 5 × 5 × 5 cm,

Fig. 2 Manufacture of RPUFC

according to JIS A 5908–2003 [17]. Determination of density was expressed in the results of the comparison between the weight and volume of the board. The MC was calculated based on the difference between the initial weight and the final weight after oven drying for 24 h at a temperature of 103 ± 2 °C. The WA test was conducted by measuring the difference in weight before soaking and weight after soaking for 24 h. The TS test was conducted by measuring the difference in thickness before soaking and thickness after soaking for 24 h The IB test was conducted by samples that glued to two iron blocks with epoxy glue and allowed to dry for 48 h. The two iron blocks were then pulled perpendicular to the sample surface using universal testing machine (UTM, 50 kN, Shimadzu, Japan) until the maximum load with a loading speed of 2 mm/min. The FS test was carried out using three point bending (40 × 5 × 5 cm) by using universal testing machine (UTM, 50 kN, Shimadzu, Japan), according with ASTM D790 [18] (Fig. 3). Data analysis in this study was used to determine the effect of oil palm trunk fiber content on a number of responses. Testing was used three replications for each test parameter.

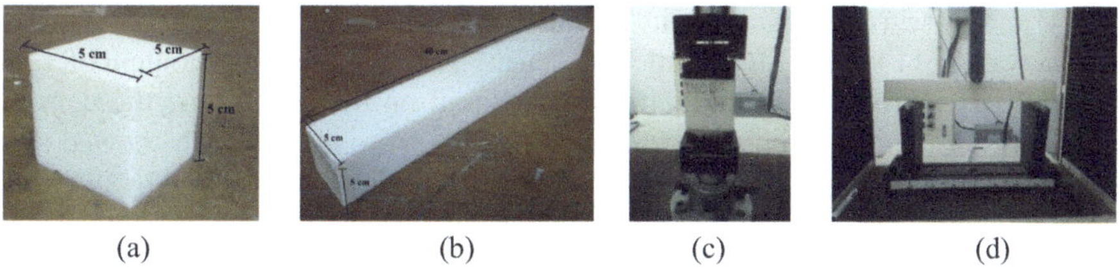

Fig. 3 Sampel size test for, **a** MC, WA, TS, IB, and **b** FS, and testing samples of **c** IB, and **d** FS

Table 2 Tack free time of RPUF

Particle content (%)	Tack free time (s)
0	324
2.5	316
5	302
7.5	296
10	279

3 Results and Discussion

3.1 *RPUF Reactivity*

RPUF tack free time values are shown in Table 2. Tack free time of RPUF in this study were ranged from 279 to 324 s, with the shortest foaming time indicated in the 10% particle content addition (279 s) and the longest time indicated in the RPUF control (324 s). Table 2 showed that increasing ramie particles in the RPUF has resulted in shorter foaming times. Previous study has stated that the addition of natural fibers in the RPUF was caused an increase in the viscosity of RPUF, which has an impact on the acceleration of foaming time [19]. Another study showed that the tack free time decreased with the addition of kenaf fiber in the RPUF [16].

3.2 *Density*

Density is an important parameter that can affect the mechanical properties of RPUF [20]. Another research was showed that the higher density of PUF with natural fiber hybrids can increase its strength [21]. The density of polyurethane depends on the pore size. The enlarged pore size was caused a decrease in foam density [22].

The density average value of RPUFC were ranged from 42.11 to 45.40 kg/m^3 (Fig. 4) and close to the target density (50 kg/m^3). The RPUFC density at 0% particle content was 42.11 kg/m^3. Meanwhile, the density value with particle content of 2.5–7.5% increased from 44.99 to 45.40 kg/m^3. Based on the results that the density value tends to increase with the addition of particle. Another research showed that the addition of particle fraction, the pores of the PUF matrix were filled by the fiber particles themselves [23]. This has an impact on the pore size of the composite which shrinks, resulting in an increasing density value. More porous in the RPUFC at 10% particle content was decreased a density. However, the density value of the RPUFC in this study tends to be uniform, because based on statistical tests, the addition of ramie particle to the RPUFC does not have a significant effect.

Fig. 4 Density of RPUFC

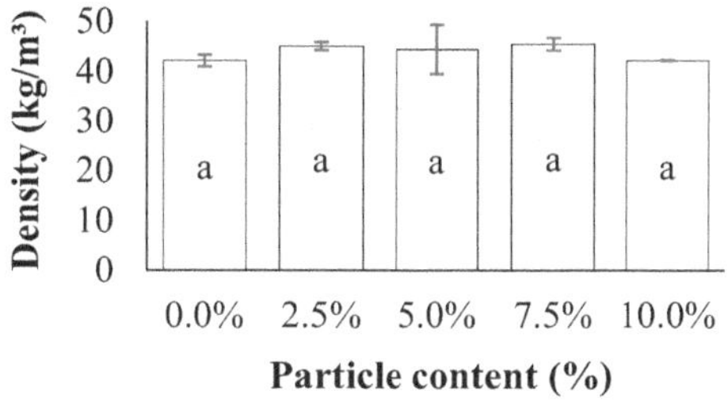

Fig. 5 Moisture content of RPUFC

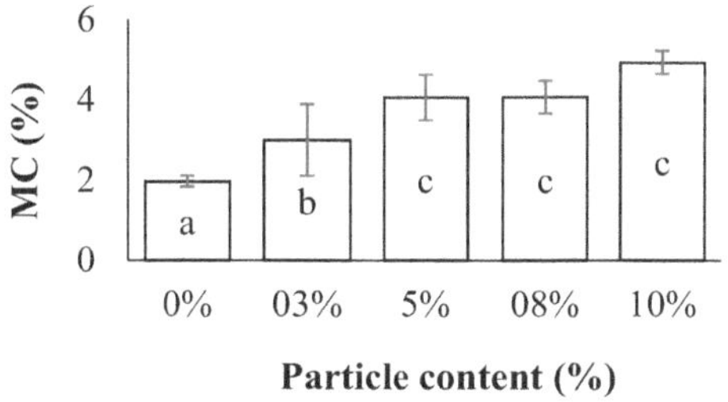

3.3 *Moisture Content*

The MC average value of RPUFC were ranged from 1.99 to 4.70% (Fig. 5). The results showed that MC was increased by increasing of ramie particle in the RPUFC. The same thing happened in other studies that the MC was increased by increasing the natural particle content in PUF [24]. This is due to the presence of –OH groups in the fiber. The higher particle content resulted a greater number of free –OH groups derived from cellulose and hemicellulose, which causes an increase in the MC. The highest value of MC against RPUFC without particles was 2.5 times. The addition of ramie particle had a significant effect on the MC of RPUFC, but the content of 5, 7.5, and 10% did not give a statistically significant effect.

3.4 *Water Absorption*

Rapid WA causes low dimensional stability and foam is damaged. Water resistance is an important characteristic of RPUFC, especially for outdoor applications [25]. The WA average value of RPUFC were ranged from 25.95 to 85.78% (Fig. 6). This is comparable to other study [26] that resulted the WA of bamboo charcoal PUF composites were in the range of 36.1–88.6%.

The results of this study indicated that WA was increased by increasing rami particle content in the RPUFC. The increase in WA is due to the presence of –OH groups in the fiber which are able to bind water molecules [24]. In addition, the

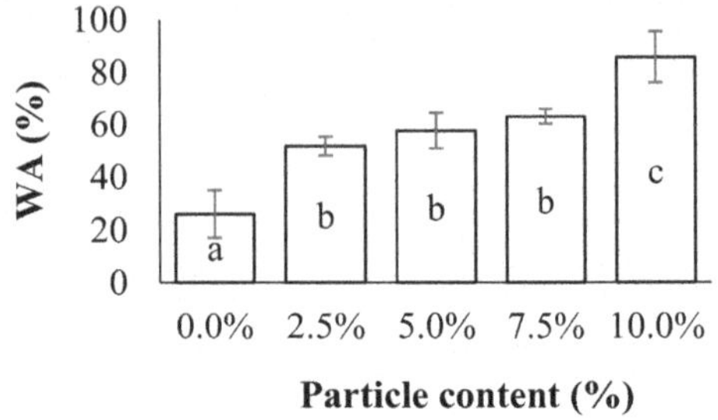

Fig. 6 Water absorption of RPUFC

increase particle content can also cause the foam cell structure to be more heterogeneous. Another research showed that the increasingly heterogeneous foam cell structure can increase the WA of PUF [27]. The highest value of WA against RPUFC without particles was 4 times. In this study, the ramie particle content had a statistically significant effect on the WA capacity of RPUFC. However, the ramie particle content of 2.5, 5, and 7.5% did not give a statistically significant effect.

3.5 Thickness Swelling

TS properties determined the dimensional stability of the RPUFC. Determination of dimensional stability is also very important for RPUFC applications in various fields [28]. The TS average value of RPUFC were ranged from 0.51 to 1.36% (Fig. 7). The TS of RPUFC tends to increase with increasing particle content. The TS of RPUFC with particle content of 7.5% was higher and significantly different, while those of 2.5, 5, and 10% were not statistically significant. The highest value of TS against RPUFC without particles was 2.5 times. These results are different from the results of other studies using treated fiber [26], that showed a TS value were decreased by increasing bamboo charcoal content in the RPUFC. It was caused by the stability of bamboo charcoal was better than untreated fiber. The high value of TS of 7.5% particle content caused by increasingly heterogeneous foam cell structure.

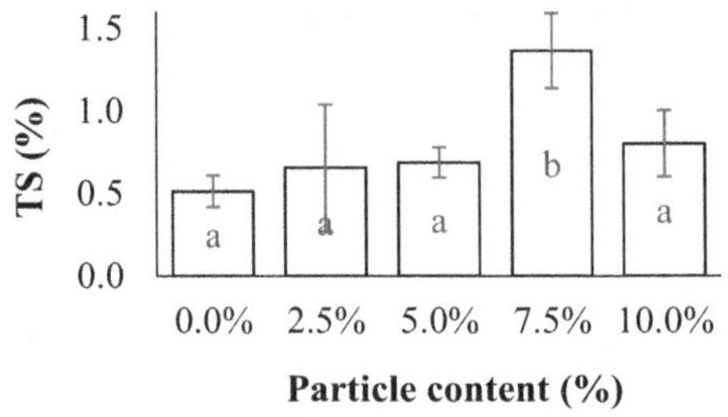

Fig. 7 Thickness swelling of RPUFC

3.6 Flexural Strength (MOE and MOR)

The MOE average value of RPUFC were ranged from 4.85 to 6.00 N/mm^2 (Fig. 8). Meanwhile, the average MOR value were ranged from 0.34 to 0.45 N/mm^2 (Fig. 9). The MOE and MOR values of the control RPUFC with the RPUFC containing ramie particles were almost the same. In fact, the MOR value of RPUFC with 2.5% ramie particles was greater than that of the control and other treatments. These results are also the same as the results of previous studies [9]. On the other hand, the addition of particle did not give a statistically significant effect on the MOE and MOR of the RPUFC. This is due to the presence of free –OH groups from lignin which causes competition with –OH from polyols to interact with NCO from isocyanates. This will trigger the formation of an irregular cell structure in the foam composite as the fiber increases. Irregular cell structure is associated with poor load transfer, thereby reducing its strength [29].

The highest MOE was found on RPUFC with 7.5% (6.00 N/mm^2) and 0% (5.82 N/mm^2) of particle content, while the highest MOR was found on RPUFC with 2.5% (0.45 N/mm^2) and 0% (0.42 N/mm^2) of particle content. This could be due to the regular arrangement of foam cells. This regular arrangement of foam cells causes better mechanical strength [30]. Another research also showed a lower MOE value of the composite with added fiber than the control [31]. This is because the addition of fiber to the PU matrix can damage the structure of the foam. The presence of fillers or the addition of fiber tends to modify the PU microstructure into smaller cells and become non-uniform was found in previous research [32].

Fig. 8 MOE of RPUFC

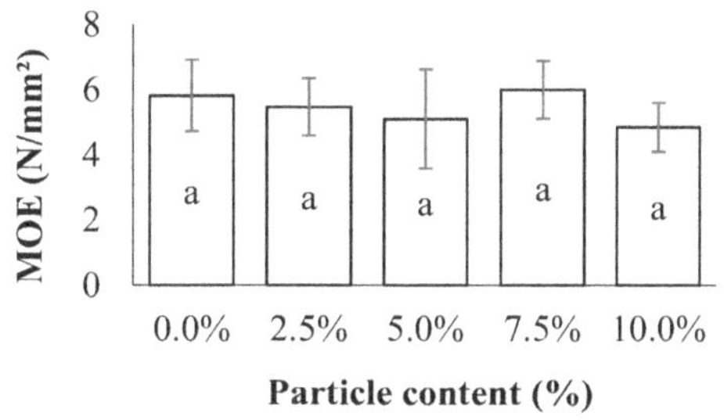

Fig. 9 MOR of RPUFC

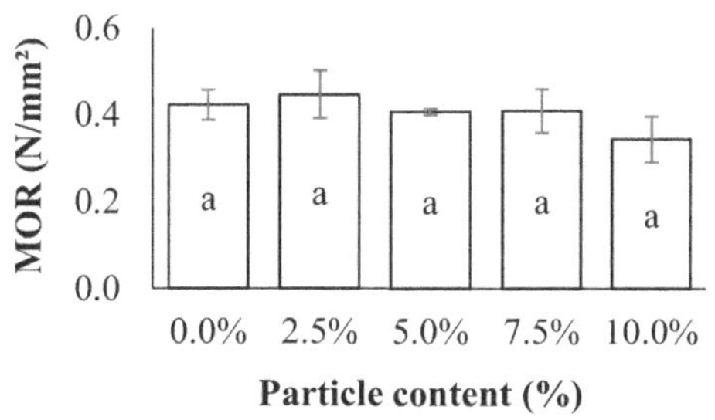

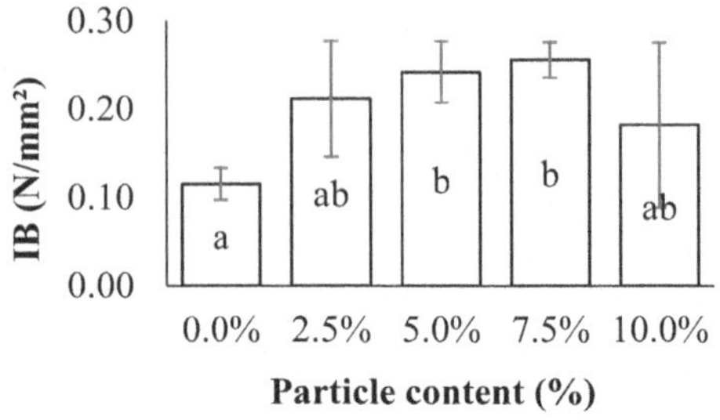

Fig. 10 Internal bond of RPUFC

3.7 Internal Bonding

IB test was carried out to provide an overview of the adhesive properties between the foam matrix and the fiber filler. The average value of RPUFC bonding strength were ranged from 0.11 to 0.26 N/mm^2 (Fig. 10). Based on the data, the IB value were increased by increasing rami particle up to 7.5% and was decrease at 10% or ramie particle. The 5 and 7.5% particles were produced the highest IB strength values of RPUFC and were statistically significantly different, but the particle content of 2.5 and 10% were not statistically significant. MC can affect to IB of RPUFC. The higher MC could lead to the reduction of adhesive crosslinking and accordingly weaken the bonding performance [33]. RPUFC with 10% particle content showed a highest MC value, and it was a reason that IB strength value was decreased. However, the IB value in this study is still lower than the IB value of previous studies, which had an IB value in the range of 0.3–0.6 N/mm^2 [34].

4 Conclusion

The addition of ramie particles in the RPUF has accelerated the free rise foaming. The physical properties (MC, WA, TS) were increased with increasing ramie particles in the RPUFC. The highest value of MC, WA, TS against RPUFC without particles were 2.5–4 times. The IB value was increased by increasing the ramie particles in the RPUFC. The MOR and MOE values of the RPUFC between the addition and without the addition of ramie particle were not significantly different. In fact, with the addition of 2.5% ramie particles, the MOR value of RPUFC was higher than RPUFC without using ramie particles. The addition of 2.5–5% ramie particles into the RPUFC showed optimum results in mechanical properties. Thereby, the addition of ramie particles in the RPUFC means reducing the use of polyurethane, hence the product is more environmentally friendly and can reduce production costs.

Acknowledgements Authors wishing to acknowledge the National Research and Innovation Agency of the Republic of Indonesia for its financial support through the National Research Program on Earthquake-Resistant, Fire-Resistant, Quick-Build and Affordable Building Technology, and JSPS Alumnae Association Indonesia (JAAI) for publication support.

References

1. Badan Pusat Statistik (BPS) (2020) Potret Sensus Penduduk 2020: Menuju Satu Data Kependudukan Indonesia. BPS RI, Jakarta
2. Giri PJP (2019) Design of building structure of the Faculty of Economics and Business, University of Muhammadiyah Prof. Dr. Hamka (Uhamka) uses precast concrete [Thesis]. Universitas Atma Jaya Yogyakarta, Yogyakarta
3. Putra RGRB, Susanto D (2017) Prefabricated house in real estate business development in Jabodetabek. IOP Conf Series Earth Environ Sci 99(0):1–9
4. Usman R (2019) Improving the production quality of wall plate and roof sandwich panels using the Six Sigma method. Spektrum Industr 17(1):1–91
5. Nugroho N, Bahtiar ET (2010) Sandwich panel performance analysis for pre-fabricated houses. Jurnal Ilmu Pertanian Indonesia 15(3):158–162
6. Silva MC, Takahashi JA, Chaussy D, Belgacem MN, Silva GG (2010) Composites of rigid polyurethane foam and cellulose fiber residue. J Appl Polym Sci 117:3665–3672
7. Saliba CC, Ore´fice RL, Carneiro JRG, Duarte AK, Schneider WT, Fernandes MRF (2005) Effect of the incorporation of a novel natural inorganic short fiber on the properties of polyurethane composites. Polym Test 24(7):819–824
8. Li P, Tao Y, Sheldon QS (2014) Modal analysis of upright piano sound board by combining finite element analysis and computer aided design. Bio Resour 46(3):376–384
9. Chang L (2014) Improving the mechanical performance of wood fiber reinforced bio-based polyurethane foam [Tesis]. University of Toronto, Toronto
10. Tao Y, Li P, Cai L (2016) Effect of fiber content on sound absorption, thermal conductivity, and compression strength of straw fiber-filled rigid polyurethane foams. Bioresources 11(2):4159–4167
11. Munawar SS, Umemura K, Tanaka F, Kawai S (2008) Effects of alkali, mild steam, and chitosan treatments on the properties of pineapple, ramie, and sansevieria fiber bundles. J Wood Sci 54(1):28–35
12. Suryana S, Rochana A, Susilawati I, Indiriani NP (2017) Ramie (Boehmeria nivea) plant nutrient quality as feed forage at various cutting ages. Anim Prod 19(2):111–117
13. Haroen WK (2016) Diversifikasi Serat Pulp Untuk Produk Inovatif. J Lignocellulose Technol 1:15–25
14. Guo Y, Xia Z, Xu J, Dai X (2018) Steam injection-pressing preparation and forming mechanism of binderless ramie stem particleboard for modern furniture. In: International conference on advanced chemical engineering and environmental sustainability (ICACEES 2018) ISBN: 978-1-60595-571-1
15. Kraitapea N, Thongpina C (2016) Influence of recycled polyurethane polyol on the properties of flexible polyurethane foams. Energy Procedia 89:186–197
16. Soberi NSM, Rahman R, Zainuddin F (2017) Effect of kenaf fiber of morphology and mechanical properties of rigid polyurethane foam composite. Mater Sci Forum 888:188–192
17. Banik I, Sain MM (2008) Role of refined paper fiber on structure of water blown soy polyol based polyurethane foams. J Reinf Plast Compos 27(14):1515–1524
18. Japanese Standard Association [JSA] (2003) JIS A 5908: 2003. Particleboards. Japanese Standard Association, Japan
19. American Society for Testing Materials (2002) ASTM D790-02. Standard test methods for flexural properties of unreinforced and reinforced plastics and electrical insulating materials. American Society for Testing Materials, Philadelphia
20. Kim SH, Park HC, Jeong HM, Kim BK (2010) Glass fiber reinforced rigid polyurethane foams. J Mater Sci 45(10):2675–80
21. Sridhar BS, Shashikala AR, Vishnu PK (2020) Mechanical characterization of polyurethane foam and hybrid natural fibre based sandwich composite. Palarch's J Archaeol Egypt 17(9):5588–5604

22. Yusuf S (2016) Effect of alkali process and mass fraction on morphology, bending strength and sound absorption coefficient of polyurethane/coir fiber composites on Muffler components [Thesis]. Institut Teknologi Sepuluh Nopember, Surabaya
23. Rosamah E, Hossain MS, Abdul Khalil HPS, Nadirah WOW, Dunganid R, Amiranajwa ASN, Suraya NLM, Fizree HM, Omar AKM (2016) Properties enhancement using oil palm shell nanoparticles of fibers reinforced polyester hybrid composites. Adv Compos Mater 1–14
24. David NV, Azlan M (2018) Moisture absorption properties and shock cushioning characteristics of bio-based polyurethane foam composites. J Mechan Eng 5(2):157–168
25. Yusuf AK, Mamza PAP, Ahmed AS, Agunwa U (2016) Physico-mechanical properties of rigid polyurethane foams synthesized from modified castor oil polyols. Int J Sci Res Publ 6(7):548–556
26. Chen YC, Tai W (2018) Castor oil-based polyurethane resin for low-density composites with bamboo charcoal. Polymers 10:1–12
27. Marcovich NE, Kuranska M, Prociak A, Malewska E, Bujok S (2017) The effect of different palm oil-based bio-polyols on foaming process and selected properties of porous polyurethanes. Polym Int 66:1522–1529
28. Ali EZ, Zubir SA (2016) The mechanical properties of medium density rigid polyurethane biofoam. In: MATEC web of conferences, pp 1–5
29. Azni MAN, Sinar AA, Firuz Z, Supri AG (2013) Characterization and mechanical properties of biomass polyurethane. J Adv Environ Biol 7(12):3696–3699
30. Legiviani R (2016) Effect of the comparison of the composition of the polyurethane and the mass fraction of coconut fiber on the sound absorption coefficient and the flexural strength of the composite in the Muffler application [Thesis]. Institut Teknologi Sepuluh Nopember, Surabaya
31. Sismantoro A (2017) Characterization of oil palm shell particle reinforced polyurethane acoustic material [Thesis]. Institut Teknologi Sepuluh Nopember, Surabaya
32. Buyukakinci BY, Sokmen N, Kucuk H (2011) Thermal conductivity and acoustic properties of natural fiber mixed polyurethane composites. Tekstil ve Konfeksiyon 21(2):124–132
33. Roffael E (1993) Formaldehyde from particleboards and other wood based panels. Forest Research Institute Malaysia (FRIM), Kuala Lumpur, Malaysia
34. Chaydarreh KC, Shalbafan A, Welling J (2017) Effect of ingredient ratios of rigid polyurethane foam on foam core panels properties. J Appl Polym Sci 1–8

Preparation of Hydrogel-Based Chitosan Using Citric Acid as Cross-Linker and Dialdehyde Cellulose from Sugarcane Trash (*Saccharum officinarum* L.) for Matrices Drug Delivery

A. S. Hanifah, D. Sondari, S. Hermanto, and A. Ahmad

Abstract Hydrogel was a three-dimensional network that has a hydrophilic group, that can swell, and hold a large amount of water that can be applied in biomedical such as matrices of drug delivery. Preparation dialdehyde cellulose from sugarcane trash leaves (*Saccharum officinarum* L.) using Microwave-autoclave pretreatment by used oxalic acid and sodium hydroxide followed by oxidation using periodate was the experimental works effort. Dialdehyde cellulose (DAC) was used as hydrogel reinforcement with the Schiff base reactions towards chitosan. The citric acid as a cross-linker agent was used 10% (w/w). Their structural and functional properties were characterized by Fourier Transform Infrared (FTIR). This study found that the hydrogel can swell up to 3567% after 15 min in water and their application as a drug delivery matrix. Encapsulation efficiency of the hydrogel towards methylene blue ranged from 45 to 85%, with drug loading range from 22 to 23 mg/g and releasing for five hours.

Keywords Hydrogel · DAC · Drug delivery · Schiff base reaction · Cross-linked

---

A. S. Hanifah (✉) · D. Sondari
Research Center for Biomass and Bioproducts, National Research and Innovation Agency, Cibinong, Bogor, West Java, Indonesia
e-mail: Sabilaarzqa1610@gmail.com

A. S. Hanifah · S. Hermanto
Department of Chemistry, Science and Technology Faculty, State Islamic University Syarif Hidayatullah Jakarta, Ciputat, Banten, Indonesia

A. Ahmad
Department of Chemistry, Mathematics and Natural Science Faculty, Hasanuddin University, Tamalanrea, Makassar, Indonesia

M. A. Rahandi Lubis (ed.), *Proceedings of the 1st International Conference of Lignocellulose*, Green Energy and Technology,
https://doi.org/10.1007/978-981-19-1596-3_16

1 Introduction

In Indonesia especially on Java and Sumatra island, sugarcane was a plant that was widely cultivated for commercial needed as a consumption sugar. Which produces waste in the form of unused stems (*baggase*) and leaves [25]. Sugarcane trash leaves contain much cellulose and hemicellulose, which can be the sources de-polymerized by chemicals or enzymes enhanced into sugars monomer [19]. Sugarcane trash leaf and baggase were cellulose sources because they still have approximately 37.65% of cellulose [25]. The pretreatment effective to expose the cellulose by removing the lignin or hemicellulose was hydrolysis [11]. The acidic hydrolysis of lignocellulose substrates degrades the hemicellulose fraction into its sugar monomers, and alkali-based pretreatments were leaving the cellulose and hemicellulose into its residues and remove the lignin [15, 24].

An three-dimensional network of hydrophilic polymer that have ability to swell and hold a large amount of water called as a Hydrogel. For a substance to be classified as a hydrogel, it must contain at least 10% water by weight (or volume). The presence of hydrophilic groups such as $-NH_2$, $-COOH$, $-OH$, $-CONH_2$, $-CONH-$, and $-SO_3H$ contribute to the network's hydrophilicity [3].

The hydrogels from chitosan/cellulose composite were reported have gelation driven by Schiff base reactions are incredibly porous materials that are capable of loading numerous potential therapeutic agents. Including as drug delivery carriers of a malignant tumor drug, drug loading, and emotional are often modulated by standardization the composition of the composite hydrogels [17]. Controlling two parameters, the degree of cross-linking in the matrices and the affinity of the hydrogel to the aqueous environment in which swelling occurs can result in high porosity hydrogel structures. Hydrogels are highly permeable to many types of medicines due to their porous architecture, allowing pharmaceuticals to be loaded and released under the right conditions [3].

Another research reported that carboxymethyl cellulose (CMC)/polyethylene glycol (PEG)/citric acid (CA) qualified as superabsorbent hydrogel (SAP) has a typically ranged of swelling degree from 100 up to 5000%. These CMC-based hydrogels were considered the in vitro cell viability responses over 95% towards human embryonic kidney cells used as model cell lines [5]. In comparison, other research used NaCMC-based hydrogel by adding chitosan as hydrogel structure reinforcement. The addition of chitosan in the synthesis of NaCMC-based hydrogel increased absorption and showed 350% swelling power [1].

Methylene blue was chosen in this study as a drug model, because MB was known as an antimalaria drug, and because it is a highly ionized drug. Methylene blue was reported as an electron carrier, in injection form MB was used as a drug for methemoglobinemia, septic shock, and vasoplegia syndrome [8, 14].

This research aims to make hydrogel-based chitosan (Ch) using dialdehyde cellulose (DAC) as reinforcement, which will be used as a drug delivery matrices using the cross-linking method. The cross-linker used in this study was citric acid (CA).

2 Materials and Methods

2.1 Materials

The material used in this research is Sugarcane trash (PG. Rajawali II, Kalijati, Subang, Jawa Barat, Indonesia), chitosan (Sigma-Aldrich, Germany), NaOH (2% w/v, Merck, Germany), HCl (0.1 M, Merck, Germany), $C_6H_8O_7$ (Merck, Germany), $C_2H_2O_4$ (0.7% and 1.8% Wako, Japan), $NaIO_4$ (Merck, Germany), H_2O_2 (30% Merck, Germany), CH_3COOH (98%, Merck, Germany), methylene blue (Merck, Germany), and PBS (Phosphate buffered saline) are from Merck, Germany and used without purification. Distilled water was used for the whole process.

2.2 Procedures

The procedure to make a hydrogel was using several stages. First, the sugarcane trash was pretreatment to isolate the cellulose (washed, delignification, and drying). The pretreatment was using microwave reactor (START D Microwave Disgestion System, Milestone, Italia) with oxalic acid (0.7 and 1.8%) solution (5 min at 180 °C) followed by autoclave using NaOH 2% solution (1 h at 121 °C) [4, 9], holocellulose and α-cellulose analyzed using ($NaClO_2$ and NaOH solution) [18, 22] and bleach using peracetic acid (15% v/v H_2O_2 in acetic acid) [13, 21]. Cellulose was oxidized using $NaIO_4$ (0.8% w/v) to make DAC [10]. Chitosan was dissolved to 1% acetic acid and stirred with DAC the ratio (3:1) by adding CA 10% (w/w) at 60 °C in 90 min. The hydrogel was dried at 40 °C in the oven. Then the hydrogel was the characterization and being conducted in vitro test as drug delivery.

2.3 Characterization

2.3.1 Swelling Degree

The hydrogel was cut into size 10 × 10 mm^2 sample was weighed before (Wd) [5].Then soaked with distilled water (10 ml) for 15 min at room temperature. The blotting process was also carried out using filter paper attached to the hydrogel surface. Then the final weight was reweighed after immersion (Wd). The swelling power was calculated using the following equation:

$$\%\,S = (Ws - Wd)/Wd \times 100\% \quad (1)$$

2.3.2 Fourier Transform Infrared Spectroscopy

Structural and Functional group analysis on hydrogel was carried out using FTIR (IR prestige-21, Shimadzu, Japan) and recorded over a range of wavenumbers from 400 to 4000 cm^{-1}.

2.3.3 Encapsulation Efficiency and Drug Loading

Methylene Blue (MB) was used as the model drug for drug loading. MB was dissolved in different dissolution mediums, including HCl (0.1 M) and phosphate-buffered saline (PBS, pH 7.4). The hydrogel was weighed and was immersed in 0.05 mg/mL MB solution for 24 h in darkness [23]. Drug contents in the hydrogel were calculated according to the original concentration of MB and the concentration of MB in the solution after loading, which could be quantified using a Microplate reader Spectro-Nano Uv–Vis (Tecan, Tecan Infinite Nano 200) at wavelength 666 nm. The drug encapsulation efficiency and drug loading were calculated using [6]:

$$\mathrm{DL(mg/g)} = (\mathrm{C0} - \mathrm{C1.V1})\mathrm{V0/w} \tag{2}$$

$$\mathrm{EE}(\%) = ((\mathrm{C0} - \mathrm{C1}))/\mathrm{C0} \times 100\% \tag{3}$$

2.3.4 In Vitro Drug Releases

In vitro Drug release experiments were performed in a shaker (Taitec, Bio-Shaker BR-300) at a speed of 50 rpm at 37 °C at the medium solution. The drug-loaded hydrogels were placed in a buffer solution, and 25 mL of solution was collected at one-hour intervals to determine drug content using a Microplate reader spectro-Nano Uv–Vis (Tecan, Tecan Infinite Nano 200). An equal volume of the same solution medium was added back to maintain a constant volume. The calibrated standard curves calculated the concentration of the drug in the different buffer solutions.

3 Results and Discussion

The utilized lignocellulosic resources from sugarcane trash to make derivatives of cellulose, namely dialdehyde cellulose, proven to be effective. There are so many stages to make the derivatives of cellulose. The first step is to isolate α-cellulose from sugarcane trash using the microwave-autoclave pretreatment, followed by bleach and oxidation. The different pretreatment affected the α-cellulose content that was isolated from sugarcane trash leaf.

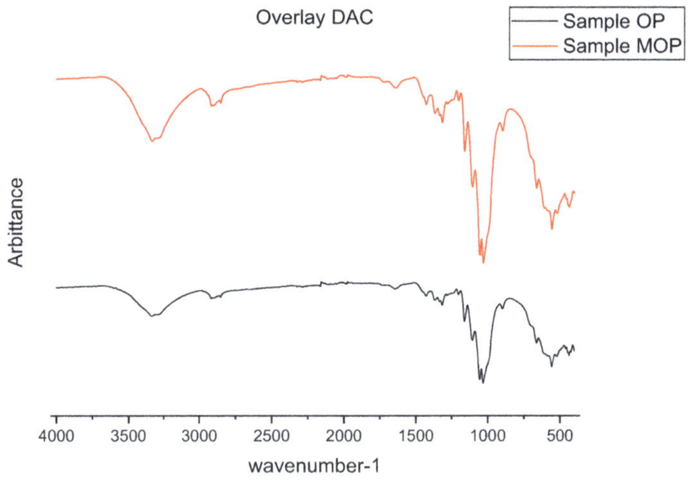

Fig. 1 FTIR spectrum from DAC

DAC (dialdehyde cellulose) was produced after the oxidation process using 0.8% periodates solution in dark conditions. DAC was made through the cleavage mechanism of the C2–C3 bonds in the glucose residues to form an aldehyde group on the backbone of the cellulose structure [7]. DAC was characterized using FTIR, which shows the chemical groups in DAC samples from sugarcane trash. FTIR spectrum of DAC shown in Fig. 1.

The spectrum IR at 3335.56 and 3334.87 cm^{-1} shows –OH group; 2915.88 cm^{-1} shows $-CH_2$ group; 1644.47 and 1641.81 cm^{-1} show group C=O aldehyde; 1104.54 cm^{-1} shows the C–O–C group which is bonds between C1–C5 and β-1,4-glycosidic bonds. Sample OP was DAC using microwave oxalic acid 0.7%—Autoclave with 2% NaOH while sample MOP was DAC using microwave oxalic acid 1.8%—Autoclave with 2% NaOH. A gel is formed when chitosan, citric acid, and dialdehyde cellulose were heated in the aqueous medium at 60 °C after 90 min stirred. The gel was then dried in a 40 °C oven to make it firm as hydrogel to eliminate water, as seen in Fig. 2.

The extension of cross-linking from hydrogel networks was determined using swelling measurements. Table 2 shows the swelling degree of hydrogel-based chitosan with DAC as reinforcement and CA as cross-linker. It can be observed that the hydrogel qualifies as superabsorbent hydrogels (SAP). SAPs are hydrophilic networks with a high water absorption capacity that absorb, swell, and retain aqueous solutions up to hundreds of times their weight (dry sample) and may be customized for various biological applications [12].

The swelling power showed that the optimum DAC was in the OCC sample. The pretreatment has affected the quality of DAC as can be seen in Table 1; by using the different concentrations of oxalic acid, the optimum ones are 0.7%, but the intensity of spectrum FTIR DAC showed 1.8% concentration of oxalic acid more intense. It turns that the OCC sample had a remarkable ability to swell.

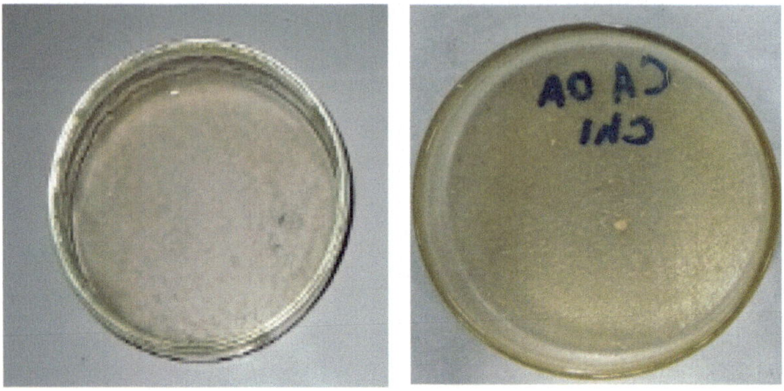

Fig. 2 Hydrogel in Petri disk

Table 1 Pretreatment holocellulose and α-cellulose content

Sample	Holocellulose content (%)	α-cellulose content (%)
OP	83.85	61.41
MOP	79.71	59.33

Table 2 Swelling degree of OCC and MACC

Sample	Swelling degree(%)
OCC	3567
MACC	440

Their possible reaction was that the free hydroxyl group and the amine group of chitosan were cross-linked with the hydroxyl group from ester in Citric Acid [16]. However, The cross-linking reaction of the hydrogel was made by the two reagents. DAC and Ch made the Schiff base reaction that made by the Imine group as their cross-linked [23]. The Schiff base reaction was made as reinforcing hydrogel. It shows the possible reaction in Fig. 3 [16, 23]. The IR spectrum from hydrogel showed the chemical groups of the hydrogel can be seen in Fig. 4.

The spectrum IR from hydrogel showed a board peak at 3374 cm^{-1} which shows –OH and –NH group; 2904 cm^{-1} were shows C–H bending; 1720 cm^{-1} shows corresponding to C=O ester stretching band [2] the higher intensity shows the more C=O group in the sample, 1640 cm^{-1} shows –C=N group from the Schiff base reaction between DAC and Ch, 1590 cm^{-1} shows the –C=O bending vibration of amide I band and –NH bending vibration [23]. According to Xu et al. [23], reported the 1730 cm^{-1} from in sample OCC and MACC the peaks shows low intensity, shows 1182 and 1052 cm^{-1} shows the correspond to C–O from chitosan ring structure.

According to the standard curves of MB absorbance in the two solutions as can be seen at Fig. 5 were showed $y = 0.128x + 0.0374$ ($R^2 = 0.9689$) for HCl medium and $y = 0.0882x + 0.09$ ($R^2 = 0.9898$) for PBS pH 7.4 medium. Drug loading and

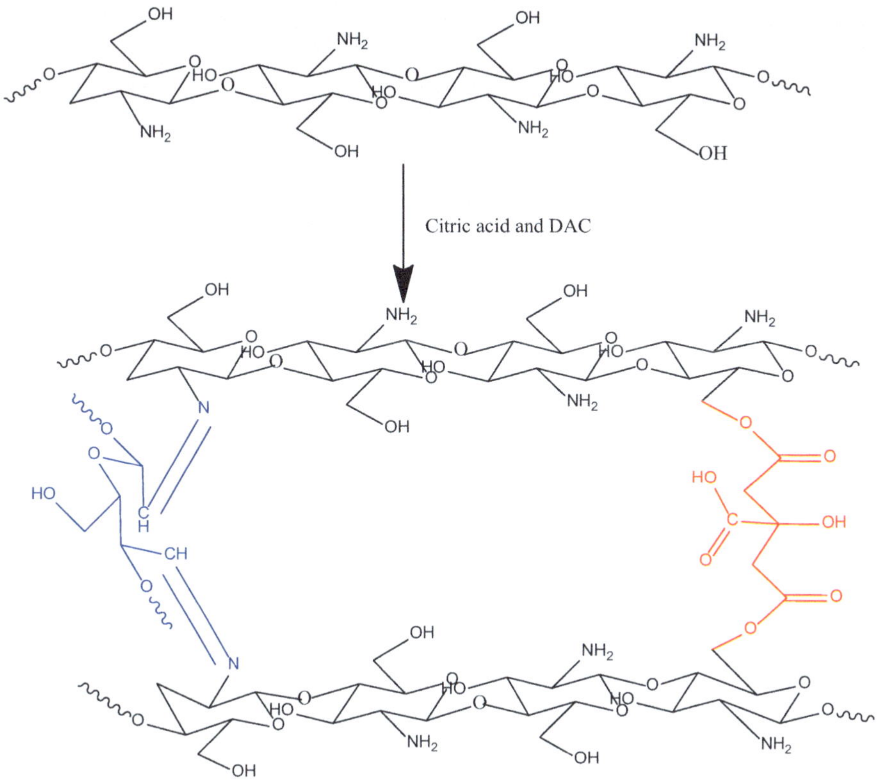

Fig. 3 Possibly cross-linking reaction between Ch/DAC/CA

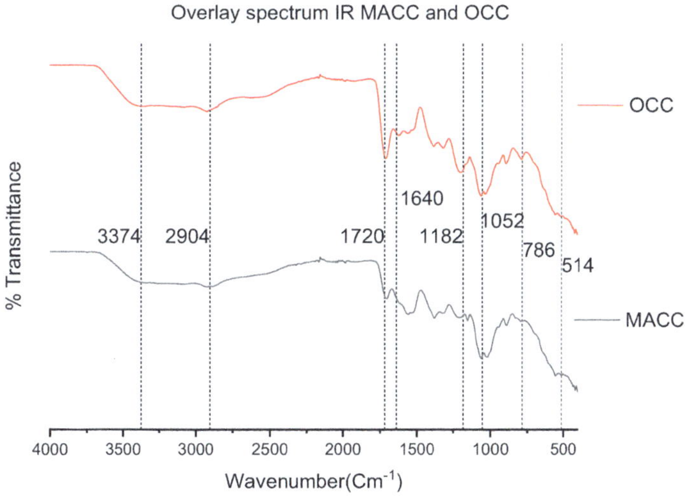

Fig. 4 IR spectrum of Ch/DAC/CA hydrogels

encapsulation efficiency are shown in Figs. 6 and 7 that sample MACC (Ch/DAC/CA from DAC sample MOP) and OCC (Ch/DAC/CA from DAC sample OP) have higher efficiency in the medium PBS rather than in the medium HCl. The highest efficiency was MACC with 85.2507%, and Drug loading was 23.1056 mg/g in PBS medium. In acidic conditions, the hydrogels are nearly 5–40% less MB than the neutral solution.

Those showed that encapsulation efficiency is directly proportional to drug loading. The more drugs were encapsulated shows, the more also drugs were loaded. The medium shows an impact that is not significantly drug-loaded. In the HCl medium

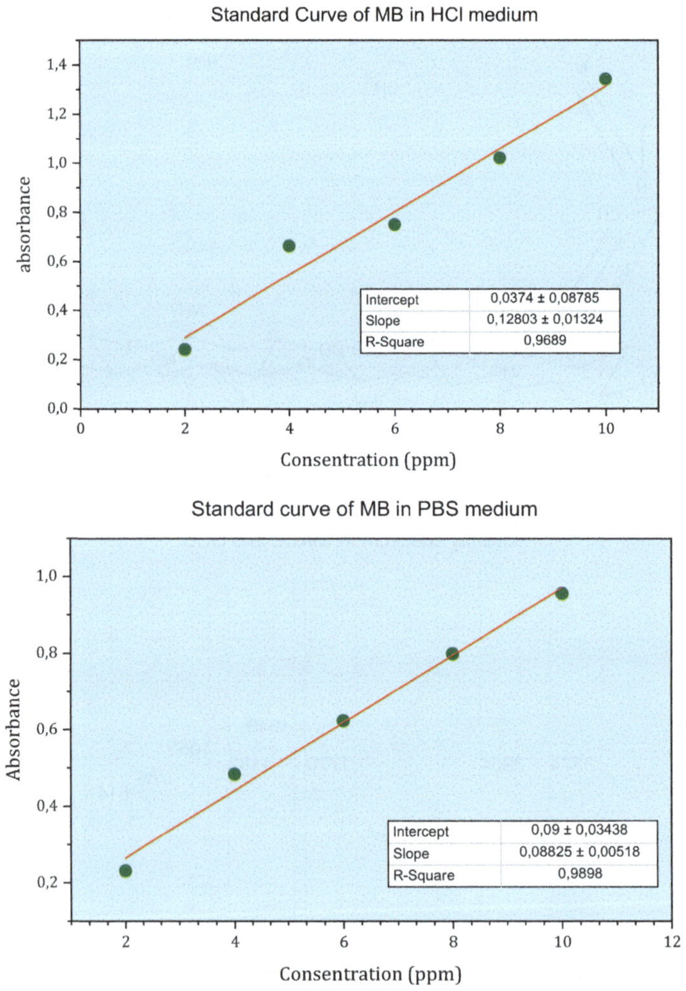

Fig. 5 Standard curves of MB in HCl medium and PBS medium

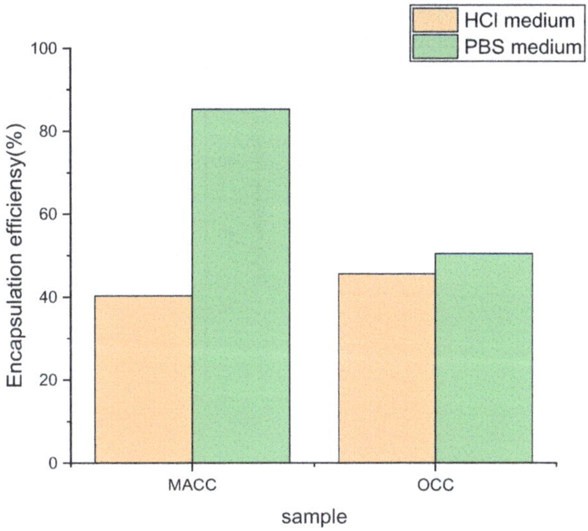

Fig. 6 Graph of Encapsulation efficiency (%) of hydrogel in the medium solution

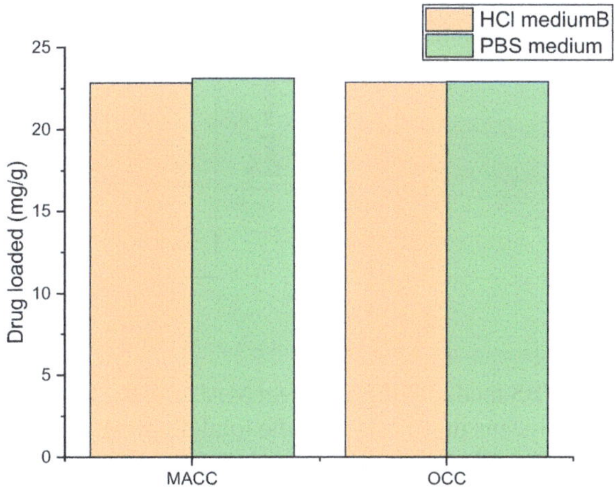

Fig. 7 Drug loading (mg/g) of hydrogel in the medium solution

sample of hydrogel, MACC was 22.84 mg/g. That means 22.84 mg of the drug were loaded per 1 g of hydrogel while sample OCC shows 22.87 mg/g in HCl medium. In PBS medium sample MACC shows 23.11 mg/g while sample OCC shows 22.9 mg/g higher than in HCl medium.

Drug release rate (%) is the percent value obtained from the ratio between the amount of drug release and the amount of drug that has been successfully encapsulated in an interval [20]. The in vitro releases of MB as a drugs model were shown in Figs. 8 and 9. The drug release profiles of the hydrogel showed burst release of the drug from 1 to 5 h.

Fig. 8 Drugs release rate (%) in PBS medium

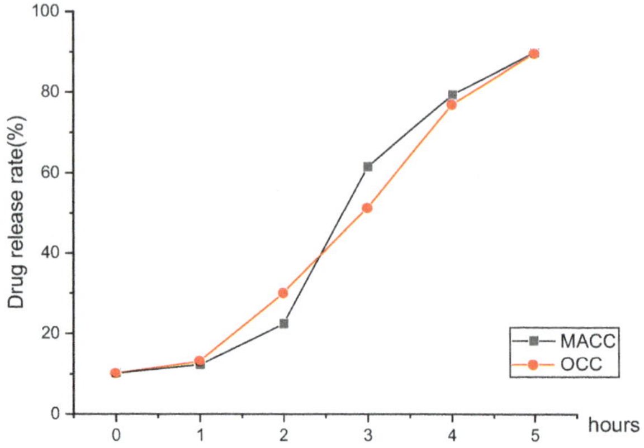

Fig. 9 Drugs release rate (%) in HCl medium

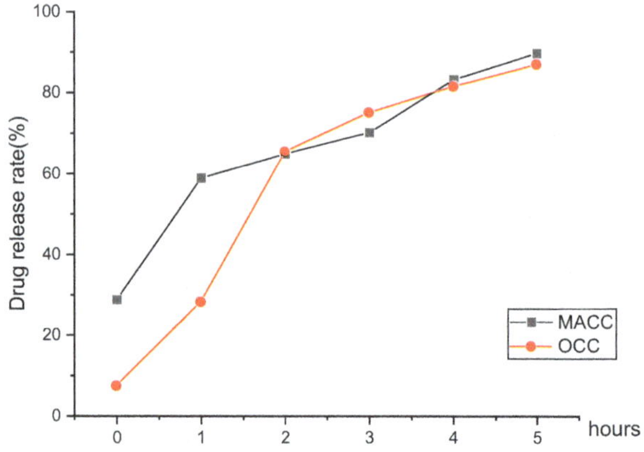

In PBS medium, the hydrogel MACC in the first hour had a release rate of 12.34%, which means that 12.34% of the total drug loaded was successfully released into the medium, and so on until released 89.84% after 5 h releasing. Higher than the release rate in HCl medium that shows 89.80% released but not significantly. However, the hydrogel OCC in PBS medium shows the release rate was 89.82% and in HCl, the medium shows drug release rate was 87.26%. After 5 h releasing. While Xu et al. [23] reported the releasing of theophylline using hydrogel from fabricated dialdehyde-cellulose nanocrystal and Chitosan show optimum release in pH 1.5 medium than pH 7.4 medium.

The cumulative drug release at PBS medium was greater than that achieved at HCl the medium. In the HCl medium, the release rate of the drug showed stability controlled by releasing the drug. The hydrogels showed great drug-controlled release characteristics and might be used as a potential matrices of drug delivery.

4 Conclusion

Based on the research conducted, it can be concluded that The hydrogel-based-chitosan crosslinked with citric acid was made by addition dialdehyde cellulose as reinforcement. The MACC and OCC hydrogel from Ch/DAC/CA showed potential carriers for drug delivery. The experimental shows that MACC hydrogel can swell up to 440% in water after 15 min and OCC hydrogel can swell up to 3567%. Both sample hydrogel shows great Encapsulation efficiency, drug loading, and drug release in PBS medium. The hydrogel can release the optimum drug for 5 h.

Acknowledgements The authors are grateful for the funding research through National Research Program grant funding 2020-2021 from The Ministry of Research and Technology/National Research and Innovation Agency of The Republic of Indonesia and LPDP. Integrated Laboratory of Bioproducts (iLaB) of research Center for Biomaterials-Indonesian Institute o Science for its equipment research support, as well as analysis support from E-Layanan Sains (ELSA), LIPI, Indonesian Institute of Science.

References

1. Adi SH, Heryani N (2019) Synthesis and optimization of sodium carboxymethyl cellulose and chitosan-based hydrogel using a hydrothermal method. https://doi.org/10.14203/widyariset.5.1.2019.1-10
2. Alonso D, Gimeno M, Olayor R, Vázquez-Torres H, Sepúlveda-Sánchez JD, Shirai K (2009) Cross-linking chitosan into UV-irradiated cellulose fibers for the preparation of antimicrobial-finished textiles. Carbohyd Polym 77(3):536–543. https://doi.org/10.1016/j.carbpol.2009.01.027
3. Bahram M, Mohseni N, Moghtader M (2016) An introduction to hydrogels and some recent applications. InTech. https://doi.org/10.5772/64301
4. Cao W, Chen S, Ronghou L, Renzhan Y, Xiaowu W (2012) Comparison of effects of five pretreatment methods on enhancing the enzymatic digestibility and ethanol production sweet sorghum bagasse. Biores Technol 111:215–221
5. Capanema NSV et al (2017) Superabsorbent cross-linked carboxymethyl cellulose-PEG hydrogels for potential wound dressing applications. Elsevier. https://doi.org/10.1016/j.ijbiomac.2017.08124
6. Fang et al (2008) Preparation of cross-linked starch microspheres and their drug loading and releasing properties. Elsevier. https://doi.org/10.1016/j.carbpol.2008.03.005
7. George D, Maheswari PU, Sheriffa Begum KMM, Arthanareeswaran G (2019) Biomass-derived dialdehyde cellulose cross-linked chitosan-based nanocomposite hydrogel with phytosynthesized zinc oxide nanoparticles for enhanced curcumin delivery and bioactivity. J Agric Food Chem 67(39):10880–10890. https://doi.org/10.1021/acs.jafc.9b01933
8. Ginimuge et al (2020) Methylene blue: revisited. J Anaesthesiol Clin Pharmacol 26(4):517–20
9. Hermiati E, Budi RP, Fatriasari W, Nia L, Thontowi A, Arnieyanto DR (2020). Microwave-assisted acid pretreatment for enhancing enzymatic saccharification of sugarcane trash@@. https://doi.org/10.1007/s13399-020-00971-z
10. Hõglund E (2015) Production of dialdehyde cellulose and periodate regeneration: towards feasible oxidation processes
11. Kumar P, Barrett DM, Delwiche MJ, Stroeve P (2009) Methods for pretreatment of lignocellulosic biomass for efficient hydrolysis and biofuel production. Ind Eng Chem Res 48:3713–3729

12. Ma J, Li X, Bao Y (2015) Advances in cellulose-based superabsorbent hydrogels. RSC Adv 5:59745–59757
13. Maharani DH (2012) Application of peracetic acid to remove the dye in the jeans fabric. University Airlangga, Thesis for bachelor
14. Meissner PE et al (2006) Methylene blue for malaria in Africa: results from a dose-finding study in combination with chloroquine. Malar J 5:84
15. Parawira W, Tekere M (2011) Biotechnological strategies to overcome inhibitors in lignocellulose hydrolysates for ethanol production: review. Crit Rev Biotechnol 31:20–31
16. Priyadarshi R, Sauraj, Kumar B, Negi YS (2018) Chitosan film incorporated with citric acid and glycerol as an active packaging material for extension of green chili shelf life. In: Carbohydrate polymers. https://doi.org/10.1016/j.carbpol.2018.04.089
17. Qian L (2018) Cellulose-based composite hydrogels: preparation, structures, and applications. Springer. https://doi.org/10.1007/978-3-319-76573-0_23-1
18. Rowell RM (2005) Handbook of wood chemistry and wood. CRC Press, Florida
19. Singh P, Suman A, Tiwari P, Arya N, Gaur A, Shrivastava AK (2008) Biological pretreatment of sugarcane trash for its conversion to fermentable sugars. World J Microbiol Biotechnol 24:667–673
20. Subedi G, Shrestha AK, Shakya S (2016) Study of effect of different factors in formulation of micro and nanospheres with solvent evaporation technique. Open Pharmaceut Sci J 3(1):182–195. https://doi.org/10.2174/1874844901603010182
21. Tripathi S, Mishra SP, Mishra OP, Bajpai P, Kumar S, Bajpai PK (2007) Application of Peracetic acid in chemical pulp bleaching. IPPTA: J Ind Pulp Pap Tech Assoc
22. Wise LE, Murphy M, D'Addieco AA (1946) Chlorite holocellulose, its fractionation, and bearing on summative wood analysis and studies on the hemicelluloses. Paper Trade J. 122:11–19
23. Xu Q et al (2019) Fabrication of cellulose nanocrystal/chitosan hydrogel for controlled drug release. MDPI. https://doi.org/10.3390/nano9020253
24. Zhao X, Peng F, Cheng K, Liu D (2009) Enhancement of the enzymatic digestibility of sugarcane bagasse by alkali-peracetic acid pretreatment. Enzyme Microb Technol 44:17–23
25. Zulharmita Z, Dewi SN, Mahyuddin M (2012) Pembuatan Mikrokristalin selulosa dari ampas tebu (*Saccharum officinarum* L.). Jurnal Sains dan Teknologi Farmasi 17:158–163

Analysis of High Magnesium Nickel Slag Powder as a Supplementary Cementitious Materials for High Strength Concrete

L. Oksri-Nelfia, Bunga Yubi Nabiilah, and Sotya Astutiningsih

Abstract Concrete demand is increasing along with the development of infrastructure that is emerging in Indonesia. The development of science in the sector of concrete technology allows the use of waste as a concrete material by utilizing industrial waste, as nickel slag due to the smelting process of nickel ore after going through the combustion and screening process. In this study, nickel slag will be used as a supplementary cementitious material to evaluate result on high strength concrete. Samples were made with water cement ratio 0.31 and will be compared with 100% OPC Type 1 concrete as a reference (Ref). Nickel slag powder (NSP) substitution of cement were 5, 10, 15, 20, 25 and 30% with the concrete age of 3, 7, 14 and 28 days. Specific gravity test for NSP and cement were done by the Le Chatelier method. Mineral characterization using Scanning Electron Microscopy (SEM) was carried out on nickel slag powder used and concrete substitution 30% NSP aged 3 days. The mechanical characteristics include ease of work in the field (workability) and compressive strength. Concrete was targeted to reach an average compressive strength of $\geq$ 72 MPa at aged 28 days using superplasticizer type Sika Viscocrete 8015. The compressive strength result of NSP concrete substitution at aged 28 days was 82.94 MPa for 5% NSP, 82.28 MPa for 10% NSP, 79.98 MPa for 15% NSP, 77.89 MPa for 20% NSP, 77.26 MPa for 25% NSP and 73.80 MPa for 30% NSP.

Keywords Compressive strength · Nickel slag powder · Supplementary cementitious materials · Scanning electron microscopy

L. Oksri-Nelfia (✉) · B. Y. Nabiilah
Universitas Trisakti, Jakarta Barat, Jakarta, Indonesia
e-mail: lisa@trisakti.ac.id

S. Astutiningsih
Departement of Metallurgical and Materials Engineering, University Indonesia, Depok, Indonesia

M. A. Rahandi Lubis (ed.), *Proceedings of the 1st International Conference of Lignocellulose*, Green Energy and Technology,
https://doi.org/10.1007/978-981-19-1596-3_17

1 Introduction

The development of concrete technology has made it possible to use waste as a concrete material, so that waste can be utilized optimally. Based on these thoughts, many studies tried to improve the quality of concrete from the materials side by substitute with natural materials, both in coarse aggregates or fine aggregates, and some are used as a supplementary cementitious material (SCMs) to increase the adhesion of concrete binders by utilizing industrial waste for concrete mixtures, for example; nickel slag powder. In the city of Southeast Sulawesi, Kabaena Island has a fairly high nickel content, it was first discovered in 1970, when PT. INCO Tbk, one of the largest nickel companies in Indonesia, sent some experts of its geologists to conduct a survey and ensure that 80% of the island has a fairly large nickel content and a fairly high level of nickel [1]. In every year, PT. INCO produces a relatively large amount of solid waste, namely nickel slag, which is around 70,000–72,000 tons (Novia Sari 2019). Much of the concrete produced (either normal or high strength concrete) made with waste has superior properties compared to conventional concrete in terms of strength, performance and durability [2].

Based on research conducted [10], it is clear that the use of nickel slag in concrete mixtures increases the mechanical properties of concrete, especially its compressive strength. The increase that occurs can reach about 45, 24 and 19%, respectively, for compressive strength, split tensile strength and modulus of elasticity. This good improvement is clearly related to the physical form of nickel slag aggregates. In research [7, 10], concluded that nickel slag has a positive impact as a substitute for concrete aggregates, and increases the compressive strength of concrete. And also in research [2, 11], concluded that slag as a substitute has a good impact on the strength and durability of high-strength concrete. This previous research has encouraged us to make high-strength concrete using nickel slag as a supplementary cementitious material. Through this approach, the consumption of natural materials, thermal energy and electrical energy, along with CO_2 emissions can be reduced [12].

2 Experimental

Nickel slag used was sifted nickel slag that have been sieved using sieve shaker machine with #200 sifter for $\pm$ 10 min, it produces very fine grain sizes (< 0.075 mm) which called nickel slag powder (NSP). NSP will be as hydraulic as cement, and based on Trace Metal Analysis results conducted by Indocement, NSP was proven not contain any dangerous heavy metal. With water cement ratio of 0.4, NSP concrete substitution can reach about fc′ 30 MPa up to a 25% substitution rate (Fig. 1).

Preparing samples and material testing were carried out at PT. Jaya Beton Indonesia. Samples were made in cylinders (Ø = 200 mm; H = 100 mm). Obtained data which conducted from the testing process at Laboratorium PT. Jaya Beton

Fig. 1 Nickel slag powder (NSP)

Indonesia Laboratory includes the test result of coarse aggregates and fine aggregates, compressive strength, and concrete mixture workability test with water cement ratio of 0.3. NSP substitution of cement were 0, 5, 10, 15, 20, 25 and 30%, where 0% NSP substitution was Ordinary Portland Cement (OPC) as a Reference (Ref). Compressive strength was performed at the concrete age of 3, 7, 14, and 28 days. Samples were put in the water (water curing) until the age of concrete has reached the age needed for the test. Total samples in this study were 91 samples, while 84 samples for compressive strength and 7 samples for scanning electron microscopy (SEM).

2.1 Materials

Cement which used as a comparison material is OPC Type 1 originated from PT. Indocement Tiga Roda. Cement has an adhesive and cohesive character, that used for binders which reacts with water, and used as a concrete mixture with coarse aggregates and fine aggregates. Water was taken from the laboratory facility of PT. Jaya Beton Indonesia. The coarse aggregates come from Gn. Sindur and fine aggregates comes from Gn. Sudamanik. Coarse aggregates maximum size of 5 mm and fine aggregates maximum sized of 20 mm. Specific gravity test of NSP and cement using Le Chatelier method refers to ASTM C.188 at saturated surface dry (SSD) condition with the result of 2.797 for NSP and 3.14 for cement.

Material testing for coarse aggregates and fine aggregates refers to Japan Industrial Standards (JIS), with the information (Table 1).

NSP substitution of 0% (Ref), 5, 10, 15, 20, 25 and 30% from the mass of cement. Concrete samples of 0% used as a reference (Ref) with a compressive strength result targeted to reach $fc' = 63$ MPa at the age of 28 days. To reach a great workability, superplasticizer type Sika Viscocrete 8015 was added as an admixture. Calculation

Table 1 Materials test result

Parameter	Coarse aggregates	Fine aggregates
Specific gravity (SSD)	2.59	2.57
Unit weight	1.49 gr/cm^3	1.66 gr/cm^3
Fineness modulus		
Water absorption	2.38%	2.17%
Water content	3.41%	9.20%
Sludge levels	–	1.95%
Abrasion test (Los Angeles)	17.76%	–

of concrete mix design refers to SNI 03-6468-2000, 2000, although this regulation only regulates the mix design of high strength concrete using fly ash, fly ash is one of a pozzolanic material, while NSP was categorized as a pozzolanic material which has been confirmed by the results of X-Ray Diffraction (XRD) analysis [5]. Based on the ratio of W/(c + p) with superplasticizer, the water cement ratio was set at 0.31 (Table 2).

Mixer capacity was up to 0.3 m^3. The mixing methods start with 100% coarse aggregates and fine aggregates in saturated surface dry (SSD) condition. Followed with 100% cementitious materials in dry condition. These materials were mixed for approximately 3 min. Next, add 50% water and 100% superplasticizer was slowly poured in. Thereafter, 50% water were added. According from the steps of the materials placed, the concrete was mixed for approximately 6 min. The workability of concrete was measured by concrete slump test. Slump was allowed to reach $\geq$ 200 mm for high strength concrete with superplasticizer according to SNI 03-6468-2000.

Table 2 Mix design

Material	Ref.	NSP 5	NSP 10	NSP 15	NSP 20	NSP 25	NSP 30
Water (kg)	96.32	84.86	85.14	85.42	85.70	85.98	86.26
Cement (kg)	591.94	562.34	532.74	503.15	473.55	443.95	414.35
NSP (kg)	0	29.60	59.19	88.79	118.39	147.98	177.58
Coarse (kg)	1109.38	1109.38	1109.38	1109.38	1109.38	1109.38	1109.38
Fine (kg)	590.85	587.54	584.22	580.90	577.58	574.26	570.94

[a] Proportion per m^3 of mixture (according to aggregate wetness condition)

2.2 Methods

Compressive strength was performed at the concrete aged of 3, 7, 14, and 28 days. Samples was put in the water (water curing) with a temperature of 20 °C until the concrete has reached the age needed for the test. At the compressive strength test, samples will be pressed with optimum strength (Figs. 2 and 3).

Properties of Concrete

Workability test using slump test with Abram's cone where the mould size is height = 150 mm and Diameter is 100 mm. Testing the workability of concrete with the aim of knowing the slump value of each concrete made in cylinder size (Ø = 100 mm and H = 200 mm). For testing the compressive strength is analyzed based on ASTM C470. Each of the concrete specimens were tested at 3, 7, 14 and 28 days using a PILOT Press digital compression device (Capacity 2000 kN). Prior to testing the compressive strength, the concrete must first soak at a water temperature 20 °C. Superplasticizers are not used in the concrete fabrication process. However, the addition of admixture in the concrete mixture will affect the concrete properties and ease of work in the field.

Characterization of Nickel Slag Powder (NSP). The density of NSP was measured using Le Chatelier procedure according to standard test method for density of hydraulic cement [4]. Scanning Electron Microscopy (SEM) conducted to determine the surface characteristics and object texture, to find out the shape and sizes of the constituent particles of the object, to observe the grains structures of the object,

Fig. 2 Concrete samples on preparation

Fig. 3 Concrete samples on water curing

and to obtain quantitatively data of the elements and particles contained in the object. SEM was performed on one sample, namely 30% NSP for the age of 7 days. For 30% NSP sample, cylindrical sample are cut using bar cutter until the sample has dimensions $\pm$ 1.5 $\times$ 1.5 cm and $\pm$ 2 mm thick. After the sample was cut, mashed the sample using abrasive paper. First using CAMI 60 abrasive paper (rough), then using CAMI 80 abrasive paper (medium), then finally using CAMI 100 abrasive paper (fine). After the surface of the sample is very smooth, the sample is ready for SEM.

Concrete samples that have been prepared are entered into a vacuum and coating system. The samples must reach 2×10^{-1} milibar for coating process. The thicker and larger the sample, the longer the vacuum process takes time. Sample should be made at least $\pm$ 2 mm with an area of 2 cm $\times$ 2 cm. After the sample finished the vacuum process, sample will be coated with gold so that the surface can be read by an electron microscope. The coating process lasts for $\pm$ 2 min. After that, the coated sample is inserted into an electron microscope. The electron microscope used is brand Inspect^{tm} F. Zoom the sample 150 times to 500 times then specify several points to view the EDS (Energy-Dispersive X-Ray Spectroscopy). EDS serves to identify the mineral content in the sample being tested. Zoom the sample 2000 times to 3000 times at the specified point (Fig. 4).

Fig. 4 30% NSP Concrete sample for SEM

3 Results and Discussion

3.1 Compressive Strength

The compressive strength test was applied after the concrete aged 3, 7, 14 and 28 days of water curing at 20 °C by using three cylinders for each test. Compressive strength data are listed in Table 3 and showed at Fig. 5.

Based on the results of the compressive strength obtained, it can be seen that the nickel slag powder gives increased strength to each of its substitutes. The compressive strength value increases until the concrete age of 14 days. At the concrete age of 28 days, 5% NSP and 10% NSP got a higher compressive strength result than the Reference. The higher strength result of the NSP substitution concrete is due to the pozzolanic reaction of Nickel Slag Powder (NSP), according to research [3] that nickel waste is a material that has a pozzolanic effect, where the composition of the chemical compound is $SiO_2 + Fe_2O_3 + Fe_2O_3 > 70\%$ as regulated in the specifications ASTM C618-93. In this case, nickel waste described was nickel slag powder which used as a partial substitution material for cement. The use of slag, can increase the compressive strength of concrete by 22%. And it can be concluded that all NSP substitution concrete samples reach the compressive strength of $fc' \geq 41.4$ MPa which can be categorized as a high strength concrete according to SNI 03-6468-2000.

Table 3 Compressive strength result

Material	Ref	NSP 5	NSP 10	NSP 15	NSP 20	NSP 25	NSP 30
fc′ at 3 days (MPa)	66.17 ± 0.63	66.04 ± 1.41	59.80 ± 0.97	58.03 ± 0.89	55.68 ± 0.98	53.68 ± 0.86	51.82 ± 3.06
fc′ at 7 days (MPa)	71.47 ± 1.16	66.55 ± 1.20	59.98 ± 0.46	59.39 ± 0.61	58.01 ± 0.84	56.33 ± 0.54	52.31 ± 1.37
fc′ at 14 days (MPa)	72.83 ± 2.21	72.52 ± 8.14	72.45 ± 4.68	65.70 ± 0.67	59.37 ± 1.78	58.91 ± 2.87	58.06 ± 10.55
fc′ at 28 days (MPa)	80.49 ± 0.26	82.94 ± 7.79	82.28 ± 0.58	79.98 ± 0.39	77.89 ± 2.45	77.26 ± 0.75	73.80 ± 6.27

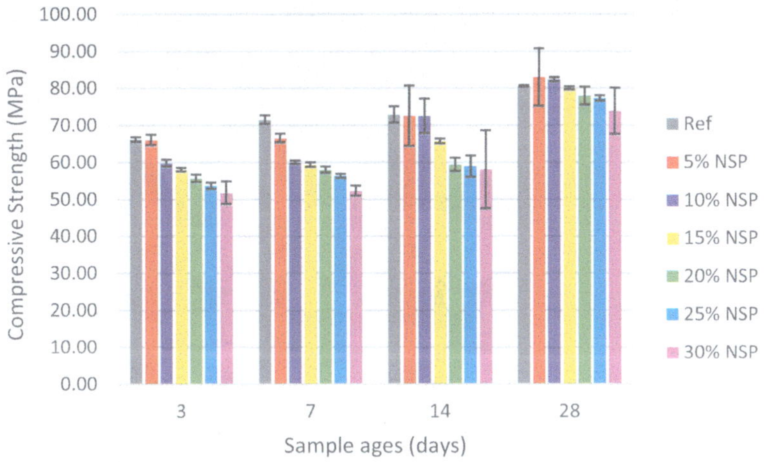

Fig. 5 Compressive Strength result

3.2 Scanning Electron Microscopy (SEM)

It is well known that for a given substitution level with mineral admixtures, the early ages characteristic of high strength concrete is influenced by the reactivity of mineral admixtures. The more reactive the mineral admixtures are, the higher the early strength of the concrete will be [6]. SEM was performed on 30% NSP concrete substitution at the age of 7 days. SEM aims to see and investigate the surface of objects directly. The magnitude for the SEM test starts from 150 to 3000 times. Later Energy-Dispersive X-Ray Spectroscopy (EDS) will identify the percentage of mineral compound content in the specimens test, where the results are strengthened by the results of the X-Ray Fluorescence (XRF) [5].

Nickel Slag Powder (NSP)

It can be observed on the 3000 times magnified nickel slag powder, that the nickel slag powder has a sharp and irregular shape. Besides, the EDS results show that nickel slag powder predominantly contains silica, calcic, magnesia and alumina (Figs. 6 and 7).

30% Nickel Slag Powder Concrete Aged 7 days

Magnification of 150 times as a global image on a sample of 30% NSP concrete that is 7 days old. Observations were also made at several points because the global image shows differences in the shape, texture, size and colour of the object's constituent particles. The solid black colour indicates voids due to air voids in the concrete. Referring to research [8], that the formation of a black gap between the fibre and matrix on the object indicates the presence of pores in the object. For the point being

Fig. 6 Magnitude 3000 times of nickel slag powder

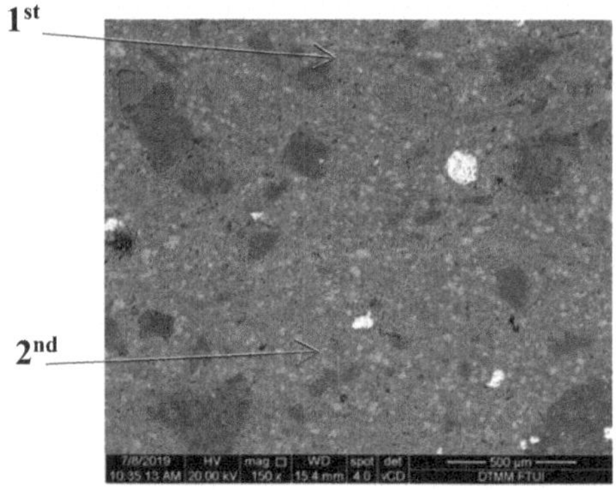

Fig. 7 Magnitude 150 times of 30% nickel slag powder concrete aged 7 days

reviewed, a magnification of 3000 times is carried out so that the particles making up the object can be seen more clearly (Figs. 8 and 9).

It can be seen from the results of the SEM test that has been carried out, the first spot is the point where the nickel slag powder gathers, indicated by the high levels of silica (Si) and magnesium (Mg). The second spot shows high silica (Si) where the compound is the same as the compound contained in cement and can form a pozzolanic reaction [13].

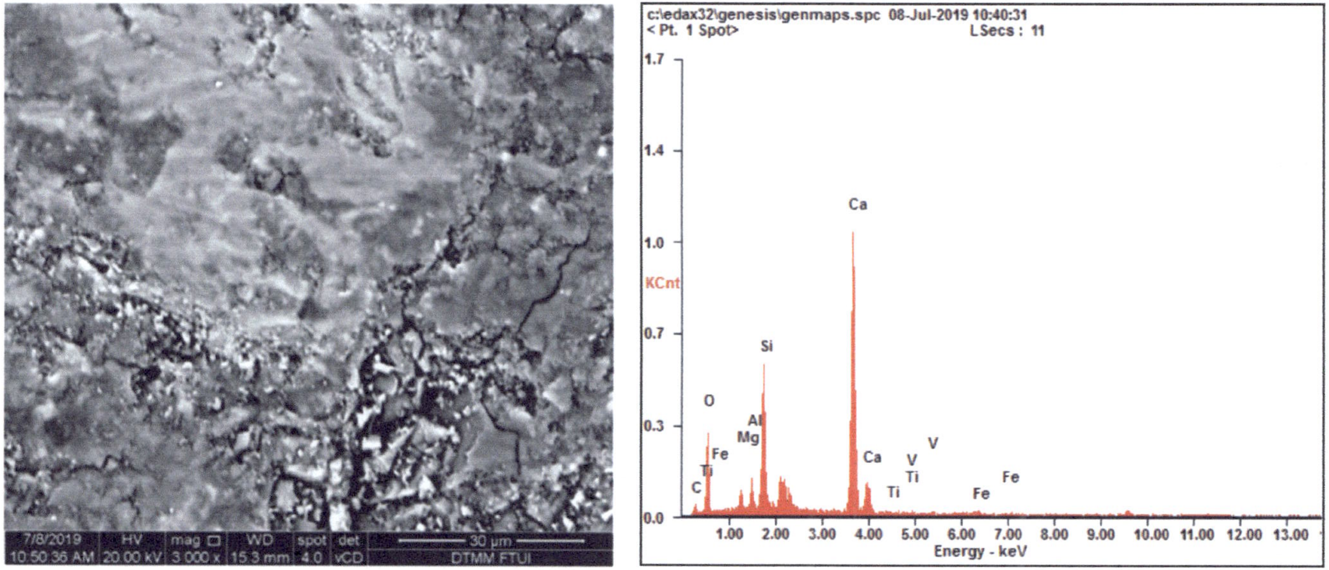

Fig. 8 Magnitude 3000 times of 30% nickel slag powder concrete aged 7 days with EDS 1st spot

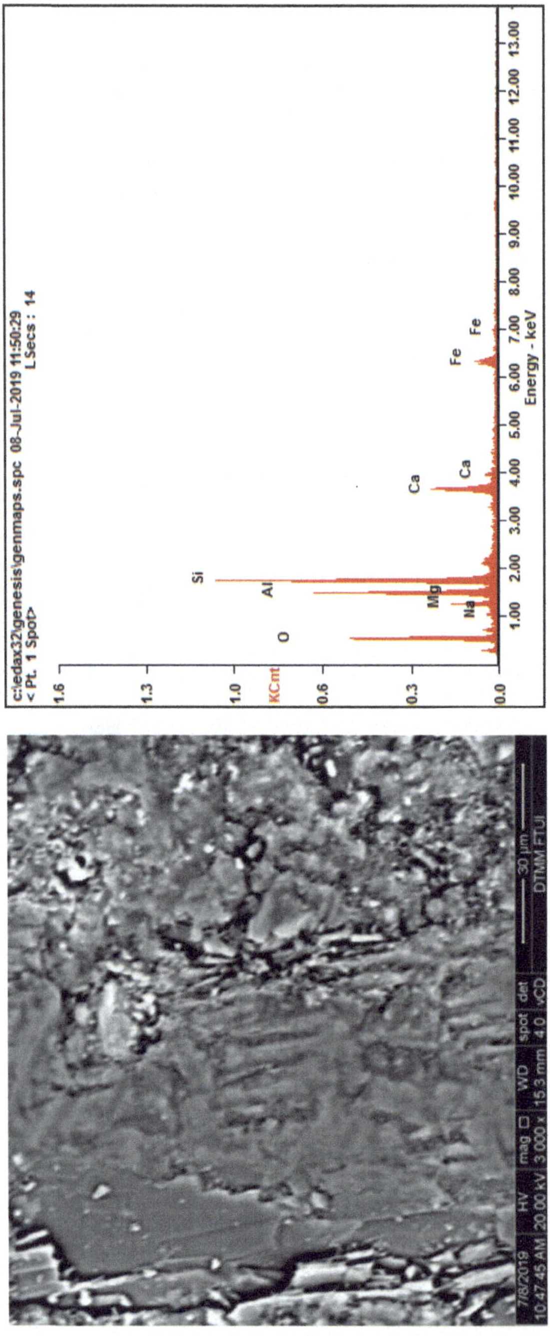

Fig. 9 Magnitude 3000 times of 30% nickel slag powder concrete aged 7 days with EDS 2nd spot

4 Conclusion

Nickel slag powder can be categorized as an alternative construction material as Supplementary Cementitious Materials (SCMs), which can minimize environmental damage due to nickel slag waste. With a water cement ratio design of 0.31 plus an admixture of superplasticizer type, the compressive strength of NSP substituted concrete at the age of 3 days is relatively high at 66.04, 59.80, 58.03, 55.68, 53.68 and 51.82 MPa. The results of the compressive strength are almost equivalent to reference concrete (Ref) at 66.17 MPa. The 30% NSP substitution concrete can be used as an alternative concrete because it can be categorized as high strength concrete and is more environmentally friendly.

It is proven that nickel slag waste can increase the result of compressive strength, the shape and texture of the slag also affects the density of concrete so that it can increase the result of compressive strength. The compressive strength of 5% and 10% NSP substituted concrete obtained higher result than 100% OPC (Ref) concrete at the age of 28 days, this was due to the pozzolanic reaction of nickel slag powder. NSP reacts in a longer period of time than cement which reacts in the short term. NSP has been shown to achieve adequate compressive strength at an early age, while maintaining higher long-term strength than OPC. This research can illustrate that high strength concrete innovation can be made from waste; it can be proven in this study with nickel slag waste. In addition, the use of nickel slag powder as a supplementary cementitious material can minimize nickel slag waste and can reduce CO_2 emissions produced by clinker from cement production.

Acknowledgements We would like to thank all contributors from Jaya Beton Indonesia Ltd for the helps in preparing samples and supporting materials, PT. Indoferro for supplying equipment and DTMM-FTUI for the helps in testing materials. We are grateful to Universitas Trisakti for their support, and generous assistance.

References

1. Ahmad F (2012) Tingkat Pencemaran Logam Berat Dalam Air Laut Dan Sedimen Di Perairan Pulau Muna, Kabaena, Dan Buton Sulawesi Tenggara. MAKARA Sci Ser 13(2):117–124. https://doi.org/10.7454/mss.v13i2.407
2. Al-Jabri KS, Hisada M, Al-Saidy AH, Al-Oraimi SK (2009) Performance of high strength concrete made with copper slag as a fine aggregate. Constr Build Mater 23(6):2132–2140. https://doi.org/10.1016/j.conbuildmat.2008.12.013
3. Ashad H (2008) Kontribusi Bubuk Slag Nikel di dalam Meminimalkan Dampak Intrusi Mikroorganisme terhadap Perilaku Fisik Material Beton 15(1):1–14
4. ASTM International, Statements B (2003). Standard test method for density of hydraulic cement: C188-95 (reapproved 2003). Chem Anal 95(Reapproved):1–2. https://doi.org/10.1520/C0187-10.2
5. Astutiningsih S, Suharno B (2019) Ni Slag Valorisation in Indonesia: a case study. In: Slag characterization, pp 4–7

6. Li G, Zhao X (2003) Properties of concrete incorporating fly ash and ground granulated blast-furnace slag. Cement Concr Compos 25(3):293–299. https://doi.org/10.1016/S0958-9465(02)00058-6
7. Mustika W, Salain IMAK, Sudarsana IK (2017) Penggunaan Terak Nikel Sebagai Agregat Dalam Campuran Beton. Jurnal Spektran 4(2):36–45. https://doi.org/10.24843/spektran.2016.v04.i02.p05
8. Rahman A, Farid M, Ardhyananta H (2016) Pengaruh Komposisi Material Komposit dengan Matriks Polypropylene Berpenguat Serat Alam Terhadap Morfologi dan Kekuatan Sifat Fisik. Jurnal Teknik ITS 5(2):209–211
9. SNI 03-6468-2000 (2000) Tata cara perencanaan campuran tinggi dengan semen portland dengan abu terbang. Standar Nasional Indonesia, p 18
10. Sudarsana K (2015) Mechanical properties of concrete using nickel slag as coarse mechanical properties of concrete using
11. Sugiri S, Saloma S, Yulianti RC (2007) Persamaan Korelasi Sifat Mekanik Beton Mutu Tinggi dengan Agregat Alami dan Slag Nikel. Jurnal Teknik Sipil 14(1):27–40
12. Sun L, Zhao Q, Xiang J, Shi J, Wang L, Hu S, Su S (2007) Effect of anions on preparation of ultrafine alpha-Al_2O_3 powder. J Cent South Univ Technol 14(6):773–778. https://doi.org/10.1007/s11771
13. Triana Dewi V, Ashari ML, Dermawan D (2017) Analisis Pengaruh Limbah Slag dan Debu EAF (electric arc furnace) 2581:292–296

The Effect of Skim Milk and Casein on the Protease Activity of *Fusarium* sp. JE-DP4a

Aerma Hastuty, Iman Hidayat, Wibowo Mangunwardoyo, and Maggy Thenawijaya Suhartono

Abstract Proteases have enormous potential in various industries, from food to pharmaceutical. Protease activity is strongly influenced by the substrate content during fermentation, especially amino acids. This study was conducted to determine the effect of skimmed milk and casein substrates on the protease activity produced by the endophytic fungus, *Fusarium* sp. strain JE-DP4a isolated from *Carica papaya* L. leaves. Selective isolation of protease-producing fungi was carried out qualitatively using skim milk agar (SMA) medium. Endophytic fungus that produced the highest protein hydrolysis index (PHI) was used for the protease fermentation assay to measure a protease activity. The fermentation was carried out on Czapek-Dox + skimmed milk and Czapek-dox + casein media at 37 °C for seven days. A total of 19 protease-producing fungal endophytes were isolated from *C. papaya* leaves. Among them, *Fusarium* sp. strain JE-DP4a exhibited the highest PHI value at 1.52. The highest protease activity was produced on Czapek-dox + casein medium with a protease activity of 46.44 U/mL, while the activity of protease enzyme in Czapek-dox + skim milk medium was 15.28 U/mL.

Keywords Protease · *Fusarium* sp. JE-DP4a · Endophyte · Skim milk · Casein

A. Hastuty (✉) · W. Mangunwardoyo
Research Center for Applied Microbiology, National Research and Innovation Agency, Cibinong, Indonesia
e-mail: aerm001@lipi.go.id

A. Hastuty
Biologi Department, Faculty of Mathematics and Natural Science, Universitas Indonesia, Depok, Indonesia

I. Hidayat · M. T. Suhartono
IPB University, Bogor, Indonesia

M. A. Rahandi Lubis (ed.), *Proceedings of the 1st International Conference of Lignocellulose*, Green Energy and Technology,
https://doi.org/10.1007/978-981-19-1596-3_18

1 Introduction

Endophytic microbes are microbes that live in plant tissues without causing obvious disease symptoms, and the host tissues remain intact and function properly. The endophytic microbes and their host plants interaction are complex and affects to the physiological process and development of their hosts [1]. The endophytic microbes, in general, are autotrophs. They produce various bioactive secondary metabolites that have potential in health, agriculture, and food industry [2–4]. The bioactive metabolites produced by the endophytic microbes is strongly influenced by environmental factors and the endophytes-hosts interaction. Several studies have reported metabolites produced by endophytic microbes including metabolites similar to those produced by their host plants, such as alkaloids, terpenes, phenolics, polyketides, antimicrobial agents such as cryptocardin, anti-cancer agents such as taxol and enzymes [5, 6]. Protease is one of the enzymes produced by the endophytic microbes that have potential in health.

Protease (E.C. 3.4.11-24) is a catalytic enzyme that catalyzes a proteolysis process and break down a protein into smaller polypeptides or amino acids [7, 8]. Microbes that produce proteases have been extensively researched and developed because they have beneficial properties such as fast growth, requiring small space for cultivation dan production, and can be manipulated genetically [9]. Among them, members of the endophytic fungi are reported potential as protease-producing enzyme to be developed to an industrial scale [10]. These include *Rhizopus*, *Aspergillus* dan *Penicillium*.

Protease activity can be influenced by the type of substrate and the concentration of substrate used in the fermentation. The common substrates used to produce protease enzymes are casein and skim milk. These substrates have significant differences in the complexity of the protein decomposition. Casein is one of the main types of protein contained in dairy products dan composed of various hydrophobic and hydrophilic regions [11]. Casein is very easy to bind with calcium phosphate, thus making the protein easily decomposed. In addition, skim milk is milk with low-fat content but has a high protein content, it thus is often used in protease research [12].

Carica papaya L. (Papaya) is the plant that produces latex proteases enzyme known as papain. The papain is commonly used in meat tenderization, therapeutics in pharmaceutical industry, dehairing or dewooling of leather, degumming process in textile, and food industry. Zhao et al. [13] reported that during the long period of co-evolution, some microbial endophytes can produce secondary metabolites similar to their host plants. Therefore, this research examined and characterized the protease enzymes produced by the fungal endophytes associated with papaya. Characterization of protease activity was carried out by testing various conditions of temperature, type, and concentration of substrate and incubation time.

2 Material and Methods

2.1 *Selective Isolation and Screening of Protease-Producing Endophytes*

Isolation and screening of protease activity were conducted on the endophytic fungi isolated from C. papaya. The fungal isolates were inoculated on Potato Dextrose Agar (PDA) medium (200 g potato, 20 g dextrose, and 16 g bacto agar in 100 mL aquadest) and incubated for 7 days at 30 °C. The mycelia were further cut using a cork borer (0.3 mm diameter), and then transferred to the Skim Milk Agar (SMA) medium (2.5 g yeast extract, 1 g D-glucose, 5 g casein, 28 g skim milk powder, 16 g bacto agar in 1000 mL aquadest) in a Petri dish to determine the potential protease activity of the fungi. The positive reaction of protease activity was indicated by the formation of a clear zone around the colony [14]. The protein hydrolysis index (IHP) was calculated by measuring the ratio of the diameter of the clear zone to the diameter of the microbial colony.

2.2 *Protease Enzyme Production*

Enzyme production was carried out using a modified method from [16]. The enzyme production was carried out on two types of media, namely Czapek-dox + casein and Czapek-dox + skim milk. A total of 7-days old 10 fungal mycelia were inoculated into 100 mL of protease enzyme production media. The cultures were incubated at a shaker incubator (30 °C, 120 rpm) for 7 days. The purpose of agitation was to obtain a proper mixing of nutrients and the addition of oxygen for fungal growth [17]. Enzyme activity was measured periodically every day. Furthermore, the filtrate and the supernatant were separated by centrifugation (5415R Refrigerated, Eppendorf, European) at 8000 rpm for 15 min. The supernatant was a crude extract of the enzyme which was then used to measure the activity of the protease enzyme.

Preparation of Bovine Serum Albumin (BSA) Standard Curve. BSA standard curves were prepared using a 50 g/mL (w/v) stock solution. The solution was prepared as follows: 0.25 mg of BSA was dissolved in 5 mL of distilled water. The concentration variations were prepared at 2, 4, 6, 8, 10, 12, 14, 16, 18, 20 g/mL The BSA solution of each concentration was 100 μL, and 5 mL of Bradford's reagent was added to each concentration followed by homogenization using a vortex. The solution was incubated at 30 °C for 30 min. The absorbance value of each final solution was measured using UV–Vis Spectrophotometer (UV-1700 PharmaSpec UV–Vis, SHIMADZU, Japan) at 595 nm. A 100 L of distilled water added 5 mL of Bradford's reagent was used as a control [15].

Preparation of Tyrosine Standard Curve. The curve was prepared based on a modification method from [16]. A standard tyrosine curve was prepared using a stock solution of 3 mg/mL, 20 mg L-tyrosine dissolved in 40 mL of distilled water. L-tyrosine was dissolved slowly in warm conditions at 37 °C until homogeneous using a hotplate magnetic stirrer. L-tyrosine was prepared with various concentrations of 0.01, 0.02, 0.03, 0.04, 0.05, 0.06, 0.07, 0.08, 0.09, and 0.1 mg/mL. The solution of each concentration was diluted with 0.1 M HCl up to 1 mL and then homogenized. The absorbance value was measured at 280 nm using a UV–Vis Spectrophotometer.

Measurement of Protease Enzyme Activity. Measurements were carried out using a modified method from [18]. A total of 2 mL of 1% (w/v) casein substrate was added with 0.5 mL of 0.2 M Tris–HCl buffer pH 7.5 in a test tube. The mixture was pre-incubated at 37 °C for 10 min. A total of 0.5 mL of the enzyme extract was added and then each tube was incubated at 37 °C, for 30 min. After that, the reaction was stopped by adding 2.5 mL of 5% trichloroacetic acid (TCA) and incubated for 20 min in an ice bath to prevent denaturation of the enzyme. The solution was further centrifuged for 5 min at 8000 rpm. After that, the enzyme activity was measured at 280 nm using a UV–Vis Spectrophotometer.

The protease activity was calculated in units of U (units) per mL of enzyme extract. One unit of protease activity is defined as the amount of enzyme that hydrolyses casein to produce 1 g of tyrosine per minute [19]. The protease activity was calculated using the following formula:

$$\text{Protease Activity (U/mL)} = \frac{\mu\text{ mol Tyrosin equivalents released } \times (A)}{(B) \times (C) \times (D)} \tag{1}$$

A = Total volume (mL) of assay

B = Time of assay (minutes) as per the Unit definition

C = Volume of enzyme (mL) of enzyme used

D = Volume (mL) used in colorimetric determination.

Measurement of Protein Content. Measurement of protein content was conducted using the [20]. A total of 100 L of enzyme solution was added to 5 mL of Bradford's reagent. The solution was further homogenized using a vortex and then was incubated at 30 °C for 30 min. After incubation, the solution was homogenized again using a vortex. The absorbance value was measured at 595 nm using a UV–Vis Spectrophotometer.

3 Results and Discussion

3.1 Selective Isolation and Screening of Protease-Producing Endophytes

A total of 19 fungal isolates were obtained from the *C. papaya* leaves. These isolates were further screened for their protease enzyme activity. Among them, four fungal isolates were exhibited (see Fig. 1).

After seven days of incubation on the SMA medium, a clear zone was formed around the colony that indicates the protease activity of the endophytic fungi. Based on the size of the isolate and the clear zone formed on the screening medium, the value of the protein hydrolysis index (PHI) of each fungal endophyte isolate was calculated. The strain of JE-DP4a known as *Fusarium* sp. exhibited the highest protein hydrolysis index value at 1.52 (see Table 1). This strain was further selected for a quantitative assay of protease activity.

If the value of the protein hydrolysis index (PHI) is > 3, it is categorized as a high value [21]. The PHI value is divided into three categories, namely the low category

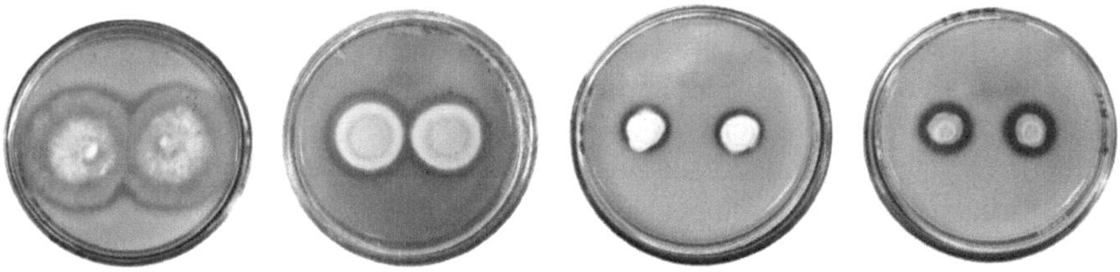

Fig. 1 Clear zones showed protease enzyme activity of four fungal isolates from *C. papaya* leaves

Table 1 Protein hydrolysis index (PHI) of fungal isolates from *C. papaya* leaves

	Protease activity	Protein hydrolysis index (PHI)
JE-BP1	–	0.00
JE-BP2	+	1.09
JE-BP3	+	1.32
JE-BP5	–	0.00
JE-DP1	–	0.00
JE-DP2	+	1.17
JE-DP3	–	0.00
JE-DP4a (*Fusarium* sp.)	+	1.52
JE-TDP2	–	0.00
JE-TDP3	–	0.00

with the PHI value < 2.1, the medium category with the PHI value 2.1–3.1, and the high category with the PHI value > 3.1 [21]. The calculation of the IHP value was obtained from the comparison of the diameter of the clear zone to the diameter of the colony. The PHI value indicates the level of protease activity produced by a microbe. The current study showed that the PHI value produced by *Fusarium* sp. strain JE-DP4a belongs to the low category (PHI value = 1.52).

3.2 Effect of Casein and Skim Milk Substrates on Protease Activity

The protease activity of a microbe can be influenced by various factors, this includes the type of substrate. This study showed that the protease activity of *Fusarium* sp. strain JE-DP4a on the medium amended with casein produced higher protease activity than the protease activity on the medium amended with skim milk (see Fig. 2). The highest activity of protease activity on the medium amended with casein was 46.44 U/mL at the seventh-day incubation, while the highest activity of protease activity on the medium amended with skim milk was 15.28 U/mL on the fifth day of incubation. On the fifth day, the protease activity on the medium amended with casein (33 U/mL) was still higher than the protease activity on the medium amended with skim milk. Based on these results, casein is a better protein substrate than skim milk in the production of protease enzyme of *Fusarium* sp. strain JE-DP4a on the CZA medium. The current protease activity has a higher activity value when compared to other studies such as Radha et al. [22] who reported *Aspergillus* sp. IS5 from soil protease activity of 0.277 U/mL, *Acremonium* sp. protease activity of 3.34 U/mL [15], *Mucor javanicus* M3 protease activity of 0.51 U/mL [23], and *A. niger* PAM18A protease activity of 0.930 U/mL [24]. However, the protease activity of the current *Fusarium* sp. strain JE-DP4a is still lower than *Aspergillus foetidus* reported by Souza et al. [8] that exhibited protease activity of 63.7 U/mL.

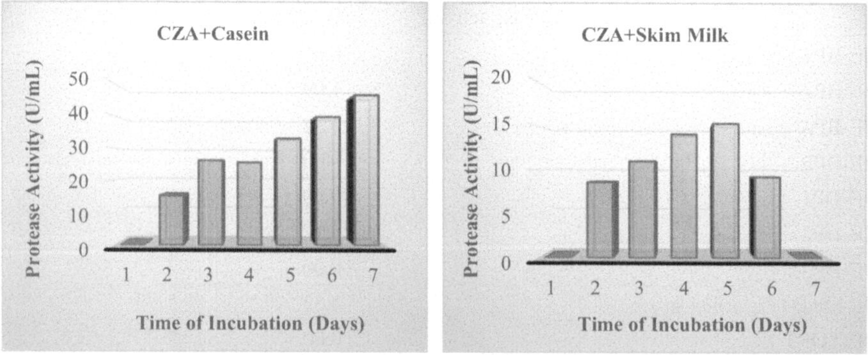

Fig. 2 Effect of substrate on protease activity of *Fusarium* sp. strain JE-DP4a

The current study showed that the protease activity of the *Fusarium* sp. strain JE-DP4a on the medium amended with casein was higher than the medium amended with skim milk. It is probably due to casein composed of various hydrophobic and hydrophilic regions, which makes it easier to bind to calcium phosphate than skim milk [12]. According to Susanti [25], casein will be hydrolyzed into peptide and amino acid compounds that are soluble in the medium. Casein protein consists of phosphoproteins that can bind to calcium to form calcium calcinate salts [26]. If casein is hydrolyzed by proteases, the white color formed in the medium will disappear and cause the colour in the liquid medium to become clear or a clear zone is around the colonies on the solid medium. The degradation of casein by protease enzyme occurs through the CO–NH peptide bond breakdown in the presence of water.

3.3 *Measurement of Protein Content*

The protein content of CZA medium containing skim milk on the 1st day was 68.17 g/mL and was completely degraded on the 4th day, while the protein content of CZA medium containing casein was 107.06 g/ml on the first day and completely degraded at the fifth day (see Fig. 3).

The value of protein content is usually inversely to protease activity. The higher the protease activity, the lower the protein content value. This is probably due to the degradation of the protein in the media by the protease enzymes which directly indicates the enzyme activity. In addition, the protein content measured from the crude extract of the protease enzyme showed the amount of substrate degraded by a proteolytic enzyme produced by the fungal endophyte. According to Ramadhani et al. [24], the protein content in the enzyme affects the catalytic power of the enzyme.

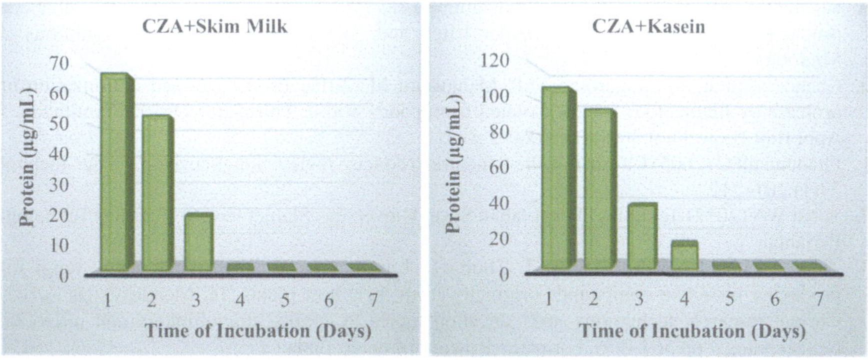

Fig. 3 The protein content of CZA medium amended with skim milk and casein substrates

4 Conclusion

A total of 19 protease-producing fungal endophytes were isolated from *C. papaya* leaves. Among them, *Fusarium* sp. strain JE-DP4a exhibited the highest protein hydrolysis index value of 1.52. The highest protease activity produced by *Fusarium* sp. strain JE-DP4a was found on the Czapek-dox medium amended with casein (46.44 U/mL).

Acknowledgements This study was supported by Hibah Riset PUTI Doktor 2020 from the University of Indonesia. The authors thank Irma Herawati for laboratory assistance. We also thank the Japan Society for the Promotion of Science (JSPS) for partial funding of the publication fee.

References

1. Tadych M, White JF (2019) Encyclopedia of microbiology, 5th edn. Academic Press, United State
2. Tan RX, Zou WX (2001) Endophytes: a rich source of functional metabolites. Nat Prod Rep 18:448–459
3. Strobel GA (2003) Endophytes as sources of bioactive products. Microb Infect 5:535–544
4. Maria GL, Sridhar KR, Raviraja NS (2005) Antimicrobial and enzyme activity of mangrove endophytic fungi of southwest coast of India. J Agric Technol 1:67–80
5. Strobel G, Daisy B, Castillo U, Harper J (2004) Natural products from endophytic microorganisms. J Nat Prod 67:257–268
6. Pavithra N, Sathish I, Ananda K (2012) Antimicrobial and enzyme activity of endophytic fungi isolated from Tulsi. J Pharm Biomed Sci 16(16):1–6
7. Alnahdi HS (2012) Isolation and screening of extracellular proteases produced by new isolated *Bacillus* sp. J Appl Pharm Sci 2(09):071–074
8. Souza PM, Filho EXF, Werneck G, Aliakbarian B, Siqueira F, Magalhães PO, Perego P, Converti A, Junior AP (2017) Production, purification and characterization of an aspartic protease from *Aspergillus foetidus*. J Food Chem Toxicol 109:1103–1110
9. Kocher GS, Mishra S (2009) Immobilization of *Bacillus circulans* MTCC 7906 for enhanced production of alkaline protease under batch and packed bed fermentation conditions. J Microbiol 7(2):1–6
10. Chandrasekaran S, Kumaresan SSP, Manavalan M (2015) Production and optimization of protease by filamentous fungus isolated from paddy soil in Thiruvarur District Tamilnadu. J Appl Biol Biotechnol 3(06):066–069
11. Phadungath C (2005) Casein micelle structure: a concise review Songklanakarin. J Sci Technol 27(1):201–212
12. Setya WA (2012) Teknologi Pengolahan Susu. Universitas Slamet Riyadi, Fakultas Teknologi Pertanian
13. Zhao J, Zhou L, Wang J, Shan T, Zhong L, Liu X, Gao X (2010) Endophytic fungi for producing bioactive compounds originally from their host plants. In: Mendez-Vilas A (ed) Current research, technology and education topics in applied microbiology and microbial biotechnology, pp 567–576. Formatex Research Center, Badajoz
14. Sharma AK, Sharma V, Saxena J, Yadav B, Alam A, Prakash A (2015) Isolation and screening of extracellular protease enzyme from bacterial and fungal isolates of soil. Int J Sci Res Environ Sci 3(9):334–340
15. Jain P, Aggarwal V, Sharma A, Pundir RK (2012) Isolation production and partial purification of protease from an endophytic *Acremonium* sp. J Agric Technol 8(6):1979–1989

16. Yusriah, Kuswytasari ND (2013) Pengaruh pH dan suhu terhadap aktivitas protease *Penicillium* sp. J Sains dan Seni Pomits 2(1):48–50
17. Kamath P, Subrahmanyam VM, Rao JV, Raj PV (2010) Optimization of cultural conditions for protease production by a fungal species. Indian J Phar Sci 72(2):161–166
18. Chow BF, Peticolas M (1948) A rapid method for the determination of proteolytic activities of enzyme preparations. J Gen Phy 32(1):17–24
19. Guangrong H, Tiejing Y, Po H, Jiaxing J (2006) Purification and characterization of a protease from *Thermophilic bacillus* strain HS08. African J Biotechnol 5(24):2433–2438
20. Bradford MM (1976) A rapid and sensitive method for the quantitation of microgram quantities of protein utilizing the principle of protein-dye binding. J Anal Biochem 72(1–2):248–254
21. Ahmad MS, Noor ZM, Ariffin ZZ (2014) Isolation and identification fibrinolytic protease endophytic fungi from *Hibiscus* leaves in Shah Alam. Int J Agric Bios Eng 8(10):1104–1107
22. Radha S, Sridevi A, Nithya VJ, Prasad NBL, Narasimha G (2012) Isolation and screening of proteolytic fungal cultures from soil contaminated with abattoir waste. Biochem an Indian J 6(7):226–230
23. Choliq A (2008) Aktivitas enzim protease dari *Mucor* javanicus yang ditumbuhkan pada media tepung singkong (*Mannihot utilissima*). J Berita Biologi 9(3):299–303
24. Ramadhani P, Rukmi MGI, Pujiyanto S (2015) Produksi enzim protease dari *A. niger* PAM18A dengan variasi pH dan waktu inkubasi. J Biol 4(2):25–34
25. Susanti E (2003) Isolasi dan karakterisasi protease dari *Bacillus subtilis* 1012M15. *Biodiversitas*. J Biol Diver 4(1):12–17
26. Choirunissa HN, Sari RY, Hastuti US, Witjoro A (2017) Identifikasi dan uji kemampuan hidrolisis pada bakteri amilolitik dan proteolitik yang diisolasi dari wadi, makanan khas Kalimantan Tengah. J Bionat 18(2):99–109

Biomass Conversion (Bioenergy, Bioproduct)

The Functional Bacteria Test as Biopriming Agents to Accelerate In-Vitro Seed Germination

Sri Widawati and Suliasih

Abstract Functional bacteria are a group of soil bacteria that serve as suppliers of Nitrogen (N) and Phosphorus (P) nutrients, growth promoters, and pathogens control. Furthermore, these bacteria can be inoculated into the seeds as biopriming to accelerate germination and optimize the potential of the seed. Therefore, this experiment aims to obtain and determine the effectiveness and activity of 5 functional bacterial as biopriming agents in accelerating the germination of 4 types of seeds in vitro. In addition, the Completely Randomized Design was used with 5 identified functional bacteria, namely *Bacillus siamensis* InaCC B1435 (A), *Chromobacterium alkanivorans* InaCC B1433 (B), *Chromobacterium alkanivorans* InaCCB1436 (C), *Stenotrophomonas maltophilia* InaCCB1434 (D), *Chromobacterium violaceum* InaCC B1437 (E), and 4 types of seeds namely *Ipomoea reptana* Poir, *Leucaena leucocephala* L., *Sorghum bicolor* L., and *Cucumis sativus* L. All treatments and control were conducted in triplicate. The result showed that the functional bacteria activity in inoculants A, B, C, D, and E can accelerate the seed germination of *I. reptana* Poir and *C. sativus* L., bacteria in inoculants B and D can accelerate the seed germination *S. bicolor* L., and inoculants B (*Chromobacterium alkanivorans* InaCCB1433) was able to accelerate the seed germination of all tested seeds (*I. reptana* Poir, *L. leucocephala* L., *C. sativus* L L., and *S. bicolor*) with the highest yield on all variables of bacterial activity as a biopriming agent, namely germination percentage (80, 50, 70, and 20%), sprouts length (9.64, 4.14, 8.89, and 9.06 cm), and vigor index (771.2, 207, 711.2, and 181.2).

Keywords Functional bacteria · Seed germination · Biopriming

All authors contributed equally to this work.

S. Widawati (✉) · Suliasih
Research Centre for Applied Microbiology, National Research and Innovation Agency, Jl Raya Jakarta-Bogor Km 46, Cibinong, Bogor 16911, Indonesia
e-mail: sriw001@brin.go.id

M. A. Rahandi Lubis (ed.), *Proceedings of the 1st International Conference of Lignocellulose*, Green Energy and Technology,
https://doi.org/10.1007/978-981-19-1596-3_19

1 Introduction

The Bacterial functional groups are a community of bacteria living in plant roots areas (Rhizobacteria), fields, sediments, roots, leaves, and plants trunks (endophytes). Furthermore, they stabilize soil fertility and facilitates the absorption of water and nutrients. These bacteria provide essential nutrients that are necessary for plant growth, such as nitrogen (N) bound in the air and soil (NFB), and elemental phosphate (P) which is released from the bonds with Al, Fe, and Ca (PSB). In addition, they produce the hormone IAA (indole acetic acid), ACC-deaminase, and siderophores, which stimulate growth, biological control (pathogens), antibiosis, and symbiosis.

In addition to plant sprouts and seedlings, bacteria functional groups can also be inoculated into plant seeds. The technique used for inoculating plant seeds is the biopriming technique, which is effective in increasing the rate and uniformity of seed growth. Furthermore, it is a seed treatment technique by inoculating seeds with beneficial bacteria for seeds protection (biological aspects) and seed hydration techniques (physiological aspects) for disease control [1]. The biopriming technique is a derivative of the priming technique, namely germination, which uses the moisture content of the seeds and temperature control to balance the metabolic activity in the seeds and maximize the potential of the seeds [2].

Some of the functional biopriming bacteria in *Zea mays* L., *Carthamus tinctorius* L., *Hordeum vulgare* L., *Solanum tuberosum* L., and *Oryza sativa* L. seeds are *Bacillus cereus, Azotobacter lipoferum, Azospirillum* sp., *Azotobacter chroococcum, Pseudomonas* sp., *Bacillus pumilus, B. furmus*, and *Bacillus polymixa* [3–7]. Inoculation of seeds can also be carried out on other types of seeds, for example on seeds of the Fabaceae, Solonaceae, Graminaceae, Poaceae, Cucurbitaceae, and Convolvulaceae families. Therefore, experiments were carried out to obtain bacterial functional groups as biopriming agents in several types of seeds such as Kangkong (*Ipomoea reptana* Poir.), Lamtoro (*Leucaena leucocephala* L.), Sorghum (*Sorghum bicolor* L.), and Cucumber (*Cucumis sativus* L.). The experiments aim to determine the activity of functional bacteria (PGPR) and to obtain the effectiveness of functional bacteria as biopriming in accelerating seed germination in vitro.

2 Materials and Methods

2.1 Liquid Inoculant

Five functional groups of bacteria were used, namely, *Bacillus siamensis* InaCC B1435 (**A**), *Chromobacterium alkanivorans* InaCC B1433 (**B**), *Chromobacterium alkanivorans* InaCC B1436 (**C**), *Stenotrophomonas maltophilia* InaCC B1434 (**D**), *Chromobacterium violaceum* InaCC B1437 (**E**). In previous studies, all bacteria were able to produce IAA, ACC-deaminase, siderophore, phosphatase, and high nitrogenase activity and therefore they have the potential to act as plant growth promoting rhizobacteria (PGPR). Each bacterial culture was rejuvenated in Nutrient Agar slant (NA = Peptone 5 g, beef extract 3 g, agar 20 g, sterile aquades 1000 mL) and incubated for 7 days at 28–30 °C, then transferred to an Erlenmeyer flask containing 250 mL of liquid NA media (without agar) and shaken at 28–30 °C for ± 3 days or until the minimum bacterial population reached 10^7 CFU mL^{-1} [8].

2.2 *Experimental Methods and Data Analysis*

This experiment applied a completely randomized design, which was organized in factorial. The first factor was liquid inoculants A, B, C, D, E, and control (no bacteria). The second factor was the seeds of *I. reptana* Poir, *L. leucocephala* L., *S. bicolor* L., and *C. sativus* L. Each treatment was in triplicate, and the observation of the effect of bacterial functional groups on the stimulation of seed germination in vitro was carried out with the variables bacterial population in the liquid inoculant, seed germination capacity, seedling length, vigor index, and bacterial colonization on sprouts. These variables were used to measure the activity of bacteria as a biopriming agent. The data obtained were analyzed using the SPSS version 13.0 with the data mean comparison through the Duncan Multiple Range (DMRT) test $P < 0.05$ (5%) [9].

2.3 *Seeds Inoculation*

The inoculation was carried out on the seeds of the *Ipomoea reptana* Poir. (Kangkong), *Leucaena leucocephala* L. (Lamtoro), *Sorghum bicolor* L. (Sorghum), and *Cucumis sativus* L. (Cucumber). In addition, it was carried out according to the following procedure: Selecting each type of seed based on the same shape (10 seeds per replication), then all seeds (separated based on species) were washed sequentially with sterile distilled water, 96% alcohol, and H_2SO_4. Furthermore, seed dormancy was triggered by immersing the seeds in warm water (temperature 60 °C) in baker glass for 24 h. About 50 swollen seeds from each species were taken out and then divided into 5 parts (10 seeds/treatment) and were put into a baker glass, containing 50% sterile distilled water and 50% liquid inoculant A, B, C, D, and E (inoculation process). The seeds in the control treatment were not inoculated (soaked with sterile distilled water). After 60 min of incubation at 28–30 °C, each type of seed was transferred into a Petridis covered with wet sterile filter paper (*Whatman*) (10 seeds/Petridis/treatment). About 1 mL of liquid inoculant (according to treatment) was added into the Petridis containing seeds and incubated for 7 days at a temperature of 28–30 °C.

2.4 Germination Percentage, Seedling Length, and Vigor Index (in Vitro)

The germination percentage (%) is the ratio of the number of germinating seeds to the total number of seeds that have germinated during a certain time in the optimal growth medium (in vitro). The capacity of germination is indicated by the presence of plumules and radicles in the seed embryo [10]. It is calculated according to the formula: Percentage of germination = Total seeds germinated/Total seeds test × 100% [11]. Meanwhile, the vigor index (physiological analysis) represents the rate of seed germination, which showed that the seed is vigorous [12]. It is calculated according to the following: Vigor Index (VI) = Germination (%) × Sprouts length (cm) [13]. The germination growth rate (length of sprouts) is measured from the bottom of the sprouts body to the tip of the sprouts.

2.5 Bacterial Colonization on Sprouts

The sprouts from each treatment were cut into pieces and placed in a test tube containing 9 mL of distilled water and homogenized on a vortex for 1 min. A series of dilutions from 10^{-1} to 10^{-5} was made in a sterile room. A total of 0.02 mL of sprouts extract from 10^{-3} and 10^{-5} dilutions was transferred to sterile Petridis and doused with NA media poured. After 3 days of incubation at 28–30 °C, the growing bacterial colonies were counted used the Vincent method [14], namely total plate count method (TPC) to obtain colony forming units (CFU).

3 Results and Discussion

3.1 The Bacterial Population on Liquid Inoculant

The results of the bacterial population with liquid inoculant were 10^8 CFU mL^{-1} for all inoculants A, B, C, D, and E (Table 1). These bacterial counts met the minimum prerequisite in the liquid inoculant. Rao (1995) stated that the carrier media in inoculants should contain a bacterial population of 10^8 CFU mL^{-1} or per gram of carrier media [15]. The bacterial population did not reach the optimum range of up to 10^{10} CFU mL^{-1}, which was predicted to be caused by inappropriate culture techniques and media that inhibit bacterial cell division [16].

Table 1 Number of functional bacterial populations in liquid inoculants

Name of functional bacteria	Treatment code	Bacterial population (CFU mL^{-1})
Bacillus siamensis InaCCB1435	A	3.15×10^8b
Chromobacterium alkanivorans InaCC B1433	B	3.90×10^8b
Chromobacterium alkanivorans InaCC B1436	C	1.30×10^8a
Stenotrophomonas maltophilia InaCC B1434	D	1.05×10^8a
Chromobacterium violaceum InaCC B1437	E	1.03×10^8a

3.2 The Test of Functional Bacteria as Biopriming to Accelerate Seed Germination in Vitro

The test of inoculants A, B, C, D, and E containing single functional bacteria (*Bacillus siamensis* InaCC B1435, *Chromobacterium alkanivorans* InaCC B1433, *Chromobacterium alkanivorans* InaCC B1436, *Stenotrophomonas maltophilia* InaCC B1434, *Chromobacterium violaceum* InaCC B1437) to accelerate germination in the seeds of 4 plant species (in vitro) resulted in statistically significant differences ($P < 0.05$) (Fig. 1a, b, Table 2).

The results of an in vitro experiment showed that a single inoculation of 5 types of functional bacteria (PGPR) was able to stimulate and increase the germination percentage, length of seedling, and vigor index compared to the control. This showed that the functional bacteria (PGPR) contained in inoculants A, B, C, D, and E are effective in stimulating and accelerating the process of seed germination (vigor index). The positive activity of these bacteria on germination was also supported by the innate nature of bacteria capable of producing IAA, ACC-deaminase, siderophore, phosphatase, and high nitrogenase activity from the results of the physiological analysis in previous studies. Furthermore, according to Suliasih and Widawati [17], inoculating PGPR on seeds stimulates and increases the percentage of seed germination and vigor index. The vigor index represents the seed germination rate, which showed that the seed is vigorous [12]. The value of the vigor index specifies if the germination is accomplished. The statement supports the results of the 5 types of bacteria inoculation in inoculants A, B, C, D, and E on 2 seed types, which were *I. reptana* Poir. and *C. sativus* L. The responses of both plant seeds were recorded as such, germination percentage (80%, 80%, 50%, 80%, 70% and 60%, 80%, 60%, 60%, 50%), sprouts length (10.3 cm, 9.64 cm, 8.3 cm, 9.8 cm, 8.51cm and 1.5 cm, 8.89 cm, 2.71 cm, 3.65 cm, 1.37 cm), and the vigor index (810, 771.2, 415, 784, 595.7 and 90, 711.2, 162.6, 219, 82.2), which were higher compared to control (Fig. 1a, b, Table 2). Similar results were observed by Shaukat et al. [18], Lenin, and Jayanthi [19] which reported that inoculating PGPR on cereal seeds showed a higher germination rate and vigor index compared to control.

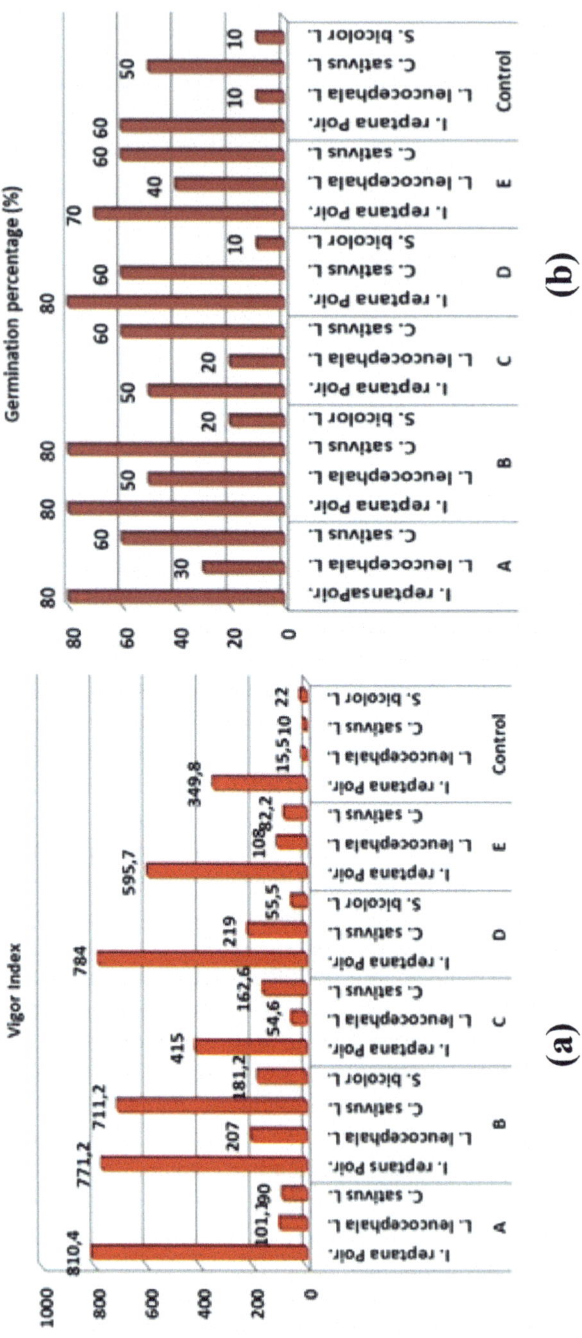

Fig. 1 Percentage of seed germination (a) and vigor index (b)

Table 2 Measurement results for the length of sprouts

Treatment code	No.	Types of plant seeds	Sprouts length (cm)
A (*B. siamensis* InaCC B1435)	1	*I. reptana* Poir.	10.13 ± 0.10^{f}
	2	*L. leucocephala* L.	3.37 ± 0.08^{abc}
	3	*C. sativus* L.	1.50 ± 0.13^{ab}
	4	*S. bicolor* L.	Did not germinate
B (*C. alkanivorans* InaCC B1433)	1	*I. reptana* Poir.	9.64 ± 0.35^{f}
	2	*L. leucocephala* L.	4.14 ± 0.77^{abcde}
	3	*C. sativus* L.	8.89 ± 0.72^{ef}
	4	*S. bicolor* L.	9.06 ± 0.77^{ef}
C (*C. alkanivorans* InaCC B1436)	1	*I. reptana* Poir.	8.30 ± 0.01^{cdef}
	2	*L. leucocephala* L.	2.73 ± 0.43^{ab}
	3	*C. sativus* L.	2.71 ± 0.13^{ab}
	4	*S. bicolor* L.	Did not germinate
D (*S. maltophilia* InaCC B1434)	1	*I. reptana* Poir.	9.80 ± 0.43^{f}
	2	*L. leucocephala* L.	Did not germinate
	3	*C. sativus* L.	3.65 ± 0.39^{abcd}
	4	*S. bicolor* L.	5.55 ± 0.35^{bcdef}
E (*S. maltophilia* InaCC B1434)	1	*I. reptana* Poir.	8.51 ± 0.82^{def}
	2	*L. leucocephala* L.	2.70 ± 0.68^{ab}
	3	*C. sativus* L.	1.37 ± 0.01^{ab}
	4	*S. bicolor* L.	Did not germinate
Control	1	*I. reptana* Poir.	5.83 ± 0.18^{bcdef}
	2	*L. leucocephala* L.	1.55 ± 0.08^{ab}
	3	*C. sativus* L.	0.2 ± 0.39^{a}
	4	*S. bicolor* L.	2.2 ± 0.12^{ab}

Notes Data were represented from average ± SD ($n = 3$). The numbers followed by the same letter were not significantly different at ($p < 0.05$) level of Duncan's test

Similarly, Hameeda [20], Niranjan [21] also described that the application of PGPR inoculants increased the germination and vigor index of pearl millet crops in vitro.

This experiment showed that 1 type of functional bacteria (PGPR), namely *C. alkanivorans* InaCC B1433 (inoculant B), was able to associate, stimulated, and accelerated the germination of 4 seed types of the plants tested. The influences of these bacteria were detected in the germination percentage (80, 50, 70, and 20%), sprouts length (9.64, 4.14, 8.89, and 9.06 cm), and vigor index (771.2, 207, 711.2, and 181.2), which were higher than the result of other bacteria in inoculants A, C, D, and E. This could be due to the influence of the IAA hormone produced by a single bacterium (*C. alkanivorans* InaCC B1433) (IAA data not shown), which stimulates bacterial activity in accelerating the germination process for the 4 types of seeds. Similar outcomes were observed by Widawati and Suliasih [22], that the single

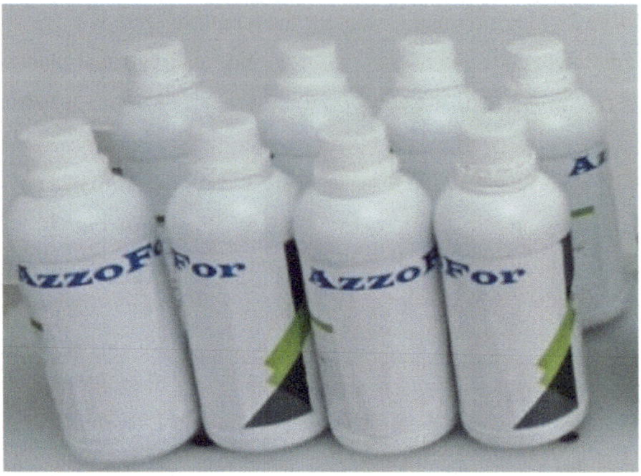

Fig. 2 AZZOFOR^{BP}: product of inoculant bacteria as biopriming agents

inoculation (*Bacillus* sp., *A. lipoferum, A. chroococcum*) on the seeds *S. bicolor* L increased sprouts length (3.60 cm, 3.59 cm, 5.54 cm) and the vigor index (444.00, 490.00, and 916.00). The results of this experiment support the research carried out by Mathivanan et al. [23] and Sengupta et al. [24] that inoculated bacteria on *Pennisetum glaucum* L. seeds and *Zea mays* L. seeds. Therefore, it can be concluded that the bacterium *C. alkanivorans* InaCC B1433 belongs to the group of biopriming agents (Fig. 2).

The inoculation of PGPR functional bacteria is supposed to promote better germination and show higher vigor index compared to control [19]. However, the *B. siamensis* InaCC B1435 (A), *C. alkanivorans* InaCC B1436 (C), dan *C. violaceum* InaCCB1437 (E) inoculation on *S. bicolor* L. seeds, and *S. maltophilia* InaCC B1434 (D) inoculation on *L. leucocephala* L promoted and stimulated the germination, which caused seed rot and eventually withered (Fig. 1a, b, Table 2). This phenomenon can be caused by several factors, such as poor seed maturity and viability, embryonic immaturity, incomplete breakdown of the physical remains of the seed, or possibly inhibitory compounds on the surface and inside the seed. Therefore the imbibition procedure (water absorption) was not optimal. As widely known, imbibition affects the germination process of the seed [25] as well as the initial germination process [26]. In brief, inoculating seeds with functional bacteria as a biofarming agent supports seed germination, because rough seeds (there is leakage due to deterioration) during the imbibition will release certain hormones that can be utilized by these bacteria [27]

3.3 *Bacterial Colonization on Sproutss*

As shown in Table 3, 5 functional bacteria (*B. siamensis* InaCC B1435, *C. alkanivorans* InaCC B1433, *C. alkanivorans* InaCC B1436, *S. maltophilia* InaCC B1434, *C.*

violaceum InaCC B1437) in inoculants B, C, and D, and E were able to live (colonize) and associate the sprouts of *I. reptana* Poir., *L. leucocephala* L., *S. bicolor* L., and *C. sativus* L., with the exception of bacteria in inoculants A, C, and E which could not colonize *S. bicolor* L., and bacteria in inoculant D that could not colonize *L. leucocephala* L. The bacterial counts colonizing the sprouts were statistically significant ($P < 0.05$).

Seven days after the inoculation of the bacteria in inoculants A, B, C, D, and E on the seeds of *I. reptana* Poir., *L. leucocephala* L., *S. bicolor* L., and *C. sativus* L., the results showed that the number of bacteria colonizing the sprouts of *I. reptana* Poir. (4.70×10^5 CFU mL^{-1}, 6.35×10^5 CFU mL^{-1}, 3.85×10^5 CFU mL^{-1}, and 3.45×10^5 CFU mL^{-1}) compared to *L. leucocephala* L., *S. bicolor* L., and *C. sativus* L. Furthermore, Gupta et al. [28] reported that the first bacterial colonies formed after 24 h of inoculation and that the number of bacteria increased significantly up to 8 days after inoculation with the number of bacteria; 1.2×10^4 dan 2.1×10^3 CFU g^{-1}. The statistically highest bacterial colonization was shown by the sprouts of *I. reptana* Poir, which was inoculated by inoculant B (6.35×10^5 CFU mL^{-1}), while the lowest colonization was shown by *S. bicolor* L. inoculated by inoculant D (1.90×10^4 CFU mL^{-1}). Similarly, research by Zaghloul et al. [29] also produced the highest bacterial colonization on roots of Faba beans inoculated with TN10 inoculant (6.4×10^5 CFU/

Table 3 Bacterial colonization on sprouts

Treatment code	Types of plant seeds	Bacterial population (CFU mL^{-1})
A (*B. siamensis* InaCCB1435)	*I. reptana* Poir.	$4.70 \times 10^5 \pm 0.00^{h}$
	L. leucocephala L.	$1.99 \times 10^5 \pm 32.04^{de}$
	C. sativus L.	$2.09 \times 10^5 \pm 35.51^{de}$
B (*C. alkanivorans* InaCCB1433)	*I. reptana* Poir.	$6.35 \times 10^5 \pm 0.00^{i}$
	L. leucocephala L.	$2.59 \times 10^5 \pm 5.20^{e}$
	C. sativus L.	$2.81 \times 10^5 \pm 61.78^{ef}$
	S. bicolor L.	$1.26 \times 10^5 \pm 20.21^{bcd}$
C (*C. alkanivorans* InaCCB1436)	*I. reptana* Poir.	$3.85 \times 10^5 \pm 0.00^{g}$
	L. leucocephala L.	$7.10 \times 10^4 \pm 3.46^{ab}$
	C. sativus L.	$1.69 \times 10^4 \pm 17.90^{cd}$
D (*S. maltophilia* InaCCB1434)	*I. reptana* Poir.	$3.75 \times 10^5 \pm 0.00^{g}$
	C. sativus L.	$2.09 \times 10^5 \pm 66.68^{de}$
	S. bicolor L.	$1.90 \times 10^4 \pm 3.46^{a}$
E (*S. maltophilia* InaCCB1434)	*I. reptana* Poir.	$3.45 \times 10^5 \pm 0.00^{fg}$
	L. leucocephala L.	$1.57 \times 10^5 \pm 15.30^{cd}$
	C. sativus L.	$1.\ 07 \times 10^5 \pm 4.91^{bc}$
Control	–	–

Notes Data were represented from average ± SD (n = 3). The numbers followed by the same letter were not significantly different at ($p < 0.05$) level of Duncan's test

root). According to the research conducted by Ji et al. [30] and Singh & Jha [31], it was reported that *Bacillus cepacia* Lu 10^{-1} and *Stenotrophomonas maltophilia* SBP-9 colonized plant roots. The bacterial ability to colonize the seeds depends on the initial period of the imbibition since it is the initial stage that determines the success of germination. The absorption of liquid by the seeds is very important for regulating and activating growth hormones. According to Dennis et al. [32], the colonization was due to the ability of bacteria to associate in the germination area and the ability of the seeds to produce exudates in the form of organic acids and amino acids in the root area.

4 Conclusion

The present study concluded that the activity of functional bacteria *B. siamensis* InaCC B1435, *C. alkanivorans* InaCC B1433, *C. alkanivorans* InaCC B1436, *S. maltophilia* InaCC B1434, *C. violaceum* InaCC B1437 in inoculants A, B, C, D, and E stimulated and accelerated seed germination of *Ipomoea reptana* Poir. Similarly, *C. sativus* L., inoculants A, B, C, and E stimulated and accelerated seed germination of *Leucaena leucocephala* L., inoculants B and D stimulated and accelerated seed germination of *S. bicolor* L., and inoculants B (*C. alkanivorans* InaCC B1433) stimulated and accelerated germination of all tested seeds with the highest yield on germination percentage, sprouts length and vigor index. Furthermore, five functional bacteria in inoculants A, B, C, D, and E, which were inoculated on 4 types of plant seeds have potential as biopriming agents.

Acknowledgements The author expresses profound gratitude to the Japan Society for the Promotion of Science (JSPS) for funding the publication fees, Ety Suryati and Nurma for their assistance in laboratory work, and Riski Mareta Ayu P (student) for technical assistance in data collection.

References

1. Reddy PP (2012) Recent advances in crop protection. Springer, New Delhi, pp 83–90
2. Rouhi HR, Surki AA, Sharif-Zadeh F, Afshari RT, Aboutalebian MA, Ahmadvand G (2011) Notulae Sci Biol 3:101–108
3. Chakraborty U, Roy S, Chakraborty AP, Dey P, Chakraborty B (2011) Rec Res Sci Technol 3:61–70
4. Sharifi RS, Khavazi K (2011) J Food Agric Environ 9:496–500
5. Sharifi RS (2012) Tech J Eng Appl Sci 2:162–166
6. Mirshekari B, Hokmalipour S, Sharifi RS, Farahvash F (2012) J Food Agric Environ 10:314–320
7. Madyasari I, Budiman C, Syamsuddin, Manohara D, Ilyas S (2017) Hort. Indonesia 8(3):192–202
8. Obaton M (1977) Ayanaba A, Dart PJ (eds) Biological nitrogen fixation in farming system of the tropics. Willey, London, pp 173–214

9. Sarwono J (2006) Analisis Data Penelitian Menggunakan SPSS-13. In: Hardjono D (eds) Andi, Yogyakarta. see http://inlislite.uin-suska.ac.id/opac/detail-opac?id=14808
10. Kuswanto H (1997) Analisis Benih. Andi, Yogyakarta, (1997), pp 1–140
11. Rini DS, Mustikoweni, Surtiningsih T (2005) Berita Biologi 7(6):307–313
12. Copeland LO, McDonald MB (2001) Seed science and technology, 4th edn. Kluwer Academic Publisher, London, p 488
13. Baki AA, Anderson JD (1973) Crop Sci 13(6):630–633
14. Vincent JM (1982) Nitrogen fixation in legume. Academic Press, London, p 288
15. Subba-Rao NS (1995) Soil microorganisms and plant growth, 3rd edn. Science Publishers, New Hampshire, p p335
16. Gaskins DS, Crosson S (2008) Biophys J 95:2063–2072
17. Suliasih, Widawati S (2017) Effect of plant growth promoting rhizobacteria and molasses on seed germination and seedling growth of sorghum bicolor L. Moench. In: Widyatmoko D et al (eds) The project for producing biomass energy and material through revegetation of Alangalang (Imperata cylindrical) fields 2016, the 1th SATREPS conference proceedings. Center for Plant Conservation Botanic Gardens Publishing, Bogor, pp 94–99
18. Shaukat K, Affrasayab S, Hasnain S (2006) Res J Microbiol 1(4):330–338
19. Lenin G, Jayanthi M (2012) Int J Res Pure Appl Microbiol 2(4):37–42
20. Hameeda B, Rupela OP, Reddy G, Satyavani K (2006) Biol Fertil Soils 43:221–227
21. Niranjan SR, Shetty NP, Shetty HS (2017) J Pest Manage 50(1):41–48. https://doi.org/10.1080/09670870310001626365
22. Widawati S, Suliasih (2018) The effect of plant growth promoting rhizobacteria (PGPR) on germination and seedling growth of sorghum bicolor L. Moench. In: The 7th international symposium for a sustainable humanosphere 2017, IOP conference series: earth and environmental science. IOP Publishing, UK, vol 166, pp 1–10
23. Mathivanan S, Chidambaram ALA, Sundaramoorthy P, Baskaran L, Kalaikandhan R (2011) Pestic Fitomed (Beograd) 26(2):141–146
24. Sengupta C, Amita B, Malusare S (2015) Effect of plant growth promoting rhizobacteria on seed germination and plants development of Zea mays. In: International Journal of Research in Advent Technology Special Issue National Conference on "ACGT 2015", pp 32–40
25. Marthen, Kaya E, Rehatta H (2013) Pengaruh perlakuan pencelupan dan perendaman terhadap perkecambahan benih sengon (*Paraserianthes falcataria* L.). Agrologi 2(1):10–16
26. Salisbury FB, Ross CW (1995) Fisiologi Tumbuhan Jilid I. ITB Press, Bandung, pp 227–234
27. Rodríguez RZ, Luis G, Hernández M, Amador BM, Puente EOR, Capistrán LL, Diéguez ET, Matson MVC (2015) Forests J 6:3109–3122
28. Gupta H, Saini RV, Pagadala V, Kumar N, Sharma DK, Saini AK (2016) J Soil Sci Plant Nutr 16(3):558–577
29. Zaghloul RA, Abou-Aly HE, Tewfike TA, Noha M, Ashry NM (2016) J Pure Appl Microb 10(1):277–290
30. Ji X, Lu G, Gai Y, Gao H, Lu B, Kong L, Mu Z (2010) Microbiology 10:243
31. Singh RP, Jha PN (2017) Front Microbiol 8:1945
32. Dennis PG, Miller AJ, Hirsch PR (2010) FEMS Microbiol Ecol 72:313–327

Production of Bioethanol from Dwarf Late Napier Grass (*Pennisetum purpureum*) Stems by Simultaneous Saccharification and Fermentation (SSF) Method Using *Clostridium acetobutylicum* Bacteria

H. Karim, D. S. T. Putra, A. Ahmad, Mahyati, A. Wadi, and D. Sondari

Abstract The cellulose of dwarf-late Napier grass belongs to the carbohydrate group that has not been utilized optimally. The cellulose can be utilized with the help of microbes to produce bioethanol. This study aims to produce bioethanol from the stems of dwarf-late Napier grass (*Pennisetum purpureum*) through the process of saccharification and simultaneous fermentation (SSF) using *Clostridium acetobutylicum* bacteria. The SSF method combines enzymatic hydrolysis and fermentation processes. As variables of this study, in the first stage, dwarf-late Napier grass stems were pre-treatment with the addition of 3% NaOH and bleaching with 3% H_2O_2 to remove lignin from the cellulose. In the second stage, the dwarf-late Napier grass stem cellulose was fermented by the SSF method. The results showed that the dwarf-late Napier grass stem cellulose after pre-treatment was 75.37%. The optimum condition of fermentation was obtained at pH 6.5 for ten days of fermentation with the use of dwarf-late Napier grass stems as much as 68 g after pre-treatment was able to produce 13.5 mL of bioethanol with a concentration of 86.25%. The results

H. Karim, D. S. T. Putra—Authors with equal contribution.

H. Karim
Department of Pharmacy, School of Pharmacy YAMASI, Makassar, Indonesia

D. S. T. Putra · A. Ahmad (✉)
Department of Chemistry, Mathematics and Natural Science Faculty, Hasanuddin University, Makassar, Indonesia
e-mail: ahyarahmad@unhas.ac.id

Mahyati
Laboratorium of Bioprocess, Polytechnic Ujung Pandang, Makassar, Indonesia

A. Wadi
Pangkep Regency, Polytechnic of Agriculture, Mandalle, Pangkep, Sulawesi Selatan, Indonesia

D. Sondari
Research Center for Biomass and Bioproducts, BRIN, Cibinong, Jawa Barat, Indonesia

M. A. Rahandi Lubis (ed.), *Proceedings of the 1st International Conference of Lignocellulose*, Green Energy and Technology,
https://doi.org/10.1007/978-981-19-1596-3_20

show that dwarf-late Napier grass has the potential to produce products in the form of bioethanol as a renewable energy source.

Keywords Bioethanol · Dwarf-late Napier grass stem · SSF

1 Introduction

The limited availability of conventional energy sources such as fossil-based fuels has become the main energy source throughout human life. Through this energy source, the world's industry develops and supports all aspects of life. Consumption patterns are increasing day by day, and environmental pollution and waste are directly proportional to global warming. This problem is certainly a serious concern for the community and the government. Thus, this also encourages academics or researchers to find alternative sources of new energy that are friendly to the environment [1]

Elephant grass, also known as Napier grass (*P. purpureum*), is generally an under-utilized plant other than being used as fodder for ruminants [2]. Several studies have reported the genotype of Napier grass, one of which is dwarf-late Napier grass [3]. The morphology of the stems of mini elephant grass with a very short distance compared to other elephant grass so that the stem texture of mini elephant grass is slightly softer and is effectively used as an animal feed with high nutrition, especially for ruminant's animals. But there are still 10–15% of the portion that is not consumed or left by ruminants. If properly processed, it can produce alternative environmentally friendly energy sources producing raw materials such as bioethanol [4]. In second-generation feedstock, raw materials for bioethanol refer to cellulosic biomass. Several studies have reported that the mini elephant grass cellulose content is quite high, which, if used properly, can become a high-value product, one of which is bioethanol [5–7]

The synthesis of bioethanol from a natural resource like cellulose includes two main stages, hydrolysis and fermentation. If done separately, the method is called Separated Hydrolysis and Fermentation (SHF) process. The widely developed method today is a method that combines stages simultaneously called Simultaneous Saccharification and Fermentation (SSF). SSF is a fermentation method using the help of microbes such as *C. acetobutylicum* bacteria, which can produce enzymes that can simultaneously hydrolyze cellulose into glucose and convert it into bioethanol [8]. Several studies on using different types of elephant grass substrates such as Sakius et al. [9] to produce bioethanol from Elephant grass (*P. purpureum* Schumach) by SSF method using *C. acetobutylicum* bacteria that the optimal fermentation condition was obtained on the pH of 6.5 in 10 days of fermentation time with the bioethanol concentration 96.24%. Stanley et al. produce bioethanol from Elephant grass (*P. purpureum*) using *Saccharomyces cerevisiae* and *Aspergillus* sp. The optimal condition of the fermentation was obtained on the pH of 6.0 in 6 days fermentation with temperature 30 °C produce an ethanol yield of 9 ml was obtained from 25 ml of plant extract [10]. This study aims to determine the optimum and the level of bioethanol that can be produced by using dwarf late napier grass (*P.*

purpureum) as a substrate through simultaneous saccharification and fermentation (SSF) using *C. acetobutylicum* bacteria to have more selling value in the community.

2 Material and Methods

2.1 Materials

Dwarf-Late Napier grass (*P. purpureum*) stems, *C. acetobutylicum* bacteria, meat extracts, and yeast extracts were used in this study. The chemicals NaOH, $(NH_4)_2SO_4$, K_2HPO_4, KH_2PO_4, $MgSO_4.7H_2O$, NaCl, $MnSO_4$, $CaCl_2$ anhydrate, $FeSO_4.7H_2O$, $CoCl_2.6H_2O$, H_2O_2, CaO, ethanol absolute, cysteine, asparagine, and casein with pro analysis grade were purchased from Sigma-Aldrich, USA.

2.2 Methods

2.2.1 Preparation of Sample

Dwarf Late Napier Grass stems cut into small pieces with a size of 1–2 cm, dried in the sun to dry and then ground with cruiser until smooth and screened with 60 mesh (0.2 mm).

2.2.2 Analysis of Lignin, Cellulose and Hemicellulose

About 1 g dry sample (**a**) was added to 150 mL of distilled water, heated at 90–100 °C with a water bath for 1 h. The result was filtered and the residue was washed with 300 mL hot water. The residue then was dried with the oven until constant weighed (**b**). The residue was added with 150 mL H_2SO_4 1 N then refluxed with a water bath for 1 h at 100 °C. The result was filtered and washed with distilled water until neutral, then dried (**c**). The dry residue was added to 10 mL H_2SO_4 72% and soaked at room temperature for four hours. Then, it was heated with 150 mL of H_2SO_4 1 N in the water bath for an hour in a cooler behind. The residue was filtered and washed with distilled water until neutral and then heated in an oven at a temperature of 105 °C and the results were weighed until the weight remained (**d**) the next residue was ached and weighed (**e**) calculation of levels of cellulose lignin and hemicellulose content as follows [11]:

$$\text{Cellulose content} = \frac{\mathrm{c} - \mathrm{d}}{\mathrm{a}} \times 100\% \tag{1}$$

Table 1 Composition of the material to the media for slanting [9]

Material	Total (g)
Glucose/corn flour	4.00
KH_2PO_4	0.08
KH_2PO_4	1.00
Yeast extract	2.00

$$\text{Lignin content} = \frac{\text{d} - \text{e}}{\text{a}} \times 100\% \tag{2}$$

$$\text{Hemicellulose content} = \frac{b - c}{a} \times 100\% \tag{3}$$

2.2.3 Pre-treatment Process

Pre-treatment was carried out by soaking Dwarf-late Napier grass stems cellulose flour (1000 g) in NaOH 3% (1:9) for 24 h. Then, the residues were bleached with a solution of H_2O_2 3% (1:9) for 24 h and then washed with distilled water until neutral and then dried in an oven at a temperature of 105 °C until constant weight [9].

2.2.4 *Rejuvenation* C. Acetobutilicum *Bacteria*

Nutrition media in Table 1 was mixed with 100 mL of meat extract and then heated while being stirred until dissolved. A few test tubes were put up and then 10 mL extract solution was pipetted into each test tube. The tubes were corked with cotton and aluminium foil and subsequently sterilized at 121 °C for 15 min. After that, the tubes were cooled in a state tilted. Pure cultures of *C. acetobutylicum* bacteria were scratched in a zig-zag pattern on Petri media using a sterilized loop scraper which was done in a sterilization cupboard (Ent case). Then, the pure cultures were grown in an incubator at 37 °C for seven days.

2.2.5 Preparation of Fermentation Media for the Determination of Optimum Fermentation Conditions on Bioethanol Production

Dwarf-late Napier grass stem was added into a 500 mL glass beaker for fermentation media beaker 5000 mL. 100 ml of water for inoculum media and 1000 mL of water was added to the fermentation medium and heated while stirring to form a gel. The above ingredients (Table 2) were added to the gel and were stirred until dissolved. Regulated pH was adjusted to pH 5.0 with phosphate buffer. Then, the mixture was diluted to 1500 mL for media fermentation and 150 mL for inoculum media.

Table 2 Composition of material for fermentation media [9, 12]

Material	Total	
	Inoculum media	Fermentation media
The cellulose of dwarf-late Napier grass	3 g	30 g
KH_2PO_4	0.112 g	1.12 g
K_2HPO_4	0.112 g	1.12 g
Yeast extract	1.5 g	15 g
$(NH_4)_2SO_4$	0.3 g	3 g
$MgSO_4.7H_2O$	0.003 g	0.03 g
$MnSO_4$	0.0015 g	0.015 g
$CoCl_2.6H_2O$	0.0015 g	0.015 g
$FeSO_4.7H_2O$	0.0015 g	0.015 g
$CaCl_2$	0.25 g	2.5 g
NaCl	0.15 g	1.5 g
Asparagine	0.3 g	3 g
Cysteine	0.0075 g	0.075 g
Casein	0.075 g	0.75 g
Meat extract	15 mL	150 mL

150 mL of fermentation medium was added to each of ten 250 mL Erlenmeyer. Each Erlenmeyer was filled with cotton and aluminium foil and sterilized in an autoclave for 15 min at 121 °C. The Erlenmeyer was moved into the sterile room and added with a stock of pure culture of *C. acetobutylicum* bacteria closed with cotton and then fermented in an incubator shaker for two days for inoculum media. Into each Erlenmeyer fermentation media is added 15 mL of inoculum media and then fermented in a shaker incubator shaker with variation times on 6, 8, 10, 12, and 12 h at a temperature of 37 °C with a speed of 150 rpm. Fermentation media was filtrated and distilled at a temperature of 78 °C to obtain pure bioethanol and analysed by a refractometer to obtain a refractive index. Next, the level of bioethanol was estimated using the comparison graph (linear regression equation) of the refractive index versus standard ethanol in percentage (v/v) [8, 13].

3 Results and Discussion

3.1 Determination of Lignin, Cellulose, and Hemicellulose

Pre-treatment in this study aims to increase the levels of cellulose that can be hydrolyzed in the cellulose fermentation process to reduce the *learning* content in the sample. The preliminary treatment of this research includes the delignification step

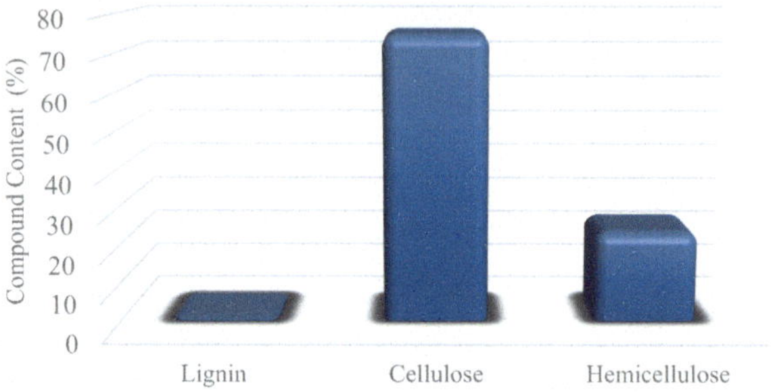

Fig. 1 Determination graph of Lignin, Cellulose and Hemicellulose

using an alkaline solvent in the form of 3% sodium hydroxide (NaOH), followed by a bleaching process using 3% hydrogen peroxide (H_2O_2) to remove residual lignin in the samples (Fig. 1).

The results showed an increase in cellulose content from 24.43 to 75.37% from 68 g biomass (dry weight) after pretreatment. The increase in cellulose content occurred because the lignin content decreased to 0.619%, which was originally 3.669%. A significant decrease in levels also occurred in hemicellulose from 71.88 to 24.01%. The decrease in levels was due to the lignin and hemicellulose having dissolved with the filtrate during washing. The levels of hemicellulose and lignin are known to represent amorphous parts that are more easily hydrolyzed in the *pre-treatment* process using sodium hydroxide (NaOH) and the bleaching process using hydrogen peroxide (H_2O_2). The decrease in levels increased the cellulose content obtained in the test. In addition, the decrease in lignin and hemicellulose levels can increase the accessibility of *C. acetobutylicum* bacteria to cellulose so that the pre-treatment can eliminate one of the inhibiting factors for bioethanol fermentation.

3.2 Determination of Optimum Fermentation Time and pH

Determination of the optimum conditions of fermentation in this study includes the time and optimum pH of fermentation. The complete data can be seen in Table 3.

The optimum time for this fermentation was shown on the ten days with a bioethanol content of 5.75%. In this phase, bacteria have reached the peak of cell division and produce higher bioethanol than other time variations. This phase is called the stationary phase, in which the bacteria living in the fermentation medium reach a constant state. In this phase, the nutrients and substrates used by bacteria begin to decrease, causing these bacteria to compete for nutrients. On the ten until the 12 days, the bioethanol content decreased. The decrease in bioethanol levels on day 12 can indicate that no cellulose in the media can be converted into bioethanol or the levels of nutrients needed by bacteria to survive are not available, causing bacteria to enter the death phase (Fig. 2).

Table 3 Effect of time and optimum pH of fermentation on refractive index and bioethanol content

pH	Day	Refractive index	Bioethanol (%)
5.0	6	1.3310	0.25
	8	1.3312	0.75
	10	1.3318	2.25
	12	1.3310	0.25
5.5	6	1.3315	1.50
	8	1.3317	2.00
	10	1.3321	3.00
	12	1.3312	0.75
6.0	6	1.3318	2.25
	8	1.3320	2.75
	10	1.3325	4.00
	12	1.3316	1.75
6.5	6	1.3320	2.75
	8	1.3325	4.00
	10	1.3332	5.75
	12	1.3318	2.25
7.0	6	1.3315	1.50
	8	1.3320	2.75
	10	1.3322	3.35
	12	1.3315	1.50

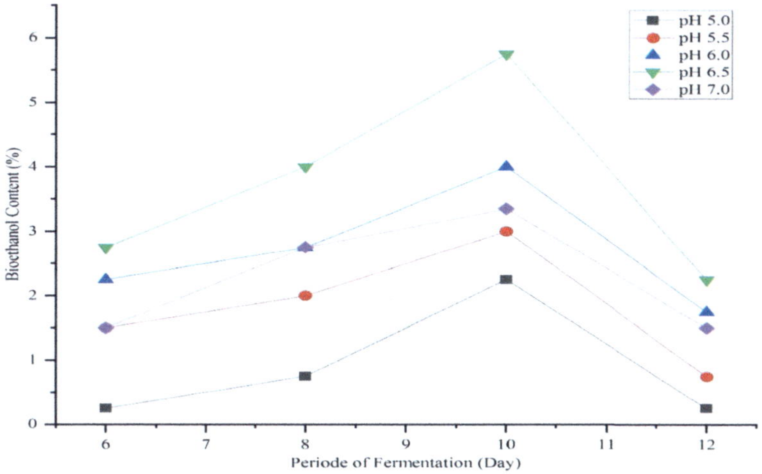

Fig. 2 Determination graph of optimum fermentation conditions of pH and fermentation time

3.3 Determination of Optimum pH Fermentation

Determination of the optimum pH of fermentation done by adding phosphate buffer in this study aims to condition the degree of acidity according to the specified variation. The pH variation in this study took a pH range of 5–7, which was fermented on ten *days* and then continued with the determination of the bioethanol content produced based on the relationship between standard ethanol content and refractive index through a linear equation. The results showed that in the pH range of 5–6 enzymes produced from bacterial activity, converting cellulose into bioethanol was not maximized. At pH 6.5, the bioethanol content was higher than other pH variations. The enzyme activity produced by bacteria in converting cellulose into bioethanol works optimally in catalyzing the formation of bioethanol so that the resulting product is also optimal. Meanwhile, at pH 7.0, the bioethanol content decreased to 3.35%. This result could be since the bacteria experienced a decrease in making bioethanol at pH 7.0. In addition, most bacteria with the genus *Clostridium sp* cannot live at neutral pH [14] (Fig. 3).

In this study, changes in pH could occur during the fermentation process. This change was because the bacteria *C. acetobutylicum* had the potential to produce other acidic compounds such as formic acid, acetic acid, butyric acid and lactic acid. The formation of this acid compound occurs in the growth phase of the bacteria *C. acetobutylicum*, known as the acidogenic phase [15]. The pH condition can affect the cell division activity of *C. acetobutylicum* bacteria. In addition, the catalytic area of the cellulase enzyme in hydrolyzing cellulose is strongly influenced by the optimum pH conditions in the fermentation medium [16]. These *compounds, which* are formed, reduce the bioethanol content because the converted cellulose is not completely converted into bioethanol. The data on the optimum fermentation conditions were then used in the bioethanol production procedure using higher levels of nutrients and substrates to see the maximum levels of bioethanol that could be produced using dwarf-late Napier grass cellulose flour as a substrate.

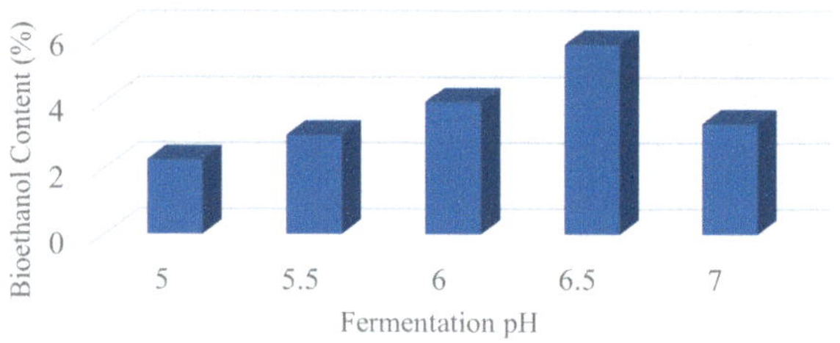

Fig. 3 Determination graph of the effect of fermentation pH bioethanol levels at the optimum time of fermentation on day 10

3.4 Production of Bioethanol on Optimum Condition Fermentation

Bioethanol production was carried out using the same steps in the fermentation media procedure by adding 68 g of cellulose substrate with a total fermentation volume of 1500 and 150 mL of *C. acetobutylicum* bacterial inoculum media then adjusting until the pH 6.5 (optimum pH) using phosphate buffer. The media was grown in a Shaker Incubator at 37 °C at a speed of 150 rpm for ten days.

3.4.1 Determination of Bioethanol Content

The stages of bioethanol production include purification of bioethanol obtained by adding calcium oxide (CaO) to the distillate, which aims to bind water to the sample, followed by repeated distillation using fractional *distillation, which* aims to increase bioethanol levels. The distillation results obtained in this study after purification were 13.5 mL. The bioethanol content was determined based on the relationship between the standard ethanol content and the refractive index of the distillate by calculating the linear equation obtained from the refractometer. *Quantitative* analysis was carried out by measuring the refractive index using a refractometer and qualitative analysis of bioethanol using gas chromatography (GC Shimadzu 2010 Plus).

From the graph of the relationship between the refractive index and the standard ethanol concentration (Fig. 4), the linear regression equation $y = 0.0004x + 1.3309$ is obtained. The refractive index of bioethanol data at the production stage obtained is 1.3654. This equation would give 86.25% as the bioethanol content of the obtained bioethanol sample. *Qualitative analysis* was performed by using gas chromatography by comparing the ethanol standard and distillates chromatogram.

The standard ethanol chromatogram based on Fig. 4 obtained a peak retention time (Rt) of 1.784. In this study, 99.8% absolute ethanol was used as the standard ethanol for comparison with the distillate chromatogram. The fermented distillate chromatogram data *are presented* in Fig. 5.

The results of the distillate chromatogram presented in Fig. 6 obtained a peak retention time (Rt) of 1.773. When compared with the peak retention time of standard 99.8% ethanol in Fig. 5, which is 1.784, the difference is not too far so that it can be obtained that the distillate contains ethanol. The difference in peak retention time in standard ethanol chromatograms and distillate results may occur because the

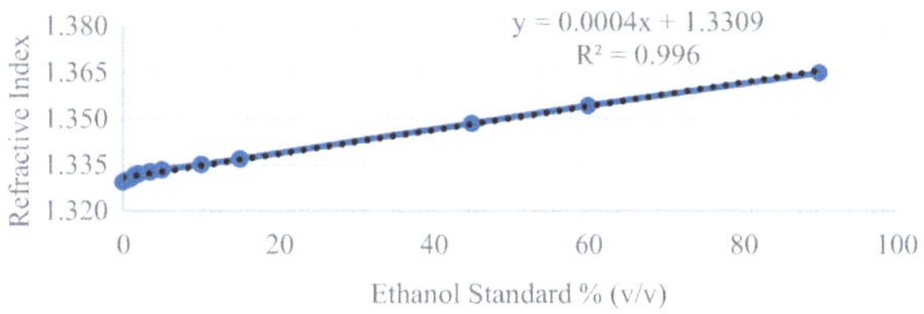

Fig. 4 Standard curve of ethanol concentration versus value of the refractive index

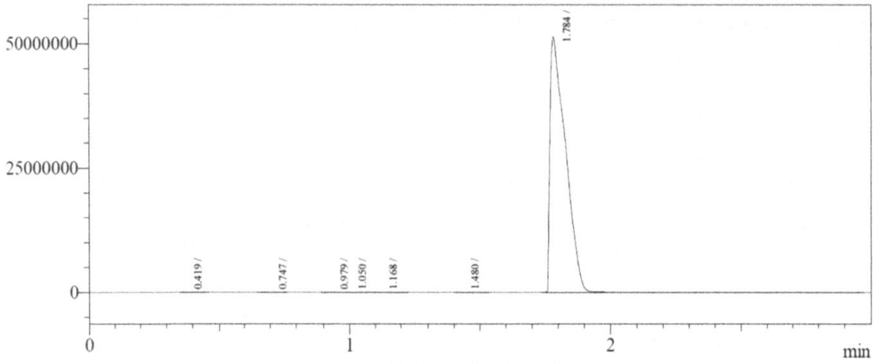

Fig. 5 Chromatogram of ethanol standard by GC method

ethanol produced is not very pure. The peaks detected are contaminants such as acetic acid and other organic compounds, both produced bioethanol from the by-product of bioethanol. Cellulose substrates from cotton waste were used by Arthe et al. [17] for bioethanol production by using SHF process through enzymatic hydrolysis from microbial extracellular and followed fermentation process with an bioethanol yield of 8.9 g/l. In another report, produce bioethanol from Elephant grass (*P. purpureum) using* yeast *Saccharomyces cerevisiae* and *Aspergillus* sp., was obtained on the optimum pH of 6.0 in 6 days fermentation with temperature 30°C to produce an bioethanol yield of 9 ml was obtained from 25 ml of plant extract [10]. The result of this work was very high bioethanol content and slightly increased compared to the other research given by bioethanol content was of 84% yield from the cellulose of banana pseudo stem [18].

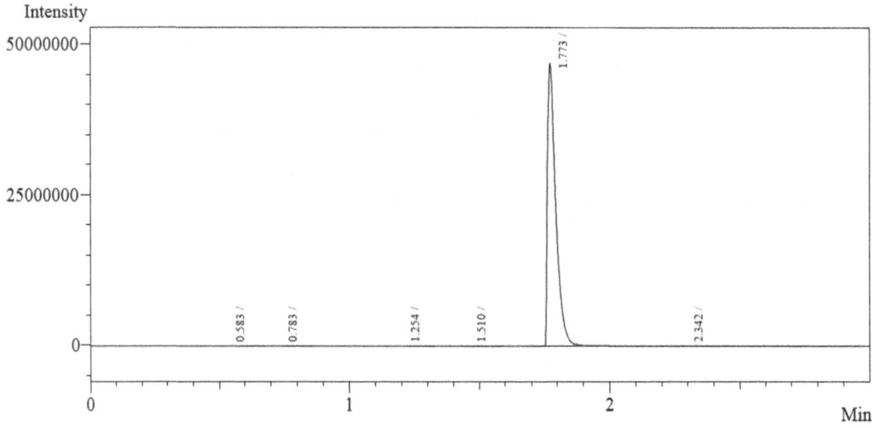

Fig. 6 Chromatogram of bioethanol sample after purification by GC method

4 Conclusion

Dwarf-late Napier grass stem with cellulose content after delignification is 75.37%; hemicellulose was 24.01% and lignin was 0.619%. The optimum fermentation conditions were obtained at pH 6.5 and fermentation time for ten days, with bioethanol production levels obtained at 86.25%. In conclusion, dwarf-late Napier grass has the potential to produce products in the form of bioethanol as a renewable energy source in the future.

References

1. Mohanty B, Abdullahi II (2016) Bioethanol production from lignocellulosic waste—a review. Biosci Biotechnol Res Asia 13(2):1153–1161
2. Desy PV, Hanafi DN, Wahyuni HT, Sadeli A (2018) Preservation technique of dwarf elephant grass cuttings (*Pennisetum purpureum CV mott*). Indonesian J Agric Res 01(3):211–217
3. Umami N, Akashi R, Gondo T, Ishigaki G, Tanaka H (2016) Study on callus induction system of 4 genotype of napiergrass (*Pennisetum purpureum*). Anim Prod 18(3):131–140
4. Heuze V, Tran G, Giger-Reverdin S, Lebas F (2016) Elephant grass (*Pennisetum purpureum*). Feedipedia, a programme by INRA, CIRAD, AFZ and FAO. http://www.feedipedia.org/node/395, Last accessed 12 Dec 2019
5. Rohadi TNT, Ridzuan MJM, Majid AMS, Khasri A, Sulaiman MH (2020) Isolation and characterisation of cellulose from cortex, pith and whole of the *Pennisetum purpureum* Effect of sodium hydroxide concentration. J Mater Res Technol 9:15057–15071
6. Fartini MS, Abdul MMS, Ridzuan MJM, Amin NAM, Gibson AG (2016) Compressive properties of Napier (*Pennisetum Purpureum*) filled polyester composites. Plast Rubber Compos 45(3):136–146
7. Indonesia Agency for the Assessment and Application of Technology (2018) Indonesia energy outlook 2018 sustainable energy for land transportation. Center of Assessment for Process and Energy Industry, Jakarta
8. Wadi A, Ahmad A, Tompo M, Hasyim H, Tuwo A, Nakasima M, Karim H (2019) Production of bioethanol from seaweed, *Gracilaria verrucosa* and *Eucheuma cottonii*, by simultaneous saccharification and fermentation methods. J Phys Conf Ser. https://doi.org/10.1088/1742-6596/1341/3/032031
9. Sakius R (2014) Production of bio-ethanol from pennisetum purpureum schumach leaves of the simultaneous fermentation system by using clostridium acetobutylicum. Hasanuddin University, Makassar, Thesis
10. Stanley HO, Ezeife CO, Onkuwe CD (2017) Bioethanol production from elephant grass (*Pennisetum purpureum*). Nigerian J Biothecnol 32:1–6
11. Datta R (1981) Acidogenic fermentation of lignocellulose-acid yield and conversion of components. Biotechnol Bioeng 23(9):2167–2170
12. Lin YY, Blaschek HP (1983) Butanol production by a butanol tolerant strain of *Clostridium acetobutylicum* in extruded corn broth. Appl Environ Microbiol 45(45):966–973
13. Ademiluyi FT, Mepba HD (2013) Yield and properties of ethanol biofuel produced from different whole cassava flours. ISRN Biotechnol 2013, ID 916481 1-7. https://doi.org/10.5402/2013/916481
14. Johnson EA (2009) Encyclopedia of microbiology, 3rd edn. Academic Press, USA
15. Buehler AE, Mezbah A (2016) Kinetic study of acetone-butanol-ethanol fermentation in continuous culture. J Plos One 11(8):1–21

16. Luo H, Ge L, Zhang J, Zhao Y, Ding J, Li Z, He Z, Chen R, Shi Z (2015) Enhancing butanol production under the stress environments of co-culturing *Clostridium acetobutylicum/ Saccharomyces cerevisiae* integrated with exogenous butyrate addition. J PLOS ONE 10(10):14–25
17. Arthe R, Rajesh R, Rajesh EM, Rajendran R, Jeyachandran S (2008) Production of bioethanol from cellulosic cotton waste through microbial extracellular enzymatic hydrolysis and fermentation. Electron J Environ Agric Food Chem 7(6):2984–2992
18. Ingale S, Joshi SJ, Gupte A (2014) Production of bioethanol using agricultural waste: banana pseudo stem. Braz J Microbiol 45(3):885–892. https://doi.org/10.1590/S1517-838220140003 00018

Influence of Simultaneous Fermentation Conditions on Production of Bioethanol from *Pennisetum purpureum* Schumach Leaves Using *Clostridium acetobutylicum*

A. Ahmad, S. Ruso, H. Karim, D. S. T. Putra, and N. La Nafie

Abstract This study aimed to identify a conversion number of celluloses of wide-leaved grass (*Pennisetum purpureum* Schumach) as the raw production of bioethanol processed through a simultaneous fermentation and a fermentation utilizing the role of *Clostridium acetobutylicum*. The cellulose of the wide-leaved grass was obtained through both fermentation processes using saccharification and simultaneously fermentation methods. The hydrolysis process during fermentation converted cellulose to glucose and simultaneously converted the glucose to bioethanol utilizing *Clostridium acetobutylicum*. The result of the research showed that the pH of 6.5 in 10 days with 6% of substrate concentration is an optimal condition. It was obtained 33.30 g ethanol at level 96.24% of 1.00 kg cellulose of wide-leaved grass (*Pennisetum pupureum*). Finally, in conclusion, wide-leafed grass has the potential to produce cellulose for conversion to bioethanol as a renewable energy source.

Keywords Bioethanol · Celluloses · Wide-leaved grass · Clostridium acetobutylicum

1 Introduction

Bioethanol is a biochemical fluid obtained from the fermentation of sugar from carbohydrates using microorganisms' assistance, followed by the process of distillation. Ethanol is produced by fermentation with the aid of microorganisms, and it

A. Ahmad (✉) · D. S. T. Putra · N. La Nafie
Department of Chemistry, Mathematics and Natural Science Faculty, Hasanuddin University, Makassar, Indonesia
e-mail: ahyarahmad@unhas.ac.id

S. Ruso
Laboratorium of Bioprocess, Polytechnic Ujung Pandang, Makassar, Indonesia

H. Karim
Department of Pharmacy, School of Pharmacy YAMASI, Makassar, Indonesia

M. A. Rahandi Lubis (ed.), *Proceedings of the 1st International Conference of Lignocellulose*, Green Energy and Technology,
https://doi.org/10.1007/978-981-19-1596-3_21

is referred to as bioethanol. In the development of bioethanol production, the raw material most widely used is many plants containing starch or sucrose, but humans consume this plant more widely.

One alternative raw material for bioethanol production is cellulosic biomass. Cellulosic biomass is an abundant natural resource and cheap, which is potent to support commercial production industry fuels like bioethanol and butanol [1]. Cellulosic biomass is obtained from agricultural waste or waste oil. One agricultural waste that can produce raw material for bioethanol is Wide-Leaved Grass.

Today the only Wide-Leaved Grass used as cattle food, sometimes even just considered a nuisance plant. However, the Wide-Leaved Grass has a high cellulose content to be used as one ingredient producer of bioethanol [2].

It is known that by hydrolysis reaction, glucose can be produced from polysaccharides of cellulose. Bioethanol production technologies in the process of hydrolysis are generally conducted by conventional methods using acid. However, this method is not environmentally friendly because it can lead the corrosive chemicals besides, the price is relatively high.

Bioprocess technology development involving enzymes during the process of hydrolysis is a process that is more environmentally friendly. In this study, it was used cellulolytic bacteria produce hydrolysis enzyme to hydrolyse cellulose to glucose bacteria used were Clostridium species in which this bacterium is a type of bacteria that can produce cellulase enzymes to hydrolyse cellulose to glucose [3–5]. Several studies using the bacterium Clostridium species for the production of some solvents, among others: Ezeji et al. in producing acetone, ethanol and butanol from corn flour with levels of 14.28 g/L [6]. Claasen et al. to produce acetone, butanol and ethanol with levels of 9.3 g/L [7]. Napoli et al. to produce ethanol with a concentration of 0.03 g/Lh^{-1} [8]. This study aims to determine the optimum pH and periodic fermentation on the level of bioethanol production using cellulose from *Pennisetum purpureum* Schumach leaves using *Clostridium acetobutylicum*.

2 Material and Methods

2.1 Materials

In this study, the wide-leaved grass stems and bacteria *Clostridium acetobutylicum* were used. Yeast extract, Bacto yeast extract, $(NH_4)_2SO_4$, K_2HPO_4, KH_2PO_4, $MgSO_4$. $7H_2O$, $MnSO_4$, NaCl, $CaCl_2$ anhydrate, $FeSO_4$ $0.7H_2O$, $Co(NO_3)_2.6H_2O$, H_2O_2, pentane, ethanol absolute, cysteine, asparagine, and casein were purchased from Sigma-Aldrich, USA.

2.2 *Procedures*

2.2.1 Preparation of Raw Material

Wide-Leaved Grass stems were cut into small pieces with a size of 1–2 cm, dried in the sun to dry and then ground with cruiser until smooth and screened with 60 mesh sieve (0.2 mm).

2.2.2 Analysis of Lignin and Cellulose

Cellulose and lignin analyses were conducted using the Chesson method [9]. A total of 1 g (**a**) dry sample was added to 150 mL of distilled water, heated at 100 °C with a water bath for 1 h. The result was filtered and the residue was washed with hot water (300 mL). The residue was then dried with a constant oven until later weighed (**b**) the residue was added 150 mL H_2SO_4 1 N then refluxed with water bath for an hour at 100 °C. The result was filtered and washed with distilled water until neutral (300 mL) and dried (**c**). The dried residue was added with 10 mL H_2SO_4 72% and was soaked at room temperature for four hours. Next, 150 mL of H_2SO_4 1 N were added and heated in a water bath for an hour in a cooler behind. The residue was filtered and washed with distilled water until neutral (400 mL) and then heated in an oven at a temperature of 105 °C and the results were weighed until the weight remained (**d**) the next residue ached and weighed (**e**) calculation of levels of cellulose and lignin content as follows:

$$\text{Cellulose content} = \frac{c - d}{a} \times 100\% \quad (1)$$

$$\text{Lignin content} = \frac{d - e}{a} \times 100\% \quad (2)$$

2.2.3 Pre-treatment Process

The preliminary experiment was performed with elephant grass flour pre-treatment process using NaOH 5–7%, performed by soaking for 24 h. Residue and the filtrate were separated. Residues bleached with a solution of H_2O_2 3% and then washed with distilled water until neutral and then dried in an oven at a temperature of 105 °C until constant weight [10].

Table 1 Composition of the material to the media for slanting

Material	Total (g)
Glucose/corn flour	4.00
KH $_2PO_4$	0.08
KH_2PO_4	1.00
Yeast extract	2.00
Meat extract	100 mL

2.2.4 Rejuvenation Bacteria with the Media for Slanting

Glucose or corn starch, KH_2PO_4, and Bacto yeast extract were mixed with 100 mL of meat extract and then heated while stirring until dissolved. A few test tubes were put up. Then 10 mL extract solution was introduced into each test tube, corked with cotton and aluminium foil, sterilized at 121 °C for 15 min and then cooled in a state tilted. Pure cultures of bacteria *Clostridium acetobutylicum* in a zig-zag scrawled on its side with the media (Table 1) for using the loop execution was done in a sterilization cupboard (Ent case) and then grown in an incubator (Memmert GmbH, Germany) at 37 °C for seven days.

2.2.5 Preparation of Fermentation Media for the Determination of Optimum Fermentation Conditions

Into a 500 mL glass beaker for media preparation beaker 5000 mL inoculum and fermentation media (Table 2) to manufacture each elephant grass-added flour, 150 ml of water for medium inoculum, and 1500 mL of water to the fermentation medium, was heated while stirring to form a gel. The above ingredients were added into the gel added and stirred until dissolved, regulated pH was adjusted to pH 5.0 with buffer phosphate, and then they were diluted to 2100 mL. Eight pieces of fruit were prepared and 300 mL solution was poured into each 500 ml Erlenmeyer stuffed with cotton and aluminium foil. Then they were sterilized in the autoclave for 15 min at a temperature of 121 °C. Erlenmeyer was moved into the sterile room, added with the stock of pure culture of bacteria *Clostridium acetobutylicum* closed with cotton and then fermented in an incubator shaker for two days for inoculum media. Into each Erlenmeyer fermentation media was added 30 mL of inoculum media and then it was fermented in a shaker incubator shaker for 2, 4, 6, 8, 10, 12, and 14 h at a temperature of 29 °C with a speed of 150 rpm. After two days, one of Erlenmeyer of the fermentation medium was taken, filtered and then distilled at 100 °C to obtain a volume of 10 mL distillates. The treatment was also made at pH of 5.5, 6.0, 6.5, 7.0, and 7.5 against the fermentation time. The production of bioethanol was the media treatment following optimum conditions obtained by fermentation. It was distilled at 78 °C temperature to obtain pure bioethanol distillate and then dried with Na_2SO_4 and subsequently analysed by gas chromatography (GC).

Table 2 Composition of material for fermentation media [11]

Material	Total (g)	
	Inoculum media	Fermentation media
Cellulose of Wide-Leaved Grass	9.000	126.0
KH_2PO_4	0.225	1.575
K_2HPO_4	0.225	1.575
$(NH_4)_2SO_4$	0.600	4.200
$MgSO_4$. $7H_2O$	0.006	0.042
$MnSO_4$	0.003	0.021
$CoCl_2$. $6H_2O$	0.003	0.021
FeSO4. 7H2O	0.003	0.021
$CaCl_2$	0.250	1.750
NaCl	0.300	2.100
Yeast extract	3.000	21.00
Asparagine	0.600	4.200
Cysteine	0.015	0.105
Casein	0.150	1.050
Meat extract	30 mL	210 mL
	Made 300 mL	Made 2100 mL

3 Results and Discussion

3.1 Determination of Optimum Fermentation Time

Data results of the influence of the fermentation period on ethanol content in some pH treatments can be seen in Table 3. Graphic determination of optimum fermentation conditions at all pH treatments can be seen in Fig. 1. From the graph, all treatments of pH on day 2 to day 6 increased levels of bioethanol are very slow. It can be seen from the graph the lag phase in which the activity of the microbes is not optimum as they are adjusting them-self to the environment. There was a minimum increase of the cell mass with no additions of the cell density number at this stage, hence it did not obtain a cell growth rate. Factors influencing the long lag phase of bacteria which varies such as the composition of media, acidity, temperature, initial inoculum number cells, and the properties physiological of microorganism in the earlier media. The cells were adapted to the environment by splitting them-self to the maximum number, this phase was generally named the logarithmic phase of the exponential phase while it continues to the exponential phase shown by the occurrence of rapid growth periods signed by splitting each cell into two cells. The degree of bacteria growth was varying and depending on the properties of genetic revelation.

Moreover, nutrients levels provided in media take an important factor with the temperature of incubating, pH and aeration. During the research, this phase lasted

Table 3 Effect of time and pH fermentation versus bioethanol content

pH	Day	Index point	Levels of bioethanol (%)
5.0	2	1.3329	2.30
	4	1.3334	2.80
	6	1.3337	3.40
	8	1.3347	5.40
	10	1.3353	6.00
	12	1.3334	2.80
	14	1.3332	2.40
5.5	2	1.3332	2.40
	4	1.3337	3.40
	6	1.3340	4.00
	8	1.3342	4.40
	10	1.3373	10.5
	12	1.3332	2.40
	14	1.3331	2.20
6.5	2	1.3332	2.40
	4	1.3337	3.40
	6	1.3341	4.20
	8	1.3376	11.20
	10	1.3398	15.60
	12	1.3347	5.40
	14	1.3338	3.60
7.0	2	1.3332	2.40
	4	1.3337	3.40
	6	1.3341	4.20
	8	1.3343	4.60
	10	1.3386	13.20
	12	1.3347	5.40
	14	1.3343	4.60
7.5	2	1.3332	2.40
	4	1.3337	3.40
	6	1.3339	3.80
	8	1.3342	4.40
	10	1.3367	9.30
	12	1.3347	5.40
	14	1.3338	3.60

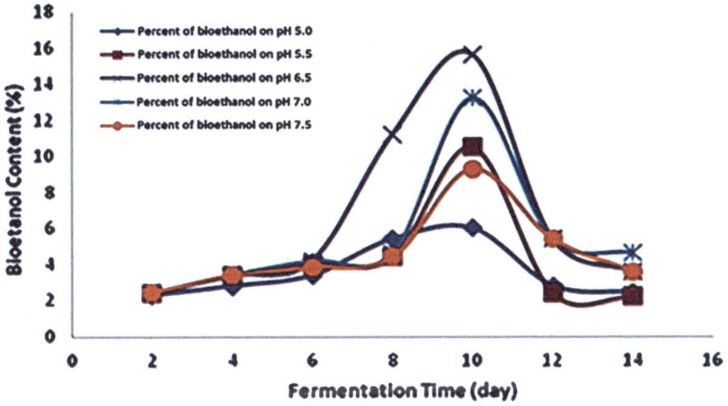

Fig. 1 Determination graph of optimum fermentation conditions of pH and fermentation time

on day eight and curry to-10 are characterized by bioethanol elevations level. It was obtained a cellular reproduction on between day 8 to day 10 with slowly increasing of biomass concentration yet the production of cell number was increased. It was also obtained a maximum point of exponential growth and reproduction of the cell. At the exponential stage, a proportional increase of cell growth with the cell concentration at the time was obtained.

On day 12 at the stationary phase, there was not optimal activity conducted by bacteria and it seemed equal between the rate of bacterial growth with the death number, it was caused by the cell division reduction degree which influenced the loss of the glucose levels and other nutrients moreover the disruption of cell division cooccurred as the accumulated toxic products. Besides, there was not an association obtained between byproducts produced by fermentation with the productivity and growth of bacteria, hence on the day enzyme was produced, it caused a decrease in bioethanol level.

On day 14 of incubation, a relative decrease in ethanol concentration caused the mortality rate to pass over the growth until the population of bacteria declined, and this happened at various times. It is concluded that day 10 is mentioned as the optimum time for *Clostridium acetobutylicum* presented in the bioethanol fermentation process with elephant grass material.

3.2 Determination of Optimum pH During Fermentation Process

The pH 5.0 to 7.5 range was used in this research while the addition of phosphate was adjusted to make optimum pH. Table 4 shows the result of the effect of pH during fermentation on the levels of bioethanol production while Fig. 2 showed by graphic related the treatment on pH and time on the level of bioethanol produced. At the pH

Table 4 Effect of pH fermentation on levels of production of bioethanol

pH	Index bias	Levels of bioethanol (%)
5.0	1.3353	6.00
5.5	1.3373	10.50
6.5	1.3398	15.60
7.0	1.3486	13.20
7.5	1.3437	9.30

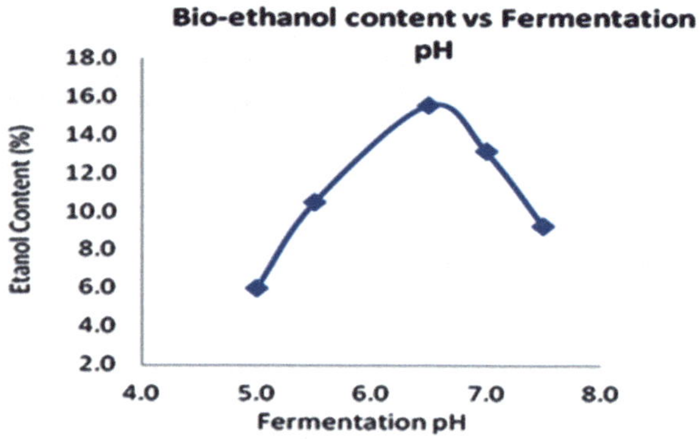

Fig. 2 The influence graph of fermentation pH on bioethanol production

of 6.5, the highest concentration of the production of bioethanol reached, which was 15.60% while the lower production was observed between pH of 5.0 to 5.5, which only 6% to 12% levels of ethanol were reached.

The range pH of 7.0 to 7.5 was mentioned as the range that produced bioethanol as much as 13% to 9%. The production of other components such as acetic acid, butyric acid and formic acid influenced the pH change. Acetic acid production was caused by a corporation of acetobacter with contaminant bacteria living together by purpose. Besides, *Lactobacillus* had a role to contaminate and convert the glucose to lactic acid by reducing the ethanol yield and inhibiting microbial growth. There was also a possibility of hydrolysing glucose had been fermented to other products as no monosaccharide was produced from the hydrolysis of polysaccharides. It is known that inhibitors presenting in biomass include lignin, a weak acid, and phenolic compounds derivatives.

The activity of enzymes produced by bacteria is a determinant of the level of bioethanol production during the fermentation process while the activity of the enzyme was influenced by several factors such as pH, ionic nature of the carboxyl group beside amino group was easily influenced by acidity. Changes in pH make any denaturation of enzyme and activity of enzymes [12]. The range pH of 6.5 was the optimum pH for *Clostridium acetobutylicum* to grow and produce maximum bioethanol compared with the other range of pH.

3.3 Determination of Levels of Bioethanol

Bioethanol was dehydrated by distillation process and returned 25.5 mL at 60 °C and it was obtained as much as 22.5 through the refining process. It was obtained only low purify. It was known that it was lower than the boiling point of ethanol, which was 78.4 °C.

After purification through distillation and drying with Na_2SO_4 salt, the sample was analyzed by using gas chromatography. The data obtained was shown in Fig. 3 that used to determine the levels of pentane hydrocarbon samples as internal standard. The internal standard was used as a correction factor to the analysis errors of the GC instrument. It is shown in Fig. 4 that the association between ethanol concentration standard with the value of A_I/A_{LS} and it is obtained regression equation y = 0.150 × − 2.241. The insertion of y value = 12.2215 of the sample was obtained at a sample rate of 96.42% of bioethanol production.

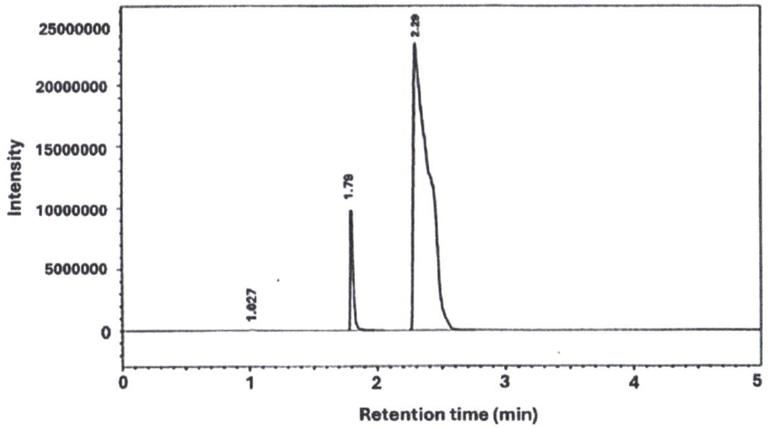

Fig. 3 Data analysis of bioethanol samples with a Shimadzu GC 2010 after purification

Fig.4 Graph showing the relationship of ethanol concentration standard with a value of A_I/A_{LS}

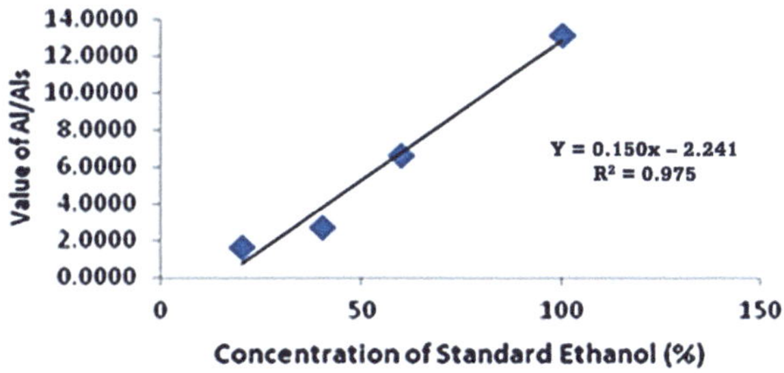

4 Conclusion

It is concluded that a pH of 6.5 at a temperature of 29 °C during ten days is the optimum condition for elephant grass in producing bioethanol through the fermentation process and one kilogram of cellulose of elephant grass can produce 3.3 g of bioethanol with a concentration level of 96.24%.

Acknowledgements We thank the Japan Society for the Promotion of Science (JSPS) for partially funding the publication fee.

References

1. Wyman CE (1999) Biomass ethanol: technical progress, opportunities, and commercial challenges. Annu Rev Energy Env 24:189–226
2. Sari NI (2009) Purification bioethanol from the grass elephant with batch distillation. www.che.itb.ac.id/sntki2009/daftar/prosiding/OTK08.pdf. Last accessed Sept 25, 2019
3. Balusu R, Paduru RMP, Seenayya G, Reddy G (2004) Production of ethanol from cellulosic biomass by clostridium thermocellum SS19 in submerge fermentation screening of nutrients using Plackett-Burman. http://www.springerlink.com/content/91323g1780726n12/fulltext.pdf. Last accessed Feb 12, 2019
4. Demain AL, Newcomb M, Wu JHD (2005) Cellulase, clostridia, and ethanol. Microbiol Mol Biol Rev 69(1):124–154
5. Riyanti EI (2009) Biomass as raw materials bioethanol. Center for Research and Development of Agricultural Biotechnology and Genetic Resources, Bogor
6. Ezeji T, Qureshi N, Blaschek HP (2007) Production of acetone–butanol–ethanol (ABE) in a continuous flowbioreactor using degermed corn and *Clostridium beijerinckii*. Process Biochem 42:34–39
7. Classen PAM, Budde MAW, Contreas AML (2000) Acetone, butanol and ethanol production from domestic organic waste by *Solventogenic clostridia*. J Mol Microbiol Biotechnol 2(1):39–44
8. Napoli F, Marzocchella GOA, Russo MEP (2005) Production of butanol in a continuous packed bed reactor of *Clostridium acetobutylicum*. P.le V. Tecchio 80–80125
9. Datta R (1981) Acidogenic fermentation of lignocellulose-acid yield and conversion of components. Biotechnol Bioeng 23(9):2167–2170
10. Selig MJ, Todd B, Selig MJ, Todd B, Vinzant TB, Himmel ME, Decker SR (2009) The effect of lignin removal by alkaline peroxide pretreatment on the susceptibility of corn stover to purified cellulolytic and xylanolytic enzymes. Appl Biochem Biotechnol 155:397–406
11. Lin YY, Blaschek HP (1983) Butanol Production by a Butanol Tolerant Strain of *Clostridium acetobutylicum* in extruded corn broth. Appl Environ Microbiol 45(45):966–973
12. Meryandini A, Widosari W, Maranatha B, Sunarti TC, Rachmania N, Satria H (2009) Isolation and characterization of bacteria cellulolytic enzyme. http://www.journal.ui.ac.id/.../07Edit1_AnjaMeryIsolasi%20BakteriLayout_new.pdf. Last accessed Sept 12, 2019

Production of Bioethanol from Elephant Grass Leaves (*Pennisetum purpureum*) by Simultaneous Saccharification and Fermentation (SSF) Method Using *Clostridium acetobutylicum*

H. Karim, N. Kurnia, Mahyati, A. Wadi, D. Sondari, and A. Ahmad

Abstract Elephant grass (*Pennisetum purpureum*) is a plant that is easy to grow in Indonesia and contains cellulose. Cellulose is one of the raw materials for bioethanol which is converted through a fermentation process. This study aims to find the potential of elephant grass leaves as a bioethanol-forming material by knowing the cellulose content of elephant grass leaves before and after delignification using NaOH 3% and H_2O_2 3%, as well as knowing the best fermentation time and pH to produce the highest concentration of bioethanol using *Clostridium acetobutylicum* bacteria. In this study, variations in fermentation time were 6, 8, 10 and 12 days, and variations in pH were 5.0; 5.5; 6.0; 6.5; and 7.0. The results of this study, the cellulose content of elephant grass leaves was 37.82% and after delignification, it became 70.24%. The best time and pH of bioethanol fermentation with *C. acetobutylicum* bacteria was for 8 days at pH 6.5 to produce a bioethanol concentration of 15.25%. Based on this research, elephant grass leaves have the potential as raw material for bioethanol production.

Keywords Bioethanol · P. purpureum · C. acetobutylicum · Fermentation

---

H. Karim
Department of Pharmacy, School of Pharmacy YAMASI, Makassar, Indonesia

N. Kurnia · A. Ahmad (✉)
Department of Chemistry, Mathematics and Natural Science Faculty, Hasanuddin University, Makassar, Indonesia
e-mail: ahyarahmad@unhas.ac.id

Mahyati
Laboratorium of Bioprocess, Polytechnic Ujung Pandang, Makassar, Indonesia

A. Wadi
Pangkep Regency, Polytechnic of Agriculture, Mandalle, Pangkep, Sulawesi Selatan, Indonesia

D. Sondari
Research Center for Biomass and Bioproducts, BRIN, Cibinong, Jawa Barat, Indonesia

M. A. Rahandi Lubis (ed.), *Proceedings of the 1st International Conference of Lignocellulose*, Green Energy and Technology,
https://doi.org/10.1007/978-981-19-1596-3_22

1 Introduction

Indonesia has begun to experience an oil balance deficit since 2003, and it has got wider every year. This issue has an impact on the needs of people who need fuel sources from petroleum, so they need alternative energy from economical materials. The emissions produced are safe for the environment and have a high octane value. These alternative fuels include bioethanol, biodiesel, biogas, and bio-briquettes [1].

Bioethanol, an alternative fuel, can be produced from plant raw materials containing lignocellulose, sugar, or starch. Cellulose in plants is the main part of the raw material to form bioethanol through the process of hydrolysis, fermentation and distillation. Elephant grass has a high cellulose content (48.05%) which can be used as a bioethanol-producing material. Another advantage is that the raw material for elephant grass is available in large enough quantities, this is because Indonesia's climate allows for the continuous growth of elephant grass. The land used can also be adjusted so that it will not compete with food land. So far, the use of elephant grass is only as cattle feed so it is still very underutilized in Indonesia [2]. In [2] using elephant grass as raw material by hydrolysis method of hydrochloric acid and S. cerevisiae produces 90–95% ethanol. Likewise in the research of [3] with elephant grass stems as raw material using the SSF method and the bacteria *C. acetobutylicum* producing 96.24% ethanol from one kilogram of raw materials.

Bioethanol production can be carried out by hydrolysis and separate fermentation *hydrolysis and fermentation* (SHF), hydrolysis and fermentation in one reactor/ Simultaneous saccharification and fermentation (SSF) or hydrolysis and fermentation in one reactor accompanied by microbial growth in the reactor/Consolidated Bioprocessing (CPB) [4]. The SSF method has the advantage that the polysaccharides that are converted into monosaccharides do not return to polysaccharides because the monosaccharides are directly fermented into ethanol [5]. This is due to the use of enzyme-producing microbes that are able to convert starch or cellulose into ethanol. One of these microbes is the bacterium *C. acetobutylicum*.

C. acetobutylicum bacteria are saccharolytic organisms (the process of breaking down sugar) and are capable of producing a number of commercially useful products, namely acetone, butanol, and ethanol [6]. Several studies using *C. acetobutylicum* bacteria, namely [7] utilizing sago pulp (*Metroxylon sp.*) produced an ethanol content of 16.5% at pH 5 with the best time of 10 days, [2] using pineapple peel produced an ethanol content of 9% at pH 6.5 with the best time of 8 days, [6] utilizing wood waste produces butanol levels of 1.88% at pH 5 with a fermentation time of 12 days, [3] using elephant grass stems to produce ethanol content of 96.24% at pH 6.5 with an optimum time of 10 days. Based on the description above, this research was carried out using elephant grass leaves as a source of cellulose with SSF methods and bacteria *C. acetobutylicum* to produce bioethanol.

2 Materials and Methods

2.1 Materials

The material used in this research is H_2SO_4 1 N, H_2SO_4 75%, NaOH 3%, H_2O_2 3%, corn flour, yeast extract, Bacto agar, K_2HPO_4, KH_2PO_4, $MgSO_4.7H_2O$, $(NH_4)_2SO_4$, $FeSO_4.7H_2O$, $CoCl_2.6H_2O$, NaCl, $MnSO_4$, $CaCl_2$, cysteine, asparagine, casein, meat extract, bioethanol absolute, CaO, distilled water, aluminum foil, gauze, cotton, *Clostridium acetobutylicum* bacteria, and elephant grass leaves.

2.2 Methods

The manufacture of bioethanol goes through several stages, pretreatment stage (washing, delignification, drying), hydrolysis (type of catalyst and catalyst concentration, pH stability), fermentation (type of yeast, yeast mass, and fermentation time), distillation (type of substance additives, temperature stability, accuracy of the distillation) greatly affect the amount/level of bioethanol and the purity of bioethanol from contaminant compounds [8]. In general, the bioethanol production process in this study can be seen in Fig. 1.

2.2.1 Sample Preparation

Elephant grass leaves were cut into small pieces 1–2 cm in size, then dried in an oven at 105 °C for 8 h, then ground with a blender until smooth and sieved through a 60 mesh (0.2 mm) sieve.

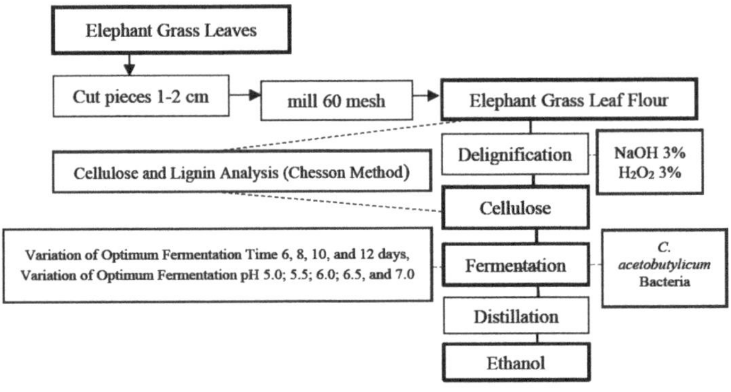

Fig.1 Schematic of work procedures

2.2.2 Sample Delignification

Elephant grass leaf flour (1000 g) were soaked with NaOH 3% (1:9) for 24 h, then filtered and washed with distilled water until the pH was neutral. The residue was soaked again with H_2O_2 3% (1:9) for 24 h, then filtered and washed with distilled water until the pH was neutral. After that, it was dried in an oven at a temperature of 105 °C, then ground with a blender until smooth and sieved through a 60 mesh sieve.

2.2.3 Analysis of Cellulose, Hemicellulose and Lignin Content (Chesson Method)

Variations of samples were analyzed before and after delignification. A total of one gram (a) of the sample was refluxed at 100 °C for 1 h in 150 mL of distilled water. The result was filtered and the residue was washed with hot water (500 mL). The residue was then dried in the oven and then weighed to a constant weight (b). The residue (b) was refluxed for 1 h at 100 °C in 150 mL H_2SO_4 1 N. The results were filtered and washed with distilled water until the pH was neutral (500 mL) then dried in an oven and then weighed to a constant weight (c). The dry residue (c) was soaked in 10 mL of H_2SO_4 72% at room temperature for 4 h. After that, 150 mL of H_2SO_4 1 N was added and refluxed for an hour. The residue was filtered and washed with distilled water until the pH was neutral (1000 mL) then dried in an oven at 100 °C and the results were weighed to a constant weight (d), then the residue (d) was smothered in the furnace and weighed to a constant weight (e). Then the lignocellulose content of the sample was calculated with the following formula (1), (2), and (3).

$$\text{Hemicellulose content} = \frac{b - c}{a} \times 100\% \tag{1}$$

$$\text{Cellulose content} = \frac{c - d}{a} \times 100\% \tag{2}$$

$$\text{Lignin content} = \frac{d - e}{a} \times 100\% \tag{3}$$

where:

a sample weight (grams).
b weight of residue on the second weighing (grams).
c weight of residue on the third weighing (grams).
d weight of residue on the fourth weighing (grams).
e weight of ash (grams).

Table 1 Composition of ingredients for fermentation of bioethanol production

Material	Total (grams)	
	Inoculum media	Fermentation media
Elephant grass leaf cellulose	3.0	60
KH2PO4	0.112	1.12
K2HPO4	0.112	1.12
Yeast Extract	1.5	15
(NH4)2SO4	0.3	3.0
MgSO4.7H2O	0.003	0.03
MnSO4	0.0015	0.015
CoC12.6H2O	0.0015	0.015
FeSO4.7H2O	0.0015	0.015
CaC12	0.250	2.5
NaCl	0.15	1.5
Asparagine	0.3	3.0
Cysteine	0.0075	0.075
Casein	0.075	0.75
Meat extract	15 mL	150 mL
Total volume	150 mL	1500 mL

2.2.4 Rejuvenation *C. acetobutilycum* Bacteria

Corn flour 4 g, KH_2PO_4 0.08 g, yeast extract 1 g, and Bacto agar 2 g were mixed and dissolved with 100 mL of meat extract then heated while being stirred until dissolved. 10 test tubes and then a 10 mL pipette were prepared and the extract solution was put into each test tube. All of the test tubes were then sterilized in an autoclave at 121 °C for 15 min and then cooled. Pure culture *C. acetobutylicum* were streaked in a zig-zag manner on the media using a sterile loop. This study was conducted inside a sterilization cabinet (entry case) before being cultured in fermentation media with the composition detailed in Table 1. The cultivation process took place in a shaker incubator at 37 °C for seven days.

2.2.5 Bioethanol Production

Elephant grass leaf flour was added into a 100 mL glass beaker to manufacture inoculum media and a 5000 mL beaker to manufacture fermentation media. It was dissolved by adding 75 mL of water for the inoculum medium and 800 mL of water for the fermentation medium and heated while being stirred to form a gel. The above ingredients were added to the gel and stirred until dissolved. The pH of the solution was adjusted to pH 6.5 with phosphate buffer, and then the mixture was diluted to

150 mL and 1500 mL. Ten pieces of 250 mL Erlenmeyer were prepared, and then the solution was added in each 150 mL, plugged with cotton and aluminum foil. The mixture was sterilized in an autoclave for 15 min at 121 °C. Erlenmeyer was transferred to a sterile room. After cooling, 15 mL of inoculum medium was added to each Erlenmeyer using a sterile measuring cup. Erlenmeyer closed again, then put in a shaker incubator for eight days at 37 °C at a speed of 150 rpm. After eight days, filtered and then distilled at 78 °C to obtain ethanol distillate.

2.2.6 Bioethanol Distillation

The distillation apparatus was properly assembled and ignited. Then the fermented liquid was put into a distillation flask. The heating temperature was kept at 80 °C. The distillation process was carried out for 8 h until the ethanol no longer dripped. The distillation results are then distilled again to a temperature of 78 °C to obtain ethanol distillate.

2.2.7 Qualitative and Quantitative Analysis of Bioethanol

The distilled bioethanol sample was added with CaO until it was saturated. Let stand overnight until clear then filtered. The bioethanol obtained was analyzed for qualitative tests using gas chromatography and quantitative with a refractometer.

3 Results and Discussion

3.1 Analysis of Hemicellulose, Cellulose, and Lignin Content

In making bioethanol, plant raw materials containing cellulose, sugar, or starch were the main components, especially cellulose. However, cellulose was not found in its pure form in plants but was associated with other polysaccharides such as lignin and hemicellulose, in the presence of other compounds. The levels of lignocellulose content can be seen in Fig. 2.

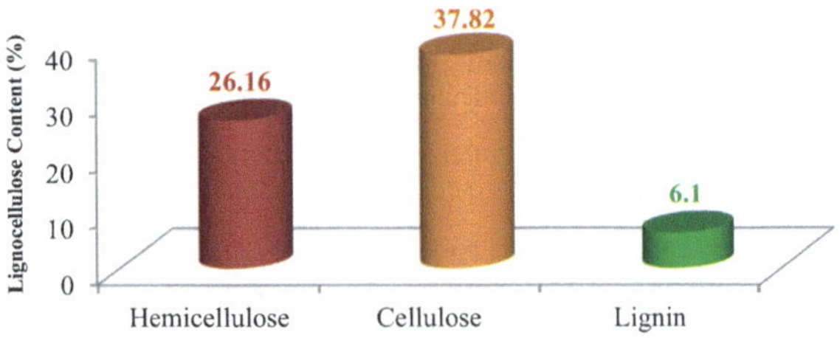

Fig. 2 Graph of lignocellulose content of elephant grass leave

Table 2 The effectiveness of delignification on elephant grass leaf samples by pretreatment and bleching

Component	Initial rate (%)	Pretreatment NaOH 3%		Bleaching H_2O_2 3%	
		Rate (%)	Effectiveness (%)	Rate (%)	Effectiveness (%)
Lignin	6.10	4.81	21.15	4.64	23.93
Hemicellulose	26.16	13.66	47.78	11.45	56.23
Cellulose	37.82	69.91	45.90	70.24	46.08

Based on the research data illustrated in the graph above, it is known that elephant grass leaves before delignification had 37.82% cellulose and 6.1% lignin content. When compared with several studies such as [9] palm leaves have 30% cellulose content and 25% lignin, [10] Sorghum leaves have 26.89% cellulose and 17.80% lignin content, [11] Elephant grass has 32.7% cellulose and 9.198% lignin content, as well as [12] elephant grass stems have a cellulose content of 24.43% and lignin 3.669%, this comparison shows that elephant grass leaves can be appropriate raw material in the manufacture of bioethanol with high cellulose content and low lignin.

Based on Table 2, NaOH 3% was able to degrade lignin as much as 21.15%, and degrade hemicellulose in the sample as much as 47.78%, resulting in cellulose as much as 69.91%, followed by the addition 3% H_2O_2 was able to degrade lignin in the sample as much as 3.61% and degrade hemicellulose as much as 16.19%, resulting in as much as 70.24% cellulose with 34.97% cellulose yield. The delignification method used in this study could increase the cellulose content up to 46% and remove lignin 24%. This shows that the delignification process used was less effective in increasing the cellulose content and reducing the lignin content in the sample. This was different from the [12] study that used the similar delignification method, which was able to degrade lignin up to 83% and increase cellulose content to 68%. This difference was due to the lignocellulose content in the elephant grass stem samples which was much less than that of the elephant grass leaves so that the effectiveness was different.

3.2 *Optimization of Fermentation Time and pH of Bioethanol Production*

3.2.1 Determination of Optimum Fermentation Time

Based on the data in Fig. 3, it can see that for all pH treatments the ideal time for fermentation of *C. acetobutylicum* bacteria with elephant grass leaf samples in the bioethanol fermentation process occurred on day 8 which marked by an increase in bioethanol levels. This results was consistent with the other research of [13] with pineapple peel as substrate whose best time of fermentation for *C. acetobutylicum*

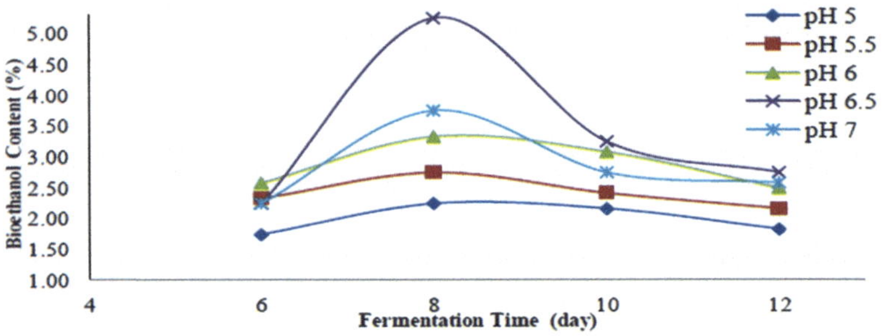

Fig. 3 Graph of determining the optimum conditions of fermentation with variations in pH and fermentation time on bioethanol levels

bacteria was found at 8 days. However, this condition was slightly different from the other research conducted by [3, 7, 11, 12, 14] using bacteria *C. acetobutylicum* which showed the bioethanol fermentation on day 10. The difference was due to the difference in the composition of the bacterial nutrient ratio, the source of cellulose, the amount and content of the substrate used so that it affected the work effectiveness of *C. acetobutylicum* bacteria.

3.2.2 Determination of Optimum Fermentation pH

Based on Fig. 4, the optimum fermentation time for the performance of *C. acetobutylicum* was on the 8th day for all pH variations. Determination of the optimum pH of fermentation on 8th day can be seen in Fig. 4. In this study, the pH conditions varied from 5.0 to 7.0. The desired degree of acidity was obtained by adding a phosphate buffer. The addition of a buffer here was intended so that the pH conditions followed the desired pH.

Based on the graph, it is known that pH 5.0–6.0 produces lower levels of bioethanol, which is 2.25%–3.33%, for pH 7 also produces lower levels of bioethanol, which is 3.75%. The highest concentration of bioethanol was produced at pH 6.5 treatment, which was 5.25%, this was under the research of [3] and [12]. The high

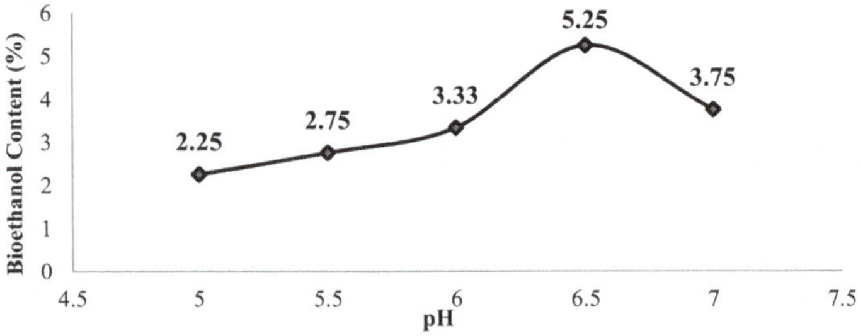

Fig. 4 Graph of the effect of pH of the fermentation medium on bioethanol levels

and low levels of bioethanol in the fermentation process can be caused by the activity of enzymes produced by bacteria. Enzyme activity was affected by pH because the ionic nature of the carboxyl and amino groups was easily influenced by pH. Changes in pH that are not suitable will cause the catalytic region and the conformation of the enzyme to change. In addition, changes in pH also cause enzyme denaturation and result in loss of enzyme activity (Meryandini et al. 2009 in [3]).

3.3 *Qualitative and Quantitative Analysis of Bioethanol Production*

3.3.1 Qualitative Analysis of Bioethanol with Gas Chromatography

Ethanol content is qualitatively determined by the retention time of the standard peak (ethanol), where if the peak retention time of the sample is close to the retention time of the standard peak, then the sample can be presumed to contain ethanol. The results of the sample chromatogram of this study can be seen in Fig. 5.

The standard ethanol chromatogram had a retention time (Rt) of 1.784, while the sample ethanol chromatogram had a retention time (Rt) of 1.792. Based on the retention time (Rt) of standard ethanol and the sample, the difference in retention time (Rt) of 0.008, it can be believed that the bioethanol distillate produced from the elephant grass leaf fermentation process contains bioethanol.

The other peaks in Fig. 5 are impurities from the distillation sample. This result was probably due to the dehydration process that was not optimal in the sample so that there was still water that had not been bound to CaO. Impurities can also cause impure samples in the form of by-products formed in the fermentation process or the bioethanol distillation process that was not good. The sample was not pure and affected the separation process of the components contained in the distillate.

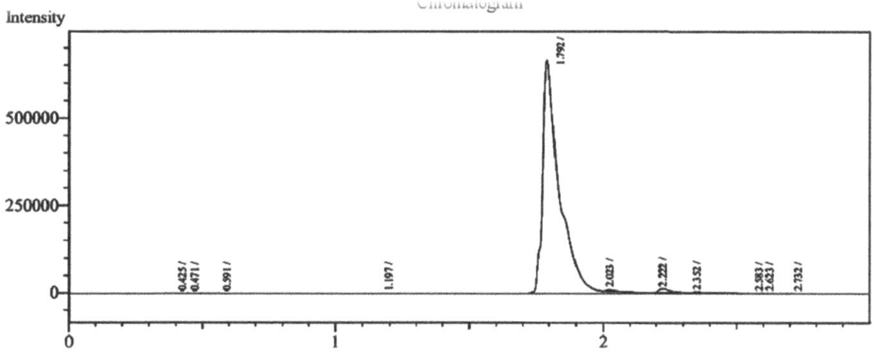

Fig. 5 Chromatogram of the analysis of bioethanol samples using GC method

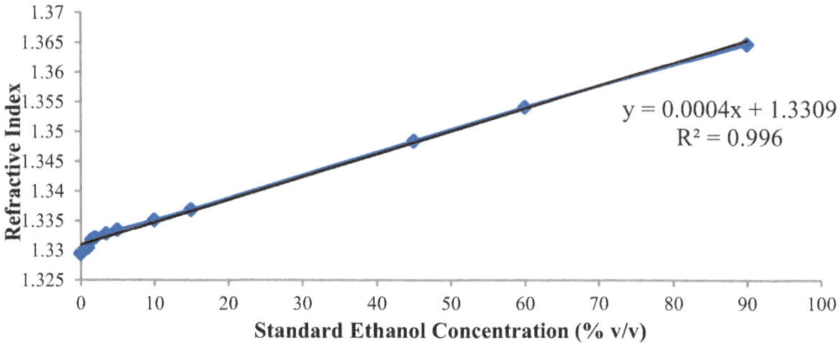

Fig. 6 Graph of ethanol standard solution

3.3.2 Quantitative Analysis of Bioethanol with Refractometer

The straight-line equation resulting from the comparison graph of the refractive index and standard ethanol concentration was $y = 0.0004x + 1.3309$ (Fig. 6) so that by using the regression formula where the refractive index value of the sample was 1.3370, the bioethanol content produced from fermented elephant grass leaves was 15.25% with a fermentation time of 8 days. The low ethanol content may be due to the non-optimal distillation process causing the sample to not be pure enough, it can be seen from the chromatogram in Fig. 5. This shows that the fermentation process does not only produce ethanol but also produces other compounds such as acetic acid, butyric acid, acid formic acid or lactic acid [7].

4 Conclusion

Based on the research conducted, it can be concluded that elephant grass leaves have a cellulose content of 37.82% with a lignin content of 6.1%. The delignification process plays a role in removing lignin to 4.64% so that the cellulose content increases to 70.24%. Elephant grass leaf cellulose fermented by *C. acetobutylicum* bacteria and the best fermentation time and pH was 8 days at pH 6.5 resulting in 15.25% bioethanol content. The results showed that elephant grass leaf cellulose can used as a raw material to produce bioethanol at the industry level in the future.

References

1. Khaidir (2016) Agricultural waste treatment as alternative fuel. Agrium J 13(2):63–68
2. Sari NM, Sri RM, Elvi Y (2018) Production of bioethanol from pineapple skin waste using *Clostridium acetobutylicum* bacteria with variations in inoculum concentration and addition of nutrients. Let's go Techn 05(1):1–6

3. Ruso S (2014) Bioethanol production from elephant grass stem (*Pennisetum purpureum Schumach*) with simultaneous fermentation system using *Clostridium acetobutylicum*, Thesis, Graduate Hasanuddin University, Makassar
4. Susmiati Y (2018) Prospects of bioethanol production from agricultural waste and organic waste. J Agroind Technol Manage 07(2):67–80
5. Agnesia VP (2017) Effect of total time and period of yeast provision on bioethanol production from cassava rubber by acid hydrolysis and modified fermentation method.Thesis, Chemical Engineering Study Program, Faculty of Engineering, University of Muhammadiyah, Surakarta
6. Fajariah HD, Wahyono H (2014) Utilization of sawdust into biobutanol by cellulase hydrolysis and bacterial fermentation of *Clostridium acetobutylicu*m. J Pomits Tech 03(2):2301–9271
7. Armidha C (2019) Production of bioethanol from sago dregs (*Metroxylon Sp.*) through process optimization of staged pretreatment and fermentation using *Clostridium acetobutylicum* bacteria and bread yeast (*Saccharomyces cerevisiae*). Thesis, Department of Chemistry, FMIPA, Hasanuddin University, Makassar
8. Bawazir I, Mokh HB, Asroful A (2021) Effect of variation in yeast mass and fermentation time of water hyacinth (*Eichhornia Crassipes*) waste on bioethanol levels. J Smart Technol 02(2):159–163
9. Manik BJR, Sri RM, Komalasari K (2019) Hydrolysis of palm leaves into glucose with variations in concentration of phosphoric acid (H_3PO_4) and substrate-solvent ratio. Let's FTEKNIK 06(2):1–7
10. Narindri B, Muhammad NC, Ria M (2016) Sorghum (*Sorghum bicolor L.Moench*) leaves bioethanol production. Biota 01(1):44−50
11. Herawati N, Pratama A, Juniar H (2019) Bioethanol production from elephant grass by delignification and hydrolysis process. AvoER National Seminar 11(1):45–51
12. Putra DTS (2021) Bioethanol production from dwarf late napier grass (*Pennisetum purpureum*) stem by with simultaneous saccharification and fermentation (SSF) method using *Clostridium acetobutylicum* bacteria. Thesis, Graduate Hasanuddin University, Makassar
13. Muria SR, Elvie Y, Mayang S (2018) Pineapple peel fermentation using *Clostridium acetobutylicum* with variations in nutrition and inoculum. J Sci Technol 17(2):56–60
14. Wadi A, Ahmad A, Tompo M, Hasyim H, Tuwo A, Nakajima M, Karim H (2019) Production of bioethanol from Seaweed, *Gracilaria verrucosa* and *Eucheuma cottonii*, by simultaneous saccharification and fermentation methods. J Phys Sci Conf Ser. https://doi.org/10.1088/1742-6596/1341/3/032031

Application of Lignocellulase-Producing Strains Actinomycetes to Decompose Plant Waste Materials

Shanti Ratnakomala, Fahrurrozi, and Puspita Lisdiyanti

Abstract Actinomycetes are common microorganisms in the soil, as they play an essential role in the degradation of plant cells walls. This study selected three isolates (ID6-455, ID6-479 and, ID6-279) from previous studies with cellulolytic, mannanolytic, and xylanolytic activities. The three isolates belonged to the genus Streptomyces and produced clear zones on media containing Carboxy Methyl Cellulose (CMC), Locust Bean, and Birchwood Xylan as substrates. Strains ID6-455 are cellulolytic actinomycetes, ID6-479 are xylanolytic actinomycetes, and ID6-279 are mannanolytic actinomycetes. In this study, the viability of the mixed actinomycetes cultures was monitored in the final compost product, studied the effect of three selected actinomycetes isolates in the application of plant waste biomass decomposition in the form of eucalyptus litter, rice straw, and corn stalk waste as substrates. The addition of mixed inoculum for biomass composting can accelerate the maturation of biomass waste compost. Compost from corn stalks degraded faster than rice straw and eucalyptus leaves. The added actinomycetes culture can survive at high composting temperatures.

Keywords Cellulase · Actinomycetes · Biomass · Compost

1 Introduction

Photosynthesis is producing millions of tons of cellulose, the main component of plant biomass. In plants, cellulose comprises almost 50% carbohydrates, while hemicellulose shall consist of 15–34%. Farmers discarded agricultural waste generated in

S. Ratnakomala (✉) · P. Lisdiyanti
Research Center for Biosystematics and Evolution, National Research and Innovation Agency, Cibinong, Indonesia
e-mail: shanti_ratna01@yahoo.com

Fahrurrozi
Research Center for Marine and Land Bioindustry, National Research and Innovation Agency, Lombok, Indonesia

M. A. Rahandi Lubis (ed.), *Proceedings of the 1st International Conference of Lignocellulose*, Green Energy and Technology,
https://doi.org/10.1007/978-981-19-1596-3_23

field and processing sites without being further used. Cellulose sources used include agricultural and forest products, waste paper, and industrial waste [18]. Cellulolytic microorganisms play an essential role in the biosphere because they recycle cellulose [5], especially in the anaerobic breakdown of cellulose waste [1].

Rice cultivation produces large quantities of straw. The most abundant components in rice straw are cellulose and hemicellulose encrusted with lignin. Besides, it contains a small amount of protein, making it a high C:N ratio [2]. Currently, to eliminate a massive amount of post-harvest rice waste in open-air field burning. One alternative that farmers often do is incorporating straw into the soil, where the microbial enzymes will convert the lignocellulosic components of the straw into compost. However, most crop straws have bulk in volume and slow bio-digestion when introduced directly into the soil. The addition of actinomycetes cultures can degrade cellulose and dissolve lignin extensively as their main metabolic activity. Thus, they are important agents of lignocellulosic decomposition in soil [17]. Compost microorganisms that have previously been identified include *Bacillus* sp., *Thermus* sp., *Streptomycetes* sp., *Aspergillus* sp. among others [8, 15]. However, actinomycetes are relatively less explored for biodegradation processes that utilize lignocelluloses for solid agro-waste management and wastewater treatment. Thus, there is also a need to search and explore novel actinomycete strains for various biodegradation applications.

Composting is a biological process in which biodegradable organic wastes are stabilized and converted by the action of some microorganisms. A humus-rich product (compost) used as a soil conditioner and organic fertilizer is made under controlled humidity, temperature, aeration, and pH conditions [14]. Compost is partially decomposed organic matter. The good compost looks like soil called humus. Compost has turned to a dark brown, easily crumbled, and has an earthy aroma when decomposition is complete. Through a biological process, soil microorganisms will break down and degrade plant tissue into simple compounds. Compost contains total NPK nutrients and contains other compounds that are beneficial to plants. So that compost is a multivitamin for the earth and plants, which can improve the physical and chemical properties of the soil.

Naturally, organic matter will decompose and be degraded into compost or humus. Still, it takes a long time, between half to a year, depending on the material and conditions. The composting process can take place more quickly by adding an activator or compost inoculum. This activator is a consortium of microorganisms that functions to accelerate the decomposition of organic matter into compost. The processes occurring in a compost pile are similar to those that break down organic matter in the soil. The decay of organic matter into mature compost depends on the ability of microorganisms to produce and secrete certain enzymes when the environment can be ideal for the microbes to do their work [11].

This study aimed to evaluate the effect of adding a mixed culture of actinomycetes, which have lignocellulolytic activity during the composting of plant waste biomass, such as eucalyptus litters, rice-straw, and corn stalk waste as substrates. In addition, the viability of the mixed actinomycetes culture and the quality of mature compost from each type of biomass were monitored in the final compost product.

2 Material and Method

2.1 Culturing Actinomycetes

The previous study screened 94 actinomycetes collected in Kutai National Park in East Kalimantan, Gili and Kuta in Lombok Island, and Cibinong in West Java [12, 3]. In this study, we selected three isolates from a previous study (ID6-455, ID6-479, and ID6-279) that have cellulolytic, mannanolytic, and xylanolytic activity to analyze their ability to decompose plant waste materials. The actinomycetes were re-cultured from glycerol stock and inoculating bacterial isolate in a plate containing yeast malt extract (YM) agar. The Petri dish has streaked then incubated at 28 °C for approximately seven days.

2.2 Detection of Actinomycetes Producing Enzymes

The three selected actinomycetes (ID6-455, ID6-479 and, ID6-279) were tested for their ability to produce cellulase, mannanase, and xylanase enzymes. The test was carried out by growing the three actinomycetes on YM agar medium containing specific substrates such as Carboxymethyl Cellulose, Locust bean gum, or Birchwood Xylan. A preliminary analysis for lignocellulolytic activity was conducted using Congo red dye. The clear zone methodology with Congo red was used to detect the strain's CMCase, Mannanase and Xylanase activity.s The plates were then incubated at 30 °C for four days to allow hydrolyzed the substrates. Plates were then flooded with 0.25% Congo red for 15–20 min, then washed with 1 M NaCl for 15 min. The clear zone formation was indicated hydrolysis of lignocellulose degradation.

2.3 Composting Methods

Biomass of domestic litter plants (eucalyptus leaf litter, corn stalk, and rice straw waste) was collect from the Wonosari area, Yogyakarta was used in these experiments. After sorting, the organic raw materials were chopped into fractions smaller than 5 cm. Each sample wastes was then mixed with 20% mixed cultured media containing three potential actinomycetes ID06-455, ID06-479, and ID06-279 (v/w). The composting with activator inoculants of growing actinomycetes was prepared by the following method.

The cultures actinomycetes from the log phase of growth were mixed according to the above treatments. Five kilograms of dried rice straw, eucalyptus leaf litter, and corn stalk were used for each compost pile. The biomass samples were divided into two treatments, with and without adding inoculum, with three replications. Water was added to the compost pile, maintained a moisture content range of 60%, and kept throughout for two months of the experiment. The compost piles were turned over weekly, and samples were taken each month for evaluation. Samples were taken at 0, 30, and 60th days of composting from both control and treatment composting systems and analyzed physically and chemically:

Physical analysis. The measurement of temperature was determined by using a hand thermometer. In addition, the naked eye observed observations of biomass degradation and changes in the color of compost in the compost pile.

Chemical analysis. pH was determined by stirred 10 g of sample with 100 ml of distilled water for 30 min, and the pH of the suspension was measured using a pH meter. Second, the total carbon content of dried and powdered compost samples was determined dry combustion method. Third, the ash content was determined after drying and burnt at 105 °C and 550 °C, respectively, in a muffle furnace for about three hours [19]. Fourth, the total nitrogen was determined by the Micro-Kjeldahl method. Fourth, the wet digestion method determined the total phosphorus content of dried and powdered compost samples. The total potassium content of dried and grinding compost samples was analyzed using a flame photometer. The determination method used is following the literature [10].

Microbiological analysis. Bacteria and actinomycetes were isolated from the composting samples. PCA and Humic acid vitamin agar media were used for total bacteria and actinomycetes. The incubation time was four days for actinomycetes and two days for total bacteria. All microbial counts were calculated based on a colony-forming unit (CFU/mg compost). The average number of microorganisms isolated on three plates was considered the viable cell number.

3 Results and Discussion

The ability to degrade specific substrates is tested on three strains (ID6-455, ID6-479, and ID6-279). As a result, the actinomycete strains ID06-279 was capable of producing mannanase and cellulase enzyme. The isolates ID6-455 and ID6-479 were able to degrade substrate showed a great activity against Locus bean gum, CMC, and Birchwood xylan (Table 1). Thus, the three actinomycetes isolates can produce lignocellulolytic enzymes that can degrade plant material waste. All the strains were identified into the genus *Streptomyces* sp.

Compost generally consists of decomposed plant or animal residues. Compost can improve physical condition, improve soil fertility when applied to garden land. The organic matter in compost can improve soil porosity by binding compost particles with the soil together, and the soil aeration and plant roots more accessible to works. The binding of soil particles with compost helps increase aeration, water infiltration, root penetration and reduces soil surface hardening. Compost also increases the activity of soil microorganisms which release enzymes, nutrients, and other beneficial growth-promoting materials into the soil. Therefore, it helps enhance soils low in

Table 1 Isolates actinomycetes that have a potential activity to degrade cellulolytic and hemicellulolytic substrates

Isolate	Diameter in 4 days colony (cm)			16S rDNA closest to	BLAST Search %
	Locust bean gum	CMC	Birchwood Xylan		
ID06-279	3.2	1.9		*Streptomyces*	97
ID06-455	2.9	4.40	3.4	*Streptomyces*	97
ID06-479	2.7	3	3.7	*Streptomyces*	92

organic matter. Indonesian Agricultural Ministry has issued a minimum standard required to qualify a compost as an organic fertilizer, as listed in Table 2.

All organic matter contains carbon and nitrogen compounds. The carbon-to-nitrogen (C:N) ratio of organic material will estimate the relative amounts of the two elements it has. Carbon is an essential component of cellulose and lignin which gives strength to plant cell walls. Nitrogen can be found in many other compounds in plant cells. The ratio is based on the dry weight percent of carbon and nitrogen in the material. Compost from a particular organic waste will have a different C:N ratio with compost from other organic materials. The ratio of about 30:1 is ideal for microbial growth and degrading plant waste into compost (Table 3) [16].

This study used three organic raw waste materials (eucalyptus leaf litter, rice straw, and corn stalk waste). These materials were abundant in the Wonosari area that has the processing plant of eucalyptus oil. In addition, dried leaves and a lot of rubbish were used only as firewood by local communities. Waste corn and rice straw are also abundant in this area because people are generally farmers. After 1 and 2 months of composting, physical and chemical analyses of the materials were determined.

Leaves of Eucalyptus represent a large percentage of total yard waste, cause the accumulation of the volume of litter, and are difficult to degrade. In addition,

Table 2 Minimum requirements of organic fertilizer (Peraturan Menteri Pertanian 2006)

No.	Parameter	Unit	Solid
1	C organic	%	> 12
2	C/N Ratio		10–25
3	Water content	%	13–20
4	pH		4–8
5	Total content P_2O_5	%	< 5
6	Total content K_2O	%	< 5
7	Microelements	ppm	
8	Zn		< 50
9	Fe		< 40
10	Heavy metal content	ppm	
11	Pb		< 50
12	Cd		< 10

Table 3 Carbon to nitrogen ratio in various materials

Organic material	C:N
Tree leaves	30–80:1
Cornstalk	60:1
Rice straw	40–100:1
Vegetable wastes	12–20:1
Cow manure	20:1

Source [16]

eucalyptus leaves can be toxic to other microorganisms because they contain aromatic compounds that can inhibit bacterial growth. There are plenty of aromatic oils in *Eucalyptus citriodora* Hook leaves, but those oils evaporate for several days. This can be seen from Table 4 that the growth of microbes from the control compost without inoculums showed a decrease in the microbial populations, both actinomycetes and other microbes.

The quality of the final products from Eucalyptus leaves compost was further assessed, as shown in (Table 4). Mixed actinomycetes inoculum did not significantly affect mineral content, micronutrients, or total phosphorous and total potassium of the composted eucalyptus leaves.

The final product of the compost Eucalyptus leaves containing mixed culture inoculum showed a little higher organic matter (43.59%) than the un-inoculated control (43.37%) (Table 5) after two months of composting without or with inoculums, no change in the texture and proximate content (Fig. 1). Eucalyptus leaves are difficult to degrade, probably because this plant contains essential oils that can inhibit the growth of most microorganisms. Although many organisms can live in such conditions, it is visible from quite many actinomycetes growth detected in the compost eucalyptus leaves. In all cases, the pH of the final product was within neutrality (pH 7.0).

Gramineae such as corn is more susceptible to actinomycete breakdown than wood [13]. Together with fungi and bacteria, actinomycetes play a significant role in mobilizing nutrients and soil humification processes associated with soils and composts [4]. As a result, the content of total phosphorus and total potassium increased after two months of composting. From our observation of the bacterial populations during composting cornstalk, it can be seen that adding a mixed inoculum actinomycetes decrease the total number of bacteria and actinomycetes after two months of composting, and this is probably because of the decrease in nutrients required for microbial growth as seen at Table 6.

Carbon is one of the most important elements required to decompose organic waste by microorganisms because it is a source of energy for microorganisms. In addition, moisture is also needed to improve soil structure [6]. Figure 2a–d showing the physical changes of corn stalks compost during the incubation period.

Characteristics of the final products from corn stalk compost were further assessed, as shown in (Table 7). The final product containing mixed culture inoculum showed higher water content (26.54%), decrease in organic matter (8.76%) compared to

Table 4 Characteristics of the final product after Eucalyptus leaves with actinomycetes inoculum for two months composting in plastic pots

Character of eucalyptus leaves		Without inoculum			With inoculum		Std
	0 month	One month	Two months	0 month	One month	Two months	
pH	7	7	7	7	7	7	< 8
Mg (%)	na	0.15	0.16	na	0.16	0.24	0.6
Na (%)	na	0.81	0.72	na	0.72	0.77	
Ca (%)	na	0.27	0.23	na	0.23	0.3	25.5
Cl (mg/100 g)	na	30.14	30.28	na	30.28	29.48	
Micronutrient							
Zn (ppm)	na	50.05	48.7	na	48.7	51.23	500
Fe (mg/100 g)	na	30.55	30.28	na	30.28	30.46	2
Total content							
N (%)	5.8	6.01	6.89	5.76	6.67	7.03	
P (%)	0.35	1.07	1.12	0.3	1.12	1.25	0.1
K (%)	0.38	2.54	2.39	0.39	2.39	2.19	0.2
Heavy metal:							
Pb (ppm)	na	2.11	1.89	na	1.89	2.17	150
Cd (ppm)	na	0.22	0.18	na	0.18	0.21	3
Total Bacteria							
Total bacteria in PCA (log CFU/ml)	na	11.14	10.83	na	9.63	8.79	
Total bacteria in HV (log CFU/ml)	7.00	7.08	7.75	9.45	8.22	10.03	
Actinomycetes (log CFU/ml)		11.45	10.68	8.86	8.97	10.08	

Inoculum = mixture culture of *Streptomyces* sp ID06-279, *Streptomyces* sp ID06-455 and *Streptomyces* sp ID06-479, na = not analysis

Table 5 Composition of Eucalyptus leaves residue composting with inoculation mixed culture actinomycetes

Analysis of composition Eucalyptus	Without inoculum		With inoculum	
	One month	Two months	One month	Two months
Water content (%)	7.06	13.13	8.31	17.76
Fiber (%)	26.11	28.22	23.57	26.11
Protein (%)	12.42	13.33	15.36	14.01
Energy (kcal/100 g)	348.88	325.32	314.4	311.12
C organic (%)	43.33	43.37	48.29	43.59

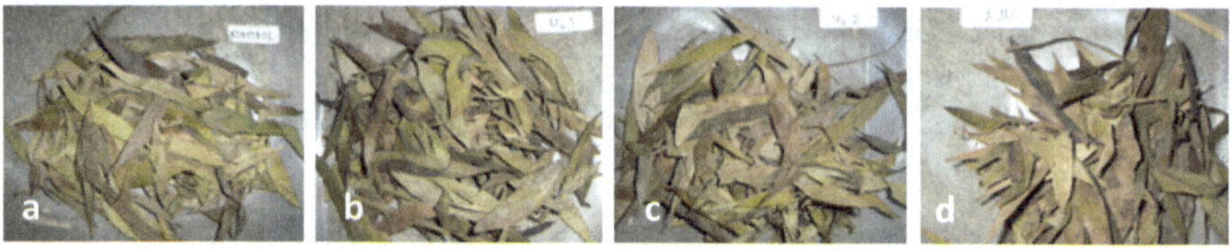

Fig. 1 The physical changes of *Eucalyptus* leave compost during the incubation period. a = control; b = one month composting; c = two months composting; d = three months composting

Table 6 Characteristics of the final product after Corn stalks with actinomycetes inoculum for two months composting in plastic pots

Character of cornstalk		Without inoculum			With inoculum		Std
	0 month	One month	Two months	0 month	One month	Two months	
pH	7	7	7	7	7	7	< 8
Mg (%)	na	0.13	0.17	na	0.23	0.2	0.6
Na (%)	na	0.77	0.75	na	0.9	0.87	
Ca (%)	na	0.25	0.29	na	0.34	0.35	25.5
Cl (mg/100 g)	na	28.74	30.56	na	31.05	29.95	
Micronutrient							
Zn (ppm)	na	49.05	48.11	na	52.07	50.76	500
Fe (mg/100 g)	na	29.05	28.9	na	29.92	31.69	2
Total content							
N (%)	1.2	1.7	2.12	1.3	1.82	2.26	
P (%)	0.74	1.15	0.98	0.78	1.3	1.5	0.1
K (%)	1.54	2.5	2.48	3.21	2.61	2.8	0.2
Heavy metal:							
Pb (ppm)	na	1.8	1.97	na	2.04	2.32	150
Cd (ppm)	na	0.15	0.17	na	0.17	0.23	3
Total Bacteria							
Total bacteria in PCA	na	11.78	9.01	na	11.57	9.65	
Total bacteria in HV	na	10.63	10.20	na	10.20	8.51	
Actinomycetes	na	11.86	10.16	na	10.82	9.47	

Inoculum = mixture culture of *Streptomyces* sp ID06-279, *Streptomyces* sp ID06-455 and *Streptomyces* sp ID06-479, na = not analysis

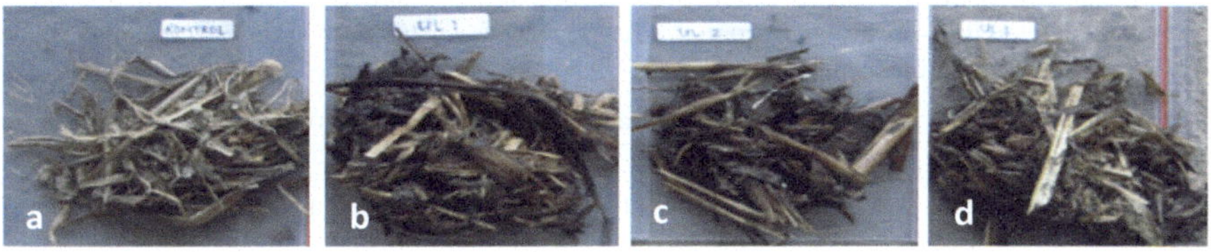

Fig. 2 The physical changes of corn stalk compost during the incubation period. a = control; b = one month composting; c = two months composting; d = three months composting

those the un-inoculated control (64.40%). Similarly, compost inoculated with this actinomycetes strain showed decreased fiber, protein, and energy content. It may be because most of the carbohydrate content in the corn stalk waste has been degraded by microbes that work during the composting. Thus, there are changes in the range of cornstalk compost after two months of composting with the provision of inoculum, but no content changes on cornstalk compost without inoculums. In all cases, the pH of the final product was within neutrality (pH 7.0).

Rice straw can be decomposed in the soil by microorganisms and transformed into a bio-fertilizer that enhances soil quality. However, the decomposition process is ultimately slow because of rice straw's characteristic comprehensive C:N ratio (Table 8). The concentrations of macronutrients (P, K, Mg, Na, Ca, Cl), micronutrients (Fe, Zn), and toxic metals (Cd, Pb) in both composts increased at the end of the process. Moreover, the pH and C/N ratio values (15.95–18) in the final composts indicated good compost development in both piles concerning organic matter biodegradation, a neutral pH, showed a high level of biodegradation and humification of fresh soil organic matter. It is suggested that the incorporation process can be potentially enhanced by controlling the type of decomposing microorganisms and certain abiotic factors such as air, temperature, moisture content, and organic amendments.

Compost rice straw pile with the addition of inoculum showed an increase in the number of actinomycetes population after two months of composting. It happens because the enzymatic activity that occurs in the compost pile has taken place. The lignocellulolytic enzymes (mannanase, cellulase, and xylanase) produced

Table 7 Composition of cornstalk composting with mix culture actinomycetes

Analysis of the composition of corn stalk	Without inoculum		With inoculum	
	One month	Two months	One month	Two months
Water content (%)	7.58	9.66	71.58	26.54
Fiber (%)	22.78	22.56	15.72	2.9
Protein (%)	23.43	20.84	11.69	10.6
Energy (kcal/100 g)	305.28	309	229.28	282.6
C organic (%)	71.53	64.40	26.80	28.76

Table 8 Characteristics of the final product after rice straw wastes with actinomycetes inoculum for two months composting in plastic pots

The character of rice straw		Without inoculum			With inoculum		Std
	0 month	One month	Two months	0 month	One month	Two months	
pH	7	7	7	7	7	7	< 8
Mg (%)	na	0.18	0.21	na	0.19	0.19	0.6
Na (%)	na	0.65	0.82	na	0.67	0.79	
Ca (%)	na	0.2	0.27	na	0.21	0.29	25.5
Cl (mg/100 g)	na	27.49	31.46	na	27.74	30.82	
Micronutrient							
Zn (ppm)	na	47.86	50.52	na	48.15	48.97	500
Fe (mg/100 g)	na	28.45	28.45	na	29.47	29.38	2
Total content							
N (%)	0.71	0.89	1.21	0.73	1.06	1.24	
P (%)	0.25	1.15	1.27	0.38	1.19	1.29	0.1
K (%)	3.6	2.04	2.5	6.13	1.92	2.48	0.2
Heavy metal:							
Pb (ppm)	na	1.68	1.85	na	1.75	2.05	150
Cd (ppm)	na	0.25	0.19	na	0.18	0.16	3
Total Bacteria							
Total bacteria in PCA	na	10.25	9.51	na	11.71	10.02	
Total bacteria in HV	10.16	7.59	9.91	10.27	8.66	9.29	
Actinomycetes	9.78	8.51	9.60	9.44	10.19	9.79	

Inoculum = mixture culture of Streptomyces sp ID06-279, Streptomyces sp ID06-455 and Streptomyces sp ID06-479, na = not analysis

from the actinomycetes consortium have degraded organic matter consist of cellulosic substrates into glucose monomers and are used to increase the actinomycetes population.

Different physical parameters influence cellulose bioconversion, and pH is an essential factor affecting cellulase production. Mrudula and Murugammal [9] had reported that the production and release of cellulase depended on the pH of the medium. Therefore, the final products from rice straw compost pots were further assessed, as shown in (Table 9).

Mixed inoculum did not significantly affect mineral content, micronutrient, heavy metal, total phosphorus, and total potassium content of the composted rice straw. However, the final product containing mixed culture inoculum showed

Table 9 Composition of rice straw residue composting with inoculation mixed culture actinomycetes

Analysis of the composition of rice straw	Without inoculum		With inoculum	
	One month	Two months	One month	Two months
Water content (%)	4.45	32.13	5.23	22.22
Fiber (%)	26.11	24.43	29.77	24.52
Protein (%)	14.01	11.42	18.15	9.67
Energy (kcal/100 g)	311.12	191.28	285.35	204.68
C organic (%)	43.59	21.84	51.79	19.79

Fig. 3 The physical changes of corn stalk compost during the incubation period. a = control; b = one month composting; c = two months composting; d = three months composting

decreased organic matter (19.79%) than the un-inoculated control (21.84%). Similarly, inoculum pots showed reduced fiber content, protein, and energy than the un-inoculated control pots. In all cases, the pH of the final product was within neutrality (pH 7.0) (Fig. 3).

Composted rice straw has a high nitrogen content. Therefore, it can represent an excellent biological fertilizer for plant growth and reduce nitrogen source chemicals because it has a slow-release character of nitrogen and phosphorus components in its residue. In principle, all materials derived from nature or organic materials such as plant litter, weeds, kitchen waste, grass, and wood scraps can be used as compost.

4 Conclusion

This study involved potential screening actinomycetes, selected three isolates with cellulolytic, mannanolytic, and xylanolytic activity. ID06-455 for cellulolytic actinomycete, ID06-479 for xylanolytic actinomycete, and ID06-279 for mannanolytic actinomycete. All the strains also have activity at least for two enzymes. The mixed inoculum for biomass composting showed that corn stalks quickly degrade than rice straw and eucalyptus leaves. Actinomycetes can survive elevated pH and temperature conditions as much as bacteria and fungi. Their specific consortia can single-handedly produce all key enzymes for composting.

Acknowledgements We thank the Japan Society for the Promotion of Science (JSPS) for partial funding of the publication fee.

References

1. Ahmed B, Aboudi K, Tyagi VK, Álvarez-Gallego CJ, Fernández-Güelfo LA, Romero-García LI, Kazmi AA (2019) Improvement of anaerobic digestion of lignocellulosic biomass by hydrothermal pretreatment. Appl Sci 9:3853. https://doi.org/10.3390/app9183853
2. Baruah J, Nath BK, Sharma R, Kumar S, Deka RC, Baruah DC, Kalita E (2018) Recent trends in the pretreatment of lignocellulosic biomass for value-added products. Front Energy Res. https://doi.org/10.3389/fenrg.2018.00141
3. Fahrurrozi, Ratnakomala S, Lisdiyanti P (2009) Screening and characterization of actinomycetes degrading hemicellulose. In: Proceeding international conference of wood science
4. Gorovtsov AV, Minkina TM, Mandzhieva SS, Perelomov LV, Soja G, Zamulina IV, Rajput VD, Sushkova SN, Mohan D, Yao J (2020) The mechanisms of biochar interactions with microorganisms in soil. Environ Geochem Health 42:2495–2518. https://doi.org/10.1007/s10653-019-00412-5
5. Jayasekara S, Ratnayake R (2019) Microbial Cellulases: An Overview and Applications. Cellulose. Pascual AR, Martín MEE (eds) IntechOpen, https://doi.org/10.5772/intechopen.84531. Available from: https://www.intechopen.com/books/cellulose/microbial-cellulases-an-overview-and-applications
6. Kulkarni NA, Shendye, Rao M (1999) Molecular and biotechnologycal aspects of xylanases. Rev FEMS Microbiol 23:411–456
7. Kumar R, Singh S, Singh OV (2008) Bioconversion of lignocellulosic biomass: biochemical and molecular perspectives. J Ind Microbiol Biotechnol 35(5):377–391. https://doi.org/10.1007/s10295-008-0327-8
8. McCaig AE, Grayston SJ, Prosser JI, Glover LA (2001) Impact of cultivation on characterisation of species composition of soil bacterial communities. FEMS Microbiol Ecol 35:37–48
9. Mrudula S, Murugammal R (2011) Production of cellulase by aspergillus niger under submerged and solid state fermentation using coir waste as a substrate. Braz J Microbiol 42:1119–1127
10. Parkinson JA, Allen SE (1975) A wet oxidation procedure suitable for determination of nitrogen and mineral nutrients in biological material. Commun Soil Sci Plant Anal 6:1–11
11. Pathma J, Sakthivel N (2012) Microbial diversity of vermicompost bacteria that exhibit useful agricultural traits and waste management potential. SpringerPlus 1:26 (2012). http://www.springerplus.com/content/1/1/26
12. Ratnakomala S, Fahrurrozi, Kartina G, Lisdiyanti P (2009) Screening and characterization of cellulase enzyme producers from indonesian actinomycetes. In: Proceeding international conference of wood science
13. Saini A, Aggarwaln NK, Sharma A, Yadav A (2015) Actinomycetes: a source of lignocellulolytic enzymes. Enzyme Res 20:1–15. https://doi.org/10.1155/2015/279381
14. Shilev S, Naydenov M, Vancheva V, Aladjadjiyan A, Aladjadjiyan A (2007) Composting of food and agricultural wastes. In book: Utilization of by-products and treatment of waste in food industry, 2007 edn. SpringerEditors, V.Oreopoulu. https://doi.org/10.1007/978-0-387-35766-9_15
15. Song J, Weon HY, Yoon SH, Park DS, Go SJ, Suh TW (2001) Phylogenetic diversity of thermophilic actinomycetes and Thermoactinomyces spp. isolated from mushroom composts in Korea based on 16S rRNA gene sequence analysis. FEMS Microbiol Lett 202:97–102
16. Starbuck C (2001) Making and using compost. Horticultural Published By University Extension, University Of Missouri-Columbia

17. Tuomela, Vikman M, Hatakka A, Itävaara M (2000) Biodegradation of lignin in a compost environment: a review. Bioresour Technol 72:169–183
18. White D (2000) The physiology and biochemistry of prokaryotes, 2nd edn. Oxford University Press, New York, NY
19. WHO (1978) International reference center for wastes disposal. In: Methods of analysis of sewage solid wastes and compost. CH-8600, Bubendorf, Switzerland

Enhancement of Enzymatic Saccharification of Oil Palm Empty Fruit Bunch by Microwave-Assisted Acid Pretreatment

Deddy Triyono Nugroho Adi, Nissa Nurfajrin Solihat, and Euis Hermiati

Abstract Pretreatment is an essential step for the bio-refinery to degrade the recalcitrance of lignocellulosic biomass structure, thus facilitating the accessibility of cellulose-degrading enzymes. In this study, pretreatment of oil palm empty fruit bunch (OPEFB) by the dilute acid (maleic and sulfuric acid) solution under microwave irradiation was investigated. The pretreatment was conducted at 160, 180, and 200 °C for 10 min, and at 180 °C for 5, 10, and 15 min. The pulp recovery, reducing sugars, and glucose released after enzymatic saccharification of pretreated OPEFB were determined. Lower pulp recovery was achieved at the higher pretreatment temperature by sulfuric and maleic acids. The reducing sugars released after enzymatic saccharification of OPEFB pretreated by maleic acid for 10 min at 160, 180, and 200 °C were 31.4, 38.7, and 40.4 g/100 g initial biomass, respectively. This result indicating the deformation of OPEFB structure by maleic acid at higher temperature successfully facilitates the enzyme to access and degrade cellulose in the pulp. However, in the case of sulfuric acid pretreatment, reducing sugars released was dropping at high temperatures. The cellulose conversion into glucose by enzymatic scarification of pretreated OPEFB was higher compared to that of untreated OPEFB. Pretreatment at 180 °C for 10 min by sulfuric and maleic acids resulted in 84.0 and 86.5% cellulose conversion, respectively. These values are about 4 times higher than the cellulose conversion from untreated OPEFB (21.8%).

Keywords Acid pretreatment · Fermentable sugars · Microwave heating · Oil palm empty fruit bunch

D. T. N. Adi · N. N. Solihat · E. Hermiati (✉)
Research Center for Biomass and Bioproducts, National Research and Innovation Agency, Cibinong, Bogor, Indonesia
e-mail: e_hermiati@yahoo.com

M. A. Rahandi Lubis (ed.), *Proceedings of the 1st International Conference of Lignocellulose*, Green Energy and Technology,
https://doi.org/10.1007/978-981-19-1596-3_24

1 Introduction

The rising demand for fuel is not in line with its availability especially that derived from fossils. This issue is enforcing researchers to find sustainable bio-based fuel from lignocellulosic biomass such as oil palm empty fruit bunch (OPEFB). In Indonesia, OPEFB is available in huge amounts as a by-product of palm oil factory which was estimated to produce 11.11 million metric tons in 2020 [1, 2]. One of the key factors to produce inexpensive and efficient biofuel from lignocellulose is the proper pretreatment method for delignification of biomass. The step of converting biomass into biofuel is divided into pretreatment of biomass, cellulose hydrolysis into sugars, sugar fermentation into bioethanol, and bioethanol purification. Increasing fermentable sugars after enzymatic hydrolysis can be achieved by an appropriate pretreatment method, which eventually will improve the whole process efficiency [3]. Pretreatment aims to obtain maximum sugars that available in biomass to be converted into bioethanol. There are some advantages and disadvantages offered by each pretreatment. A pretreatment method is considered good if it is cost-effective, safe for humans and the environment, effective in disrupting lignocellulose structure, and avoids the production of toxic substances [3–5].

Microwave is known as an environmentally friendlier pretreatment method compared to conventional heating due to the energy efficiency, shorter heating time to break down the lignocellulose matrix, reduce emission, and wastewater-free [6, 7]. Microwave heating also can lower the decomposition of some neutral monosaccharides compared to induction heating [8]. A liquid catalyst is needed to combine in microwave pretreatment to intensify the digestion of the lignocellulosic compounds. Some researchers used to dilute acid both organic and inorganic to assist microwave pretreatment where influence parameters in microwave-assisted acid pretreatment are temperature, irradiation time, pressure, and concentration of acid [7, 9]. Sulfuric acid is commonly used to hydrolyze lignocellulose compounds yet it suffers several problems such as generating the production of furfural, phenolic compounds, and other inhibitor products. The number of inhibitors produced from bamboo pretreatment using sulfuric acid is higher than those from pretreatment using organic acids, such as oxalic acid, acetic acid, and formic acid [10]. Pretreatment of lignocellulosic biomass by maleic acid has almost similar effectiveness as that by sulfuric acid for increasing enzymatic digestibility of cellulose to glucose monomers, while at the same time it produces more xylose and less furfural [11]. Therefore, the use of organic acids, such as maleic acid, was proposed to substitute sulfuric acid.

Our previous research reveals the use of pretreatment using microwave and oxalic acid has a positive effect to enhance the sugar yield from the OPEFB up to ~ 40% of initial biomass after enzymatic saccharification [12, 13]. In this study, evaluation of the efficacy of sulfuric and maleic acid to increase glucose yield was conducted to get more understanding about various effects of the use of inorganic and organic acids during OPEFB pretreatment on sugar yield after enzymatic saccharification.

2 Materials and Methods

2.1 Materials

In this study we used the same OPEFB as that reported in our previous study [13, 14]. The material was collected from Sukabumi, West Java, dried, milled and sieved to obtain samples having particle size of 40–60 mesh (250–420 μm). The OPEFB contains 8.44 ± 0.04% of moisture, 2.69 ± 0.08% of extractives, 2.91 ± 0.03% of ash, 23.62 ± 0.08% of acid soluble lignin (AIL), 41.23 ± 0.77% of cellulose, and 26.55 ± 0.57% of hemicellulosethe. Sulfuric and maleic acids as well as other chemicals used were analytical grade.

2.2 *Pretreatment of OPEFB Using Microwave and Dilute Acids*

The OPEFB fiber was dried overnight in an oven at 105 °C and cooled in a desiccator prior to the pretreatment experiment to remove the moisture. Dilute acid-microwave pre-treatment of OPEFB was performed in START D Microwave Digestion System (Milestone, Sorisole, Bergamo-Italy) such as reported in Solihat et al. 2017 and 2020 [13, 15]. Briefly, 3 g dried OPEFB was suspended in 30 mL of 1% acid (sulfuric acid or maleic acid) solution in a Teflon tube. The tube was closed tightly and put inside the microwave digester. The mixture in the reactor was stirred constantly at 195 rpm. The reactor was preheated for 12 min and the mixture was irradiated at designated temperature and duration of heating. The influence of temperature was studied at 160, 180, and 200 °C, respectively for 10 min duration of heating. The effects of the duration of heating (5, 10, and 15 min) at 180 °C were also observed. The pretreatment was terminated by cooling down the reactor immediately, and the insoluble solid, called the pulp, was separated by filtration and washed until neutral pH with distilled water. The pulp for enzymatic saccharification was stored in a sealed plastic bag and keep in a refrigerator. Pulp recovery was determined based on the weight and moisture content of the pulp.

2.3 *Determination of pH and Absorbance Value of the Pretreatment Hydrolysate*

The pH of pretreatment hydrolysate was measured using Eutech pH meter. Estimation of degradation of major chemical components of OPEFB was conducted by measuring the absorbance of brown compounds using a spectrophotometer (Hitachi U-2001 spectrophotometer, Hitachi Corp., Tokyo, Japan) at 490 nm [16].

2.4 Enzymatic Saccharification

Enzymatic saccharification was performed at 1% dry pulp loading in 0.05M citrate buffer pH 4.8 medium supplemented with 0.2 mg/mL sodium azide to prevent microbial growth. After the addition of Meicellase from *Trichoderma viridae* (Meiji Seika, initial activity 200 FPU/g) at the concentration 120 FPU/g dry pulp, the mixture was incubated at 50 °C and agitated at 150 rpm for 5 days in WIS-30 Wise Cube Incubator (witeg Labortechnik GmbH, Werthein-Germany). The enzymatic saccharification was terminated by boiling the mixture for 5 min. The enzymatic hydrolysate was isolated by filtration at room temperature and stored in a freezer for further glucose and reducing sugar analysis.

2.5 Glucose and Reducing Sugars Analyses

Glucose concentration in the enzymatic hydrolysate was determined using the Glucose CII test kit (Wako Junyaku, Co., Osaka, Japan), and the absorbance of the assay at 505 nm was measured by a Paradigm P2 Spectramax microplate reader (Molecular Devices LLC, San Jose, California, USA). The glucose yield based on the pulp and initial OPEFB biomass as well as cellulose conversion was calculated using Eqs. 1, 2, and 3, respectively.

$$Glucose\ yield(\%pulp) = \frac{(GC \times 0.1)}{ODP} \times 100 \tag{1}$$

$$Glucose\ yield(\%initialOPEFB) = glucose\ yield(\%pulp) \times PR \tag{2}$$

$$Cellulose\ conversion(\%) = \frac{(GC \times 0.1 \times 0.9)}{\left(\frac{CC}{100}\right) \times ODP} \times 100 \tag{3}$$

where GC is glucose concentration in g/L, ODP is oven-dry weight of pulp in g, 0.1 is the volume of saccharified sample in L, PR is pulp recovery in %, 0.9 is the conversion factor from glucose to cellulose, and CC is cellulose content in the initial biomass in %.

The reducing sugar concentration in the enzymatic hydrolysate was determined by spectrophotometry (Hitachi U-2001 spectrophotometer, Hitachi Corp., Tokyo, Japan) at 540 nm following the DNS method. The reducing sugar yield (RSY) was also determined based on the pulp and OPEFB biomass using Eqs. 1 and 2, where glucose was replaced by reducing sugars.

3 Results and Discussion

3.1 *Microwave-Assisted Acid Pretreatment*

The objective of pretreatment is to reduce the recalcitrance of lignocellulosic biomass structure by the aid of physical and/or chemical treatment. The combination treatment of microwave irradiation to generate high temperature in a short period and dilute sulfuric acid and maleic acid solution to degrade the structure of OPEFB were studied. At a high temperature, the presence of acid catalyst degraded some part of the OPEFB structure and dissolved it into the aqueous solvent, and finally left the degraded material known as pulp.

Figure 1 shows the pulp recovery after microwave-assisted acid pretreatment using 1% sulfuric acid and 1% maleic acid, respectively. The degradation of OPEFB by microwave irradiation combined with 1% sulfuric acid and 1% maleic acid leave the recovered pulp in the range of 36–60% and 53–73% (maleic acid) from the initial loading, respectively. At the equal heating temperature and duration, 1% sulfuric acid performed more severe degradation compared to 1% maleic acid and resulted in a lower quantity of recovered pulp. This is in agreement with a previous study that shows lower pulp recovery from sulfuric acid pretreatment compared to that from maleic acid pretreatment [17]. The influence of temperature also gave a significant impact on the degradation where the quantity of recovered pulp was reduced from 60 to 36% (sulfuric acid) and 73–53% (maleic acid) when pretreatment temperature was increased from 160 to 200 °C. However, the heating duration only gave a little impact on the OPEFB degradation. The pretreatment at 180 °C for 5, 10, and 15 min resulted in only a slight difference in the quantity of recovered pulp 50–56%, and 61–63%, from pretreatment using sulfuric acid and maleic acid, respectively. These phenomena are also shown during microwave-maleic acid pretreatment of OPEFB previously reported [17].

The difference of degradation severity between sulfuric acid and maleic acid was also studied from the aqueous solution collected by filtration after pretreatment. Table 1 compiles the value of pH and the brown compound absorbance value of the aqueous solution resulted from microwaved-assisted OPEFB pretreatment.

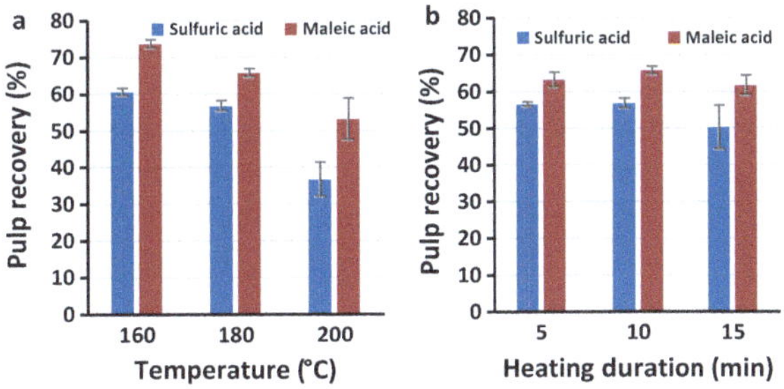

Fig. 1 Pulp recovery of microwave-assisted pretreatment of OPEFB by 1% sulfuric acid and 1% maleic acid **a** at different temperatures for 10 min heating duration and **b** at 180°C for the different heating duration

Table 1 pH and brown compound absorbance of the aqueous fraction

Parameter		pH		Brown compound abs. at 490 nm	
		Sulfuric acid	Maleic acid	Sulfuric acid	Maleic acid
Temperature (°C) for 10 min	160	0.94 ± 0.04	1.55 ± 0.04	0.869 ± 0.02	0.263 ± 0.01
	180	0.43 ± 0.04	1.76 ± 0.01	1.042 ± 0.13	0.307 ± 0.02
	200	0.85 ± 0.02	2.11 ± 0.03	1.543 ± 0.12	1.311 ± 0.26
Heating duration (min) at 180 °C	5	0.39 ± 0.06	1.94 ± 0.03	0.428 ± 0.06	0.327 ± 0.02
	10	0.43 ± 0.04	1.76 ± 0.01	1.042 ± 0.13	0.307 ± 0.02
	15	0.37 ± 0.05	1.86 ± 0.02	1.116 ± 0.06	0.401 ± 0.03

The pH of the aqueous solution indicates the concentration of hydronium (H_3O^+) ions in the aqueous which is involved in the catalytic degradation of the OPEFB structure. At all treatments, the pH value of sulfuric acid was lower than maleic acid indicating there was plenty of hydronium ions deliberated by sulfuric acid and catalyzed OPEFB degradation. The different influence of pretreatment temperature on pH value between sulfuric acid and maleic acid pretreatment was also observed. Sulfuric acid pH value remained low (< 1) with the lowest pH was observed at 0.43 after 180 °C heating. On the other side, the pH of the maleic acid aqueous fraction was getting increased from the values of 1.55, 1.76, and 2.11 by heating at 160, 180, and 200 °C, respectively. The increase of the pH due to the increase of temperature was an indication of higher consumption of hydronium ions during the heating at the higher temperature. The hydronium ion consumption in maleic acid catalysis created a limitation of the continuous degradation of the OPEFB substrate at high temperature and simultaneously recovered more pulp compared to sulfuric acid catalysis. This was a positive impact to avoid unwanted substrate degradation by the catalysis and prevent the valuable compound for the following treatment.

The appearance of brown compound dissolved in the aqueous fraction as the OPEFB degradation product was also observed. The absorbance value of 490 nm can be considered as the intensity of the brown compound of the hydrolysate [16]. The formation of brown color in the hydrolysate was caused by the polymerization of 5-HMF, furfural, and phenolics. Hence, the analysis of brown compounds can be used as a preliminary indication of the formation of inhibitor compounds, such as 5-HMF, furfural, and phenolics. As expected, pretreatment using sulfuric acid produced more quantity of brown compound during the heating at all temperatures applied compared to maleic acid. The absorbance of the former is 0.26–1.31, while the latter is 0.33–0.40. The result of absorbance of the hydrolysate from maleic acid pretreatment is in agreement with our previous results from pretreatment OPEFB using microwave-maleic acid (0.15–0.55) [17] and microwave-oxalic acid pretreatment (0.2–0.5) [9].

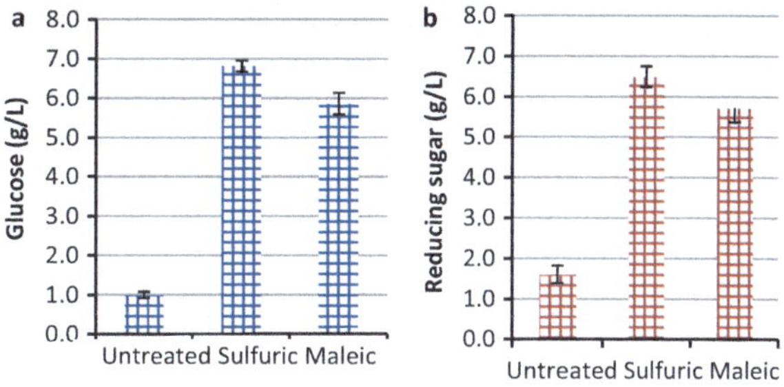

Fig. 2 The concentration of **a** glucose and **b** reducing sugar produced by enzymatic saccharification of untreated OPEFB and microwaved-assisted acid (sulfuric and maleic acid) pretreated OPEFB. Pretreatment was held at 180 °C for 10 min

3.2 Enzymatic Hydrolysis of Microwave-Assisted Acid Pretreated Pulp

The final goal of the microwave-assisted acid pretreatment was to provide better accessibility for the cellulase enzyme to reach and hydrolyze cellulose and finally produce valuable fermentable sugars. To evaluate the effectiveness of microwave-assisted acid pretreatment on OPEFB, the enzymatic hydrolysis of pretreated pulp was performed. Figure 2 shows the glucose and reducing sugar concentration after 5 days incubation of OPEFB pulp with Meicellase enzyme at 50 °C. It shows the importance of pretreatment in enhancing lignocellulosic material conversion into valuable fermentable sugars as the feedstock for the further bio-refinery process. Enzymatic hydrolysis of microwave-assisted acid pretreated pulp produced 5 times glucose concentration compared to the untreated OPEFB material.

The influence of heating temperature resulted in a different response to the reducing sugars and glucose yield by sulfuric and maleic acid catalysts (Table 2). Heating temperature from 160 to 180 °C in sulfuric acid pretreatment slightly increases the value of RSY from 36.1 to 36.7 and glucose yield from 36.4 to 38.5g/ 100g biomass corresponding to 79.6–84% cellulose conversion. At 200°C the RSY and glucose yield was dropped to only 22.4 and 19.6g/100g biomass, respectively, corresponding to 42% cellulose conversion. This situation was related to excessive degradation of OPEFB by the severe sulfuric acid that results in low pulp recovery. The excessive degradation was suspected to create an over burning of the cellulose content in the pulp. On the other hand, RSY and glucose yield obtained from the saccharification of the maleic acid pretreated pulp was increased gradually from 31.4, 38.7, and 40.4 g/100 g biomass and from 25.6, 39.6 to 40.3 g/100 g biomass, respectively, corresponding to 55.9, 86.5 and 88.0% cellulose conversion due to increase of temperature from 160, 180, to 200 °C. Different from sulfuric acid catalysis, pretreatment by maleic acid at 200 °C recovers pulp with a high cellulose conversion, indicating the successful prevention of excessive degradation of OPEFB.

Table 3 shows that prolong the heating duration until 15 min in microwave pretreatment with sulfuric acid catalyst reduces the reducing sugars and glucose yields, which corresponded with the decrease of cellulose conversion. On the contrary, the longer heating duration (15 min) in the pretreatment using maleic acid catalyst

Table 2 Reducing sugars, glucose yields, and cellulose conversion obtained after enzymatic hydrolysis of OPEFB, either untreated or pretreated by microwave and acids at the temperature of 160–180 °C

Pretreatment*		RSY (g/100 g pulp)	RSY (g/100 g biomass)	Glucose yield (g/100 g pulp)	Glucose yield (g/100 g biomass)	Cellulose conversion (%)
Untreated		–	16.6 ± 2.0	–	10.0 ± 0.7	21.8 ± 1.6
Sulfuric acid	160	59.9 ± 0.0	36.1 ± 0.7	60.4 ± 3.6	36.4 ± 1.5	79.6 ± 3.2
	180	65.2 ± 2.3	36.7 ± 0.6	68.4 ± 1.7	38.5 ± 1.7	84.0 ± 3.8
	200	59.6 ± 2.2	22.4 ± 2.8	51.6 ± 3.1	19.6 ± 4.3	42.7 ± 9.5
Maleic acid	160	42.4 ± 1.7	31.4 ± 1.8	34.5 ± 2.7	25.6 ± 2.4	55.9 ± 5.3
	180	58.6 ± 5.3	38.7 ± 2.8	60.0 ± 4.4	39.6 ± 2.2	86.5 ± 4.7
	200	71.7 ± 2.8	40.4 ± 3.3	71.5 ± 1.2	40.3 ± 2.4	88.0 ± 5.2

*Duration of heating is 10 min

increase reducing sugars and glucose yields, as well as cellulose conversion. These phenomena give more approval to the ability of maleic acid in preventing the excessive degradation of sugar components to lower molecular weight compounds, and at the same time provide better access to enzymes during the saccharification.

The maximum RSY and glucose yield obtained in this study is quite high (43.2 and 43.5 g/100 g biomass, respectively), reaching about 4 times higher than that obtained from untreated OPEFB with almost 95% cellulose converted to glucose. The RSY is higher than that obtained from microwave-oxalic acid pretreatment of OPEFB (34.60%) [9], autoclave-sulfuric acid pretreatment of OPEFB (30%) [14], microwave-alkaline pretreatment of OPEFB (17.8%) [18], and maleic acid pretreatment of sugarcane trash (38.21%) [19].

Table 3 Reducing sugars, glucose yields, and cellulose conversion obtained after enzymatic hydrolysis of OPEFB, either untreated or pretreated by microwave and acids for 5–15 min

Pretreatment*		RSY (g/100 g pulp)	RSY (g/100 g biomass)	Glucose yield (g/100 g pulp)	Glucose yield (g/100 g biomass)	Cellulose conversion (%)
Untreated		–	16.6 ± 2.0	–	10.0 ± 0.7	21.8 ± 1.6
Sulfuric acid	5	69.9 ± 2.1	39.4 ± 1.6	67.5 ± 5.4	38.1 ± 3.4	83.2 ± 7.4
	10	65.2 ± 2.3	36.7 ± 0.6	68.4 ± 1.7	38.5 ± 1.7	84.0 ± 3.8
	15	61.4 ± 2.8	32.3 ± 2.3	64.4 ± 4.8	33.8 ± 1.4	73.7 ± 3.0
Maleic acid	5	67.7 ± 1.2	43.2 ± 2.5	60.7 ± 2.4	38.7 ± 3.0	84.5 ± 6.6
	10	58.6 ± 5.3	38.7 ± 2.8	60.0 ± 4.4	39.6 ± 2.2	86.5 ± 4.7
	15	64.7 ± 2.8	39.7 ± 4.0	70.9 ± 0.8	43.5 ± 2.0	94.9 ± 4.4

*Temperature of heating at 180 °C

4 Conclusions

Pretreatment using microwave and acids has been proved could successfully increase the sugar yield from OPEFB. Pretreatment temperature dan time affects the formation of brown compounds in pretreatment hydrolysate as well as sugar yield and cellulose conversion from the saccharification of pretreated OPEFB. The highest RSY (39.4 g/ 100 g biomass) from pretreatment using sulfuric acid catalyst was obtained at 180 °C for 5 min, while that from pretreatment using a maleic acid catalyst (43.2 g/100 g biomass) was achieved at 180 °C for 5 min. It is suggested that maleic acid is a potential substitute of sulfuric acid for biomass pretreatment.

Acknowledgements The authors gratefully acknowledge JST-JICA for funding this research through Satreps Project Innovative Bio-production in Indonesia (iBioI): Integrated Bio-refinery Strategy to Promote Biomass Utilization using Super Microbes for Fuels and Chemicals Production (2013–2018). The authors also would like to acknowledge Research Center for Biotechnology, Indonesian Institute of Sciences, for facilitating glucose analysis using a microplate reader.

References

1. Kamal N (2014) Book Karakterisasi dan Potensi Pemanfaatan Limbah Sawit. ITENAS, Bandung
2. StatistaHomepage (2021). https://www.statista.com/statistics/706786/production-of-palm-oil-in-indonesia/. Last accessed July 09, 2021
3. Maurya DP, Singla A, Negi S (2015) An overview of key pretreatment processes for biological conversion of lignocellulosic biomass to bioethanol. 3 Biotech 5(5):597–609
4. Lee JW, Jeffries TW (2011) Efficiencies of acid catalysts in the hydrolysis of lignocellulosic biomass over a range of combined severity factors. Bioresour Technol 102(10):5884–5890
5. Shahriarinour M, Wahab MNA, Mustafa S, Mohamad R, Ariff AB (2011) Effect of various pretreatments of oil palm empty fruit bunch fibres for subsequent use as substrate on the performance of cellulase production by aspergillus terreus. BioResour 6(1):291–307
6. Solihat NN, Fajriutami T, Adi DTN, Fatriasari W, Hermiati E (2017) Reducing sugar production of sweet sorghum bagasse kraft pulp. In: International symposium on applied chemistry (ISAC) 2016: AIP Conference proceeding, pp 020012-1-8
7. Mikulski D, Kłosowski G (2020) Microwave-assisted dilute acid pretreatment in bioethanol production from wheat and rye stillages. Biomass Bioenerg 136:105528
8. Tsubaki S, Oono K, Onda A, Yanagisawa K, Azuma J (2013) Comparative decomposition kinetics of neutral monosaccharides by microwave and induction heating treatments. Carbohydr Res 375:1–4
9. Anita SH, Fitria, Solihat NN, Sari FP, Risanto L, Fatriasari W, Hermiati E (2020) Optimization of microwave-assisted oxalic acid pretreatment of oil palm empty fruit bunch for production of fermentable sugars. Waste Biomass Valoriz 11:2673–87
10. Li Z, Fei B, Jiang Z (2014) Comparison of dilute organic and sulfuric acid pretreatment for enzymatic hydrolysis of bamboo. BioResour 9(3):5652–5661
11. Kootstra AMJ, Beeftink HH, Scott EL, Sanders JPM (2009) Comparison of dilute mineral and organic acid pretreatment for enzymatic hydrolysis of wheat straw. Biochem Eng J 46(2):126–131

12. Hermiati E, Solihat NN, Adi DTN, Ghozali M (2018) The ultrastructure and enzyme susceptibility of oil palm empty fruit bunch pretreated by oxalic acid under microwave irradiation. In: 5th International symposium on innovative bio-production Indonesia. Bogor, pp 92–9
13. Solihat NN, Hermiati E, Adi DTN (2020) Microwave assisted dilute organic acid pre-treatment of oil palm empty fruit bunch to improve enzyme accessibility. IOP Conf Ser Mater Sci Eng 935
14. Risanto L, Fitria, Fajriutami T, Hermiati E (2018) Enzymatic saccharification of liquid hot water and dilute sulfuric acid pretreated oil palm empty fruit bunch and sugarcane bagasse. IOP Conf Ser Earth Environ Sci 141(012025):1–6
15. Solihat NN, Sari FP, Risanto L, Anita SH, Fitria F, Fatriasari W, Hermiati E (2017) Disruption of oil palm empty fruit bunches by microwave-assisted oxalic acid pretreatment. J Math Fund Sci 49(3):244–257
16. Warrand J, Janssen HG (2007) Controlled production of oligosaccharides from amylose by acid-hydrolysis under microwave treatment: comparison with conventional heating. Carbohyd Polym 69(2):353–362
17. Fatriasari W, Raniya R, Oktaviani M, Hermiati E (2018) The improvement of sugar and bioethanol production of oil palm empty fruit bunches (Elaeis guineensis Jacq) through microwave-assisted maleic acid pretreatment. Bioresour 13(2):4378–403
18. Nomanbhay SM, Hussain R, Palanisamy K (2013) Microwave-assisted alkaline pretreatment and microwave assisted enzymatic saccharification of oil palm empty fruit bunch fiber for enhanced fermentable sugar yield. J Sustain Bioenergy Syst 03(01):7–17
19. Hermiati E, Ermawar RA, Laksana RPB, Mardiah S (2021) Optimization of microwave-assisted maleic acid pretreatment of sugarcane trash by response surface methodology in Seminar Lignoselulosa 2020. Pusat Penelitian Biomaterial, Cibinong, pp 31–42

Conversion Oil Palm Empty Fruit Bunch into Reducing Sugar by Liquid Hot Water Assisted Sulfuric Acid Pretreatment

Fahriya Puspita Sari, Try Purwanti, Nissa Nurfajrin Solihat, and Widya Fatriasari

Abstract Indonesia is predicted to create 11.11 million metric tons of oil palm empty fruit bunches (OPEFB) each year as a waste. Because of the high cellulose content, this waste has the potential to be processed into a valuable product such as bioethanol. In this study, liquid hot water-assisted dilute sulphuric acid (LHW-SA) for 67.17 min at 180 °C was used as a pretreatment method to breakdown the matrix structure of OPEFB, and eventually cellulose could be converted into reducing sugar. Both filtrate and residue after pretreatment were investigated for chemical and physical properties. Analysis of liquid fraction showed that LHW-SA generates reducing sugar and a brown compound which may correlate with cellulose and hemicellulose degradation. The degradations were also shown in the chemical composition analysis of the residue as well as in functional group alteration of FTIR spectra. In the pretreated pulp, peak intensity corresponded to C–H bending of the methoxyl group at 1371 cm^{-1} was diminished. Besides, the crystallinity index analysed by XRD showed a decreasing amorphous component. LHW-SA also caused a rupture in the morphological structure of OPEFB that was attributed to increased enzyme accessibility. After enzymatic hydrolysis, as much as 40% reducing sugar yield was obtained. That proved the effectiveness of LHW-SA pretreatment to increase reducing sugar yield.

Keywords Brown compound · Physical–chemical changes · Enzymatic hydrolysis · Oil palm empty fruit bunches · Sugar production

F. P. Sari · N. N. Solihat (✉) · W. Fatriasari
Research Center for Biomass and Bioproducts, National Research and Innovation Agency (BRIN), Cibinong, Bogor, Indonesia
e-mail: nissa.nurfajrin.solihat@brin.go.id

T. Purwanti
Department of Forest Products Technology, Faculty of Forestry and Environment, IPB University (Bogor Agricultural University), Bogor, Indonesia

M. A. Rahandi Lubis (ed.), *Proceedings of the 1st International Conference of Lignocellulose*, Green Energy and Technology,
https://doi.org/10.1007/978-981-19-1596-3_25

1 Introduction

Indonesia, as the world's largest producer of oil palm, is predicted to generate 11.11 million metric tons of oil palm empty fruit bunches (OPEFB) as waste in the oil palm industry each year [1]. The number is obtained from a calculation where in 2020 the production volume of oil palm will reach 48.3 million metric tons [2]. The great quantities of OPEFB as a waste cause serious problems for the environment, and hence the utilization of OPEFB is pivotal. OPEFB has a great amount of cellulose, such as 43.75% cellulose and 26.22% hemicellulose, which is potentially converted into 2nd generation of bioethanol [3]. However, in the conversion process, the complicated linkage of lignin with cellulose in OPEFB poses a challenge. A suitable pretreatment is required to break the linkage without disrupting the cellulose structure.

One promising method is liquid hot water, also known as hydrothermal pretreatment. LHW has become a promising pretreatment technology due to its lack of need for other catalysts except water to disrupt the biomass cell wall matrix and hence can minimize degradation of xylose and be environmentally friendlier [4]. Furthermore, LHW fulfils the standard conditions for pretreatments, such as opening the structure of biomass to release high cellulose content, susceptibility to enzymatic hydrolysis, and low risk of corrosion in equipment [5]. Among chemical components in biomass, hemicellulose is the easiest to deconstruct and solubilized compound in biomass at high temperatures (200 °C) [6]. Risanto et al. compared non-and with a present acid catalyst in LHW pretreatment of OPEFB. The result showed that the presence of dilute acid, organic acid (oxalic acid) catalyst obtained higher reducing sugar than non-catalyst [7]. However, to date, the correlation between obtained reducing sugar and physical disruption of OPEFB by LHW assisted sulphuric acid (LHW-SA) pretreatment has not been discussed yet. The parallel study was also conducted to observe the effect of organic acids such as maleic acid in the optimum condition as a catalyst during LHW pretreatment on the structural changes of OPEFB.

Observing transformation of chemical and physical structure biomass after pretreatment is a critical study to evaluate the effectiveness of pretreatment [8]. Therefore, this present study was conducted to evaluate obtained reducing sugar and its correlation with physical and chemical alteration of LHW-SA pretreatment of OPEFB. The used acid concentration, temperature, and time of pretreatment were based on optimization of LHW-acid pretreatment in a previous study [9].

2 Materials and Methods

2.1 Material

In this work, the OPEFB was collected from Cibadak, Sukabumi, West Java, Indonesia in 2018. The OPEFB fiber was cut, grounded, and meshed to obtain powder with a powder size of 40–60 mesh (moisture content of 6.94%). The samples were then packed and stored in a container. All of the chemicals were of analytical grade and were not purified before use. Celluclast, a cellulase enzyme from Trichoderma reesei with a 42.3 FPU/mL initial activity, and then diluted to 2 FPU/mL with citrate buffer, was employed in the enzymatic hydrolysis of untreated and pretreated OPEFB.

2.2 Methods

Liquid Hot Water-Assisted Sulfuric Acid (LHW-SA) pretreatment of OPEFB

Five grams samples (oven-dry weight) were placed into a Teflon vessel of a hydrothermal reactor (Heating stirrer Model HHE/HM-19G-U, San-Ai Kagaku Co. Ltd., Nagoya, Japan) with a solid to liquid (S/L) ratio of 1:14 for 67.17 min at 180 °C with a sulfuric acid concentration of 1.5%. The Teflon vessel was cooled in an ice bath after the pretreatment process, and to separate the liquid and solid fractions the slurry was vacuum filtered (GAST DOA-P504-BN, USA). By using blue litmus paper, the solid fraction (pulp) was washed in distilled water until reached pH of 5–7. To calculate weight loss (Eq. 1) during pretreatment, the moisture content and the pulp weight were determined. Meanwhile, the pulp recovery was calculated by subtracting a 100%—weight loss. The pulp was collected in a sealed plastic and frozen to be used as a substrate in enzymatic hydrolysis.

$$\text{Weight loss } (\%) = \frac{W_a - W_b}{W_a} \times 100 \tag{1}$$

where w_a is pulp initial weight (g), w_b is final weight pulp (g)

Analysis of liquid Fraction

The liquid fraction was refrigerated before further analysis due to the determination of secondary products in the hydrolysate. The pH value was identified using a pH meter (Eutech, Eutec Pte Ltd, Ayer Rajah Crescent, Singapore). The brown compounds were determined using a spectrophotometer (Hitachi U-2001, Hitachi Inc., Tokyo, Japan) at a wave number of 490 nm [10], and the reducing sugar concentration was determined using the DNS method [11]. The observation was done in triplicate.

Enzymatic hydrolysis and analysis of reducing sugars

The OPEFB pulp was then hydrolysed enzymatically to observe the reducing sugar concentration. After placing 0.3 g oven dried weight (ODW) of pulp in a 50 mL centrifuge tube, the following components were added: 0.05 M citrate buffer pH 4.8 as much as 5 mL, 300 μl sodium azide 2% w/v, and 4.5 mL of cellulase enzyme, which was equivalent to 30 FPU/g enzyme loading. To take the final weight to 30 g, the citrate buffer was poured. The samples were subsequently incubated for 72 h at 50 °C and 150 rpm in a shaking incubator (Wise Cube, witeq Labortechnik GmbH, Wertheim, Germany). The mixture was boiled for 5 min in order to inactivate the enzyme, then the residues were separated from the supernatant by centrifugation. The reducing sugar yield (RSY) was calculated using the DNS technique, which uses a spectrophotometer at a wave number of 540 nm to determine the concentration of reducing sugar in the supernatant. The initial weight (g) of biomass before pretreatment was used to calculate the RSY [10].

Characterization pretreated OPEFB

The chemical component analysis

The chemical components of both before and after OPEFB were studied including moisture content [11], ash content [12], ethanol–benzene extractive content [13], acid insoluble lignin (AIL) and acid soluble lignin (ASL) [14], holocellulose content [15], α-cellulose content [16] and hemicellulose content [17]. The component loss of pretreatment was determined by considering the weight loss of biomass during pretreatment with the equation (Eq. 2) Anita et al. [3] and Hermiati et al. [18].

$$\text{Component loss} = \left(\frac{C_0 - \left(\frac{C_1}{100}\right) \times PR}{C0} \right) \times 100 \tag{2}$$

where C_0 represents the chemical component of lignin, holocellulose and hemicellulose in untreated OPEFB (%); C_1 represents the chemical component of holocellulose, cellulose, lignin or hemicellulose in pretreated OPEFB (%); and PR stands for pre-treated pulp recovery (%)

FTIR analysis

Investigating functional group of OPEFB before and after LHW-SA pretreatment was done by An Attenuated Total Reflectance-Fourier Transform Infrared (ATR)-FTIR with a spectrum two Perkin Elmer, Shelton, USA. Approximately 1 mg untreated and pretreated OPEFB was placed on a plate and then manually pressed. The IR spectra were captured by 16 scanning number and resolution 4.0 cm^{-1} within the range of wave number 4000–600 cm^{-1} by using the software of Perkin Elmer, USA. Following that, the changes of relative intensity were then analyzed by identifying the functional groups.

X-ray Diffraction (XRD) analysis

The Shimadzu XRD-700 Maxima X series (Shimadzu Corp., Kyoto, Japan) was used to characterize the crystallinity degree (Xc) of untreated and pretreated OPEFB. The samples were scanned and recorded with a 2θ angle of 10–40° every 2° per min through Cu Kα radiation ($\lambda = 0.15406$ nm) at a voltage of 40 kV and 30 mA using the Xc calculation (Eq. 3) was used the deconvolution method [19] as follow.

$$\text{Xc}\ (\%) = \frac{\text{Icr}}{\text{Ia} + \text{Icr}} \times 100 \tag{3}$$

where Icr donates the of crystalline area while Ia donates the non-crystalline area

Scanning Electron Microscopy (SEM) analysis

The morphology of samples was examined with scanning electron microscopy (JEOL JSM-6510LA, Tokyo, Japan). Samples were located on a cylinder stub that had been covered with double tape previously. The scans were carried out at a voltage of 15–20 kV with a bar size of 10 μm to capture the SEM images at 2500 × magnification.

3 Results and Discussion

3.1 Effect of LHW-SA Pretreatment on Chemical Composition of OPEFB

The pulp recovery of 38.26% was obtained after LHW-SA pretreatment. The pulp recovery achieved was slightly lower than that obtained in a prior work by Risanto et al., where pulp recovery was around 60 g/100 g biomass [7]. It may be due to different temperatures and times of pretreatment conditions. The chemical composition of OPEFB as a result of pretreatment is shown in Table 1. After pretreatment, some chemicals especially cellulose components reduced yet lignin content increased. Hemicellulose is a type of amorphous structure in biomass that is easy to hydrolyse by acid, component loss reached 71.22%. This finding is similar to the previous report where hemicellulose in treated OPEFB by microwave assisted-organic acid pretreatment was lost about 60–70% [8, 20]. However, pretreatment by organic acid preserved more hemicellulose compared to sulphuric acid due to the different strength of the acid. The same finding was also observed by comparing organic vs sulphuric acid pretreatment of bamboo, where loss of hemicellulose was higher after sulphuric acid pretreatment. However, the increase in lignin content was same the after both pretreatment [21]. In this study, AIL content increased about three times after pretreatment. That may indicate exposure of lignin after degrading amorphous cellulose content [22]. In addition, this phenomenon showed that sulphuric acid had no appreciable effect on the delignification process [21]. The trend of gaining value after pretreatment was also observed in extractive content due to re-precipitation of low molecular weight lignin fragments and polysaccharides in the pretreated fiber [23].

3.2 Liquid Fraction After LHW-SA Pretreatment

The hydrolysate was investigated, including its pH, brown compound (absorbance at 490 nm), and reducing sugar content, with the data available in Table 2. The brown compound of the hydrolysate was produced as a result of polymerization of hydroxyl methyl furfural (5-HMF) and oxidation of the phenolic structure. Some

Table 1 Chemical component of OPEFB before and after LHW-SA pretreatment

Chemical composition	Untreated OPEFB (%)	Pretreated OPEFB (%)	Component loss (%)
Moisture content	6.94 ± 0.19	3.46 ± 0.95	–
Ash content	4.26 ± 0.09	1.67 ± 0,67	–
Extractive in EtOH-benzene	1.91 ± 0.11	18.47 ± 0.87	–
Lignin	19.10 ± 0.00	61.79 ± 0.00	–
AIL	17.80 ± 0.79	60.67 ± 0.30	–
ASL	1.30 ± 0.06	1.12 ± 0.14	13.69
Holocellulose	69.74 ± 0,00	30.23 ± 1.25	83.42
α-cellulose	43.02 ± 0.23	10.13 ± 0.97	90.99
Hemicellulose	26.72 ± 0.39	20.10 ± 2.21	71.22

reports analysed brown compounds as preliminary research to investigate the effectivity of pretreatment in correlation with generating HMF and phenolic compounds [3]. The pH of the liquid fraction, 0.66, was slightly lower than in a recent work by Fatriasari et al. who employed sulphuric acid 1–3% with assisted microwave irradiation to breakdown the lignocellulose structure of OPEFB [24]. It may be due to different methods of heating processes and reaction time of pretreatment. Furthermore, the high absorbance of the hydrolysate (>1) indicated dissolving hemicellulose into furfural and decomposition of lignin [3]. However, in the future, this hypothesis should be confirmed by analysing the concentration of secondary degradation products in the hydrolysate. After pretreatment, glucose and xylose may be present in the liquid fraction as a result of hydrothermal degradation of cellulose and hemicellulose. Martín-Lara et al. employed LHW to improve the reducing sugar content of pepper waste and assessed its efficiency by analysing sugar content in the liquid and solid fraction of pretreated OPEFB. The results showed that increasing time and temperature of LHW pretreatment typically increased the recovery of glucose and xylose in the hydrolysate [25]. In this study, reducing sugar content of 0.13 g/mL was measured in the hydrolysate, which may be derived from cellulose and hemicellulose [26].

Table 2 pH, absorbance, and RSY concentration of hydrolysate after LHW-SA pretreatment

Parameter	Liquid fraction after pretreatment
pH	0.66 ± 0.01
Absorbance	1.01 ± 0.02
Reducing sugar content (g/mL)	0.13 ± 0.00

3.3 *Effect of Enzymatic Hydrolysis Solid Fraction on Reducing Sugar Yield*

As presented in Fig. 1, RSY per pulp of pretreated OPEFB after enzymatic hydrolysis increased by about 25% from untreated OPEFB (15.61%). This result proved that LHW-SA pretreatment has a positive effect on increasing enzyme accessibility and thus gained RSY. LHW increased physicochemical changes in biomass, such as reducing recalcitrance through disrupting the cell wall matrix of biomass with lower production of inhibitor compounds. Li et al. investigated the chemical structure change of poplar after LHW pretreatment, where after pretreatment the xylan solubilized about 70%, the hemicellulose of poplar reduced by 75% and the cellulose crystallinity increased by 9%. The removal of the acetyl group resulted in a slight change in lignin linkage, probably −O−4 ether bonds [4]. These alterations are important in the success of pretreatment for RSY improvement. Temperature, duration, solvent ratio, particle size, and pH of the solution are all critical factors in hydrothermal pretreatment [27, 28]. In this study, a temperature of 180 °C for 67.17 min was used for LHW-SA pretreatment according to optimization results in a parallel study. However, some reports have similar temperatures to gain higher RSY. For instance, Perez et al. optimized temperature and time for wheat straw LHW pretreatment and the optimal condition to achieve the highest hemicellulose derived sugar was at 188 °C for 40 min [29]. Martin-Lara et al. also reported that LHW pretreatment at 180 °C for 40 min obtained a maximum glucose yield of 61% with low phenolic compounds, which generally act as inhibitors. In addition, Muhd Ali et al. observed that maximum soluble sugar (68.18 mg glucose/g OPEFB) was obtained after catalyst-free hydrothermal pretreatment at 175 °C with a reaction time of 20 min [28]. The resulted RSY in the current investigation, 40%, was lower compared to our earlier study that employed organic acid for OPEFB LHW pretreatment [30]. It may be due to the different catalytic rate and pH of pretreatment [4] which are attributed to the severity factor. Lee and Jefferies found that pretreatment of oxalic acid resulted in higher glucose and xylose content compared to sulphuric acid at low severity factor because dicarboxylic acid exhibited higher hydrolytic efficiency than sulphuric acid [5].

3.4 *Effect of LHW-SA on Functional Group of OPEFB*

According to the chemical composition result, LHW-SA pretreatment removed hemicellulose and changed the chemical composition of OPEFB. This finding is proven by the identification of functional group changes in OPEFB before and after pretreatment. The FTIR spectra is shown in Fig. 2. Despite the fact there is no new functional group after LHW-SA pretreatment, it did change the band intensity and shift the band. This similar result was obtained after OPEFB microwave assisted oxalic acid pretreatment [8, 22]. The O–H stretching was ascribed to the absorption bands

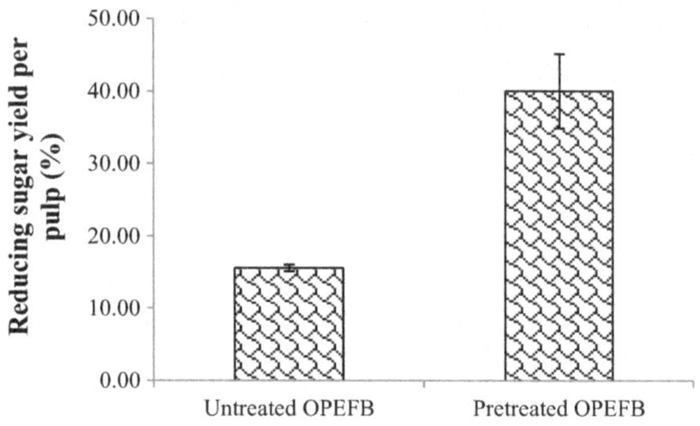

Fig. 1 Obtained reducing sugar yield per pulp of OPEFB before [9] and after LHW-SA pretreatment

between 3450 cm^{-1} and 3400 cm^{-1}. Primary and secondary alcohols, absorbed water, aliphatic and other O–H groups can be found in lignocellulosic materials (in cellulose, lignin, hemicellulose, and extractives). O–H groups in lignocellulosic material include absorbed water, primary and secondary aliphatic [25]. In the pretreated OPEFB, the peak intensity of hydroxyl group (no 1) was slightly decreased. That was confirmed as a result of chemical composition where cellulose loss increased after pretreatment. The stretching vibrations of alkyl groups (no 2–3) and C–O out of plane bending (no 19) in cellulose seemed to be stronger than in untreated OPEFB, which could be attributed to lignin and carbohydrate bond breaking [8]. In this study, sharper peaks attributed to lignin were observed at 1641 (4), 1595 (5), 1509 (6), 1460 (7), 1269 (11), 1103 (13) and 1032 (14) cm^{-1}, which correspond to stretching of C–O, C = C, C–C in the aromatic ring, aromatic ring vibration and C–O, guaiacyl ring, stretching of C–O in secondary alcohol, and C–O in primary alcohol. This observation is consistent with the chemical component result of increased lignin content following pretreatment. This phenomenon occurred as a consequence of exposing lignin linkage after cellulose degradation [31]. Meanwhile, peaks at 1371 cm^{-1} that corresponded to C–H bending of the methoxyl group (no 9) disappeared as hemicellulose and cellulose were degraded by LHW-SA pretreatment [25]. Diminished peak intensity at 897 cm^{-1}, which represents as -glycosidic linkage means hydrolysing hemicellulose content by acid pretreatment [8]. According to these findings, LHW-SA pretreatment was more severe on cellulose degradation than on lignin. A similar result was also reported for hardwood and softwood hydrothermal pretreatment [6] and sugarcane bagasse after LHW pretreatment [32].

3.5 SEM Images of Untreated and Pretreated OPEFB

In this SEM investigation, the contribution of LHW-SA pretreatment to the morphological and physical characteristics of surface OPEFB fibers was evaluated. The comparisons of SEM images before and after pretreatment are shown in Fig. 3. There are significant changes to the OPEFB structure after LHW-SA pretreatment.

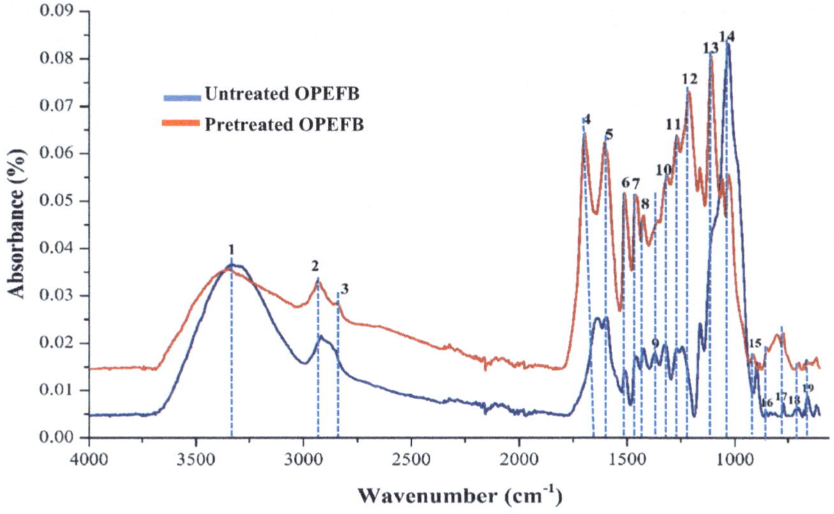

Fig. 2 FTIR spectra of untreated OPEFB [9] and pretreated OPEFB by LHW-SA

The silica bodies (red arrow) were attached to the surface of OPEFB fiber in circular craters and the surface of the OPEFB fiber was dense, smooth, without any cracks (Fig. 3a). Yunus et al. reported that the presence of silica is extra protection and resistance after lignin on biomass [33]. After treatment, silica bodies were eroded but still embedded in the fiber. This result is similar to our previous study that used microwave-oxalic acid pretreatment. The silica bodies remained intact with circular craters at low temperatures. Pretreatment at 190 °C and 200 °C can remove the silica bodies from the OPEFB fiber strand [8]. Figure 3b also revealed that LHW-SA pretreatment caused cracks and fractures (green arrow) in the surface of the OPEFB fiber. The cracks and fractures on the surface of OPEFB fiber indicate the interruption of the cell wall in the fiber structure [34]. This disruption can render enzymes more accessible to penetrate, leading to a higher RSY during the enzymatic hydrolysis process [35]. The changes in morphological structure can also be inferred because of the loss of lignin and hemicellulose caused by pretreatment.

3.6 Crystallinity Degree by XRD

XRD analysis was completed to investigate the crystallinity index of untreated and pretreated OPEFB using LHW-SA pretreatment. The chemical composition of biomass influences its crystallinity degree. Lignin and hemicellulose are amorphous fractions, while cellulose has crystalline and amorphous regions. Two strong peaks at 15°–17° and 22°–23° are visible in the diffraction spectra (Fig. 4). The crystallinity degree of the OPEFB is enhanced by these peaks. The spectra present that the peak

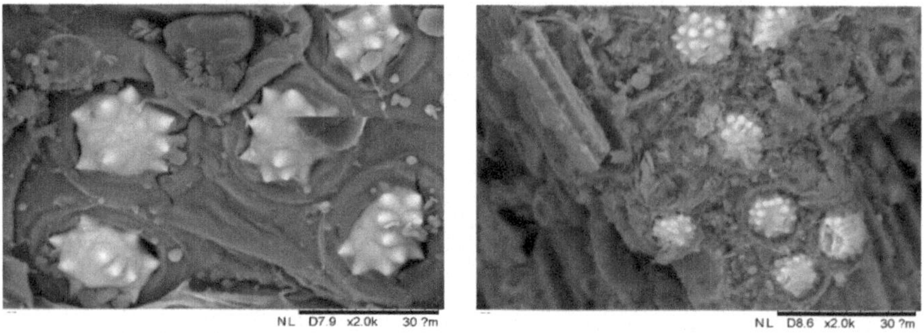

(a) (b)

Fig. 3 Morphological image of untreated **a** and pretreated **b** OPEFB by LHW-SA

in untreated OPEFB at 2 of 15°–17° is only shifted and a small peak appears in the preateated OPEFB. Untreated OPEFB has a higher peak at 2 of 22°–23° than pretreated OPEFB. After LHW-SA pretreatment, the crystalline area reduced, but the amorphous region increased (Table 3). Table 3 further reveals that untreated OPEFB has a higher crystallinity degree (23.03%) than treated OPEFB (17.76%). After pretreatment, the Xc decreases due to crystalline lattice disruption [36]. The crystalline structure of cellulose can also be turned into an amorphous structure partially by physical processes (ball milling, heating) and chemical processes (acid, alkaline pretreatment) [37, 38]. The crystallinity of cellulose plays a significant role in enzyme accessibility. There are two categories of accessibility in biomass: micro-accessibility and macro-accessibility. The amorphous region of cellulose represents the highest degree of micro-and macro-accessibility that enzymes are capable of attaining all the cellulose chains. The accessibility of enzymes to their specific substrate and the effectiveness of enzymes when adsorbed on the surface are two main factors that control enzymes' ability to hydrolyze cellulose [39]. Several features in biomass also affect enzymatic hydrolysis, such as particle size, porosity, pore-volume, and surface area [37, 39, 40].

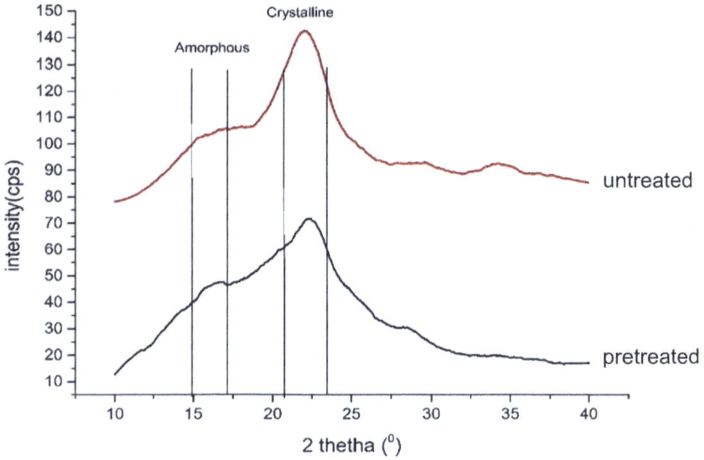

Fig. 4 XRD patterns before and after LHW-SA pretreament of OPEFB

Table 3 The crystallinity index of before and after LHW-SA pretreatment of OPEFB

Sample name	Icr (kcps.deg)	Ia (kcps.deg)	Xc (%)
Untreated [9]	0.2348	0.7850	23.03
Pretreated OPEFB	0.1853	0.8583	17.76

4 Conclusion

LHW-SA pretreatment caused alteration of chemical and physical properties of OPEFB where it was more severe on hemicellulose degradation than that of lignin. After pretreatment, the FTIR spectra show diminished peak intensity of C–H deformation in hemicellulose and cellulose at 1371 cm^{-1}, yet peaks corresponding to lignin at a range of 1450–850 cm^{-1} are more pronounced. Besides, the crystallinity degree in XRD also shows degradation of the amorphous component in pretreated OPEFB. This pretreatment successfully disrupts the surface structure of OPEFB, which opens enzyme accessibility to break down cellulose and reduces sugar. Pretreated OPEFB releases up to 40% more RSY per pulp after the enzymatic hydrolysis process, whereas untreated OPEFB only obtains 15.61%. However, further conversion of reduced sugar into bioethanol is needed to be investigated in the next study.

Acknowledgements This research was supported by DIPA 2020 funding of Research Center for Biomaterials, Indonesian Institute of Sciences (LIPI). The authors also acknowledge the facilities, scientific, and technical assistance provided by the Advanced Characterization Laboratories Cibinong—Integrated Laboratory of Bioproducts, LIPI, via E-Layanan Sains Lembaga Ilmu Pengetahuan Indonesia. Author also would like to thank to JSPS Alumni Association of Indonesia (JAAI) for publication fee.

References

1. Kamal N (2014) Karakterisasi dan Potensi Pemanfaatan Limbah Sawit. ITENAS, Bandung
2. Statista Homepage, https://www.statista.com/statistics/706786/production-of-palm-oil-in-indonesia/. Last accessed 2021/08/30
3. Anita SH, Fitria Solihat NN, Sari FP, Risanto L, Fatriasari W, Hermiati E (2020) Optimization of microwave-assisted oxalic acid pretreatment of oil palm empty fruit bunch for production of fermentable sugars. Waste Biomass Valorization 11:2673–2687
4. Li M, Cao S, Meng X, Studer M, Wyman CE, Ragauskas AJ, Pu Y (2017) The effect of liquid hot water pretreatment on the chemical-structural alteration and the reduced recalcitrance in poplar. Biotechnol Biofuels 10:237
5. Lee JW, Jeffries TW (2011) Efficiencies of acid catalysts in the hydrolysis of lignocellulosic biomass over a range of combined severity factors. Bioresour Technol 102:5884–90
6. Nitsos CK, Choli-Papadopoulou T, Matis KA, Triantafyllidis KS (2016) Optimization of hydrothermal pretreatment of hardwood and softwood lignocellulosic residues for selective hemicellulose recovery and improved cellulose enzymatic hydrolysis. ACS Sustain Chem Eng 4:4529–44
7. Risanto L, Fitria Fajriutami T, Hermiati E (2018) Enzymatic saccharification of liquid hot water and dilute sulfuric acid pretreated oil palm empty fruit bunch and sugarcane bagasse. In: ICBBogor 2017, IOP Conference series: earth and environmental science 141 pp 1–6
8. Solihat NN, Sari FP, Risanto L, Anita SH, Fitria Fatriasari W, Hermiati E (2017) Disruption of oil palm empty fruit bunches by microwave-assisted oxalic acid pretreatment. J Math Fund Sci 49:244–57
9. Fatriasari W, Ulwan W, Aminingsih T, Sari FP, Fitria Suryanegara L, Iswanto AH, Ghozali M, Kholida LN, Hussin MH, Fudholi A, Hermiati E (2021) Optimization of maleic acid pretreatment of oil palm empty fruit bunches (OPEFB) using response surface methodology to produce reducing sugars. Ind Crops Prod
10. Fatriasari W, Syafii W, Wistara NJ, Syamsu K, Prasetya B (2014) Digestibility of Betung bamboo fiber following fungal pretreatment. Makara J Technol 18
11. TAPPI (1997) Test Method T 264 cm-97 Preparation of wood for chemical analysis
12. TAPPI (2002) Test method T 211 om-02 Ash in wood, pulp, paper and paperboard: combustion at 525 °C
13. TAPPI (1997) Test Method T 204 cm-97 Solvent extractives of wood and pulp
14. Sluiter A, Hames B, Ruiz R, Scarlata C, Sluiter J, Templeton D, Crocker D (2012) Determination of structural carbohydrates and lignin in biomass—laboratory analytical procedure (LAP). In: National renewable energy laboratory technical report NREL/TP-510-42618
15. Wise LE, Murphy M (1946) Addieco: chlorite holocellulose, its fractionation and bearing on summative wood analysis and on studies on the hemicelluloses. Pap Trade J 122:35–43
16. Rowell RM, Pettersen R, Han JS, Rowell JS, Tshabalala MA (2005) Handbook wood chemistry and wood composites, ed R M Rowell. CRC Press, Boca Raton
17. Punyamurthy R, Sampathkumar D, Bennehalli B, Srinivasa CV (2013) Influence of esterification on the water absorption property of single abaca fiber. Chem Sci Trans 2:413–22
18. Hermiati E, Laksana RPB, Fatriasari W, Kholida LN, Thontowi A, Yopi Arnieyanto DR, Champreda V, Watanabe T (2020) Microwave-assisted acid pretreatment for enhancing enzymatic saccharification of sugarcane trash. Biomass Convers Biorefinery
19. Wibowo ES, Park BD (2020) Determination of crystallinity of thermosetting urea-formaldehyde resins using deconvolution method. Macromol Res 28:615–24
20. Fatriasari W, Raniya R, Oktaviani M, Hermiati E (2018) The improvement of sugar and bioethanol production of oil palm empty fruit bunches (Elaeis guineensis Jacq) through microwave-assisted maleic acid pretreatment. bioResour 13, 4378–403
21. Li Z, Fei B, Jiang Z (2014) Comparison of dilute organic and sulfuric acid pretreatment for enzymatic hydrolysis of bamboo. BioResour 9:5652–61

22. Solihat NN, Hermiati E, Adi DTN (2020) Microwave assisted dilute organic acid pre-treatment of oil palm empty fruit bunch to improve enzyme accessibility. In: International conference of forest product, IOP conference series: materials science and engineering 935
23. Carvalho DM, Sevastyanova O, Penna LS, Silva BP, Lindström ME, Colodette JL (2015) Assessment of chemical transformations in eucalyptus, sugarcane bagasse and straw during hydrothermal, dilute acid, and alkaline pretreatments. Ind Crops Prod 73:118–26
24. Fatriasari W, Anita SH, Risanto L (2016) Microwave assisted acid pretreatment of oil palm empty fruit bunches (EFB) to enhance its fermentable sugar production. Waste Biomass Valorization 8:379–91
25. Martin-Lara MA, Chica-Redecillas L, Perez A, Blazquez G, Garcia-Garcia G, Calero M (2020) Liquid hot water pretreatment and enzymatic hydrolysis as a valorization route of Italian green pepper waste to delivery free sugars. Foods 9(11):1–19
26. Eisenhuber K, Krennhuber K, Steinmüller V, Jäger A (2013) Comparison of different pretreatment methods for separating hemicellulose from straw during lignocellulose bioethanol production. Energy Proced 40:172–81
27. Grénman H, Eränen K, Krogell J, Willför S, Salmi T, Murzin DY (2011) Kinetics of aqueous extraction of hemicelluloses from spruce in an intensified reactor system. Ind Eng Chem Res 50:3818–28
28. Muhd Ali MD, Tamunaidu P, Nor Aslan AKH, Morad NA, Sugiura N, Goto M, Zhang Z (2016) Hydrothermal pre-treatment of oil palm empty fruit bunch into fermentable sugars In: International conference on chemical engineering and bioprocess engineering, IOP Conference Series: Earth and Environmental Science 36 pp1–7
29. Pérez JA, Ballesteros I, Ballesteros M, Sáez F, Negro MJ, Manzanares P (2008) Optimizing Liquid Hot Water pretreatment conditions to enhance sugar recovery from wheat straw for fuel-ethanol production. Fuel 87:3640–7
30. Sari FP, Solihat NN, Anita SH, Fitria Hermiati E (2016) Peningkatan produksi gula pereduksi dari tandan kosong kelapa sawit dengan praperlakuan asam organik pada reaktor bertekanan. Reaktor 16:199–206
31. Li X, Li M, Pu Y, Ragauskas AJ, Klett AS, Thies M, Zheng Y (2018) Inhibitory effects of lignin on enzymatic hydrolysis: the role of lignin chemistry and molecular weight. Renew Energy 123:664–74
32. Yu Q, Zhuang X, Lv S, He M, Zhang Y, Yuan Z, Qi W, Wang Q, Wang W, Tan X (2013) Liquid hot water pretreatment of sugarcane bagasse and its comparison with chemical pretreatment methods for the sugar recovery and structural changes. Bioresour Technol 129:592–8
33. Yunus R, Salleh SF, Abdullah N, Biak DR (2010) Effect of ultrasonic pre-treatment on low temperature acid hydrolysis of oil palm empty fruit bunch. Bioresour Technol 101:9792–6
34. Raina N, Slathia PS, Sharma P (2020) Response surface methodology (RSM) for optimization of thermochemical pretreatment method and enzymatic hydrolysis of deodar sawdust (DS) for bioethanol production using separate hydrolysis and co-fermentation (SHCF). Biomass Convers Biorefinery
35. Sari FP, Falah F, Anita SH, Ramadhan KP, Laksana RPB, Fatriasari W, Hermiati E (2021) Pretreatment of oil palm empty fruit bunch (OPEFB) at bench-scale high temperature-pressure steam reactor for enhancement of enzymatic saccharification. Int J Renew Energy Dev 10:157–69
36. Hall M, Bansal P, Lee JH, Realff MJ, Bommarius AS (2010) Cellulose crystallinity–a key predictor of the enzymatic hydrolysis rate. FEBS J 277:1571–82
37. Karimi K, Taherzadeh MJ (2016) A critical review of analytical methods in pretreatment of lignocelluloses: composition, imaging, and crystallinity. Biores Technol 200:1008–18
38. Xiao LP, Sun ZJ, Shi ZJ, Xu F, Sun RC (2011) Impact of hot compressed water pretreatment on the structural changes of woody biomass for bioethanol production. BioResour 6:1576–98

39. Kumar R, Wyman CE (2013) Aqueous pretreatment of plant biomass for biological and chemical conversion to fuels and chemicals. (ed) Wyman CE. John Wiley & Sons, Ltd. pp 281–310
40. Chandra RP, Esteghlalian AR, Saddler JN (2009) Characterization of lignocellulosic materials. (ed) Hu TQ. Blackwell Publishing Ltd, New Jersey pp 60–80

Screening of Pectinolytic Activity from Bacillus Strains Isolated from Indonesian Fermented Cacao Beans

Fahrurrozi, Miranti Nurindah Sari, Saesar Pambudi Utomo, Urip Perwitasari, Ario Betha Juanssilfero, and Puspita Lisdiyanti

Abstract Before being used as raw material for chocolate production, cacao beans subjected to a fermentation process. The most important part of the fermentation process is microbes and enzymes that play a role in the fermentation process in this case are pectinase enzymes. The presence of pectinase enzyme is very important in the cocoa bean fermentation process, due to pectinase plays a role in the degradation of pectin contained in the pulp, which generate faster fermentation process. In this study, out of 137 isolates (yeast and bacteria) were screened as pectinolytic producer and found 4 isolates as a pectinolytic producer from bacteria, namely IDI-001, IDI-002, IDI-004 and IDI-008. However, none of yeast to produce pectinolytic. Isolate IDI-001 has the highest clear zone and pectinolytic activity with 1.06 mm and 1.80 U mL^{-1}, respectively, followed by IDI-002, IDI-004, and IDI-008 with clear zone diameter and enzyme activity 1.00 nm, 0.84 nm, and 0.81 mm and 0.95 U mL^{-1}, 0.62 U $^{-1}$ mL, and 0.43 U mL^{-1}, respectively. Based on 16S rRNA identification, isolate IDI-001 has 100% similarity to *Bacillus subtilis*. Characterization of pectinase activity has shown that *Bacillus subtilis* IDI-001 produced pectinase enzyme with higher activity 1.80 U mL^{-1} at 48 h and pH 7.0. Further investigations are needed to characterize the pectinase enzyme and its ability to cacao beans fermentation process.

Keywords *Bacillus* · Cacao bean · Fermentation · Pectinase

Fahrurrozi (✉) · M. N. Sari · U. Perwitasari
Research Center for Marine and Land Bioindustry, National Research and Innovation Agency, Cibinong, West Java, Indonesia
e-mail: fahrurrozi@brin.go.id

S. P. Utomo
Department of Food Technology, Faculty of Life Sciences, Surya University, Pinang Kota Tangerang, Indonesia

A. B. Juanssilfero
Research Center for Environmental and Clean Technology, National Research and Innovation Agency, Cibinong, Indonesia

P. Lisdiyanti
Research of Centre for Biosystematics and Evolution, National Research and Innovation Agency, Cibinong, Indonesia

M. A. Rahandi Lubis (ed.), *Proceedings of the 1st International Conference of Lignocellulose*, Green Energy and Technology,
https://doi.org/10.1007/978-981-19-1596-3_26

1 Introduction

Indonesia as one of biggest cacao production in the world is still face with low quality of cacao bean. It is because almost 90% of cacao bean in Indonesia is unfermented bean. Fermentation of cacao bean is one of important step to produced premium cacao bean. Fermentation of cacao bean causing the death of bean embryo, and starting the enzymatic process inside the bean causing biochemical reaction and changes the color and forming precursor chocolate aroma [1, 2].

The key players during cacao bean fermentation are microbes and enzymes activate during cacao bean fermentation. During the fermentation microbes turnout compounds that are acidic and alcoholic which can cause necrobiosis if the quantity is simply too a lot of. Enzymes such as pectinase are important player during cocoa bean fermentation. These enzymes degrade pectin in pulp, causing liquefaction of pulp. The death of seed embryo causing by fermentation activated indigenous enzymes to catalyze biotransformation reactions to express the potential flavor of cacao beans [3].

Cocoa pulp contains of pectin (1–1.5%), citric acid (1–3%), sugar (10–15%), the rest is water as much as 82–87%. The sugar content in cocoa pulp consists of sucrose (60%) and a mixture of glucose-fructose (39%). A small amount of proteins, amino acid, vitamin and mineral are essential require for microbial growth are also present [2, 4, 5]. Cocoa pulp is sticky, lumpy, and cohesive derived from pectin. Pectin and pectic acid is the natural substrates of pectic enzymes [6, 7]. Two main groups of pectinase enzymes, (1) pectinemethyesterases (PME), this enzyme remove the methoxyl groups from the esterified acid, and (2) polygalacturonases (PG) and pectin lyases (PL) that working on β-(1,4)-glycosidic bond or by trans-elimination [2]. Several yeast species has been reported to produced pectinolytic enzymes e.g. *Candida, Picia, Saccharomyces, Zygosaccharomyces, Torulopsis* [8–12]. Furthermore bacteria strain *Bacillus* sp. e.g. *Bacillus cereus, Bacillus thuringensis, Bacillus subtilis, Bacillus fusiformis, Bacillus sphaericus, Bacillus pumilus* was reported have pectinolytic activities [13, 14]. Aimed of the study was characterization the pectinolytic enzymes and molecular identification of isolate which has screened from cacao bean fermentation microbe even yeast nor bacteria.

2 Methods

2.1 Microorganism Isolation

All microbes used in this study (yeast and bacteria) were provided from Laboratory of Applied Microbiology, Research Center for Biotechnology, Indonesian Institute of Sciences [15]. All yeast strains were rejuvenated from glycerol stock onto malt extract agar (MEA) (Merck, Darmstadt, Germany) containing 100 mg L^{-1} chloramphenicol (Sigma, St. Louis, Missouri, USA). All bacterial isolates were rejuvenated from glycerol stock onto plate count agar (PCA) (Merck, Darmstadt, Germany) 0.1% cycloheximide (Merck, Darmstadt, Germany).

2.2 Screening Pectinolytic Enzyme

For screening of pectinolytic yeast, all yeast strains were grown on MEA containing 1% of citrus pectin (Sigma, St. Louis, Missouri, USA), and incubation for 2 days at 30 °C. Subsequently, the yeast on MEA plates was flooded with iodine-potassium iodide solution for 30 min. The isolate yeast generating large halos were selected [14, 16]. For pectinolytic bacterial, all bacterial were grown on PCA containing 1% of citrus pectin (Sigma, St. Louis, Missouri, USA), and incubation for 2 days at 30 °C. Afterward, the bacterial isolate on PCA plates was flooded with iodine-potassium iodide solution for 30 min. The bacterial result in large halos were selected [14, 16].

2.3 Production of Pectinase Enzyme

Inoculum preparation. The fresh culture of *Bacillus subtilis* strain IDI-001 was inoculated into sterilized MEB (malt extract 0.3%, yeast extract 0.3%, peptone 0.5%, glucose 1%) with pH of 7.0 ± 0.5. Culture was grown in 5 mL MEB media in 15 mL test tube. The isolate was incubated at 30 °C on a rotary shaker (KS 3000, IKA, Germany) at 120 rpm for 24 h.

Production of pectinase. Crude pectinase enzyme was produced in 100 mL MEBP containing (MEB medium plus 1% citrus pectine) in 300 mL Erlenmeyer flasks. Sterilized MEBP medium was added with 1% of *Bacillus subtilis* strain IDI-001 inoculum. The medium plus inoculum was incubated in rotary shaker at 30 °C on a rotary shaker (KS 3000, IKA, Germany) at 120 rpm for 48 h. Afterward, the mixture was centrifuged (MX-307, Tomy, Japan) at 10.000 rpm, for 15 min. at 4 °C. Supernatant was collected as pectinase crude enzyme [14, 17].

2.4 Determination of Enzymes Activity

Pectinase activity was measured using DNS method [18] with modification. Crude enzyme extract 0.5 mL was mixed with 0.5 mL substrate (1% citrus pectin in 0.05 M phosphate buffer, pH 7.0). The mixture was incubated at 30 °C for 30 min. in Incubator (IN55, Memmert, Germany). Afterward, the mixture was added with 1.5 mL of DNS, followed by incubation at 100 °C for 15 min. to stop enzymatic reaction. As blanks, solution contain the substrate were also set up. All samples was analyzed by spectrophotometer (UV-1800, Shimadzu, Japan) at 540 nm wavelength. Galacturonic acid was used as standard [14, 19].

2.5 Characterization of Crude Enzyme Activity

To determined the optimum pH of pectinase enzyme. The enzymatic assay was conducted by measuring enzyme activity at various pHs 5–7 in phosphate buffer [14, 16].

2.6 Molecular Identification

The methods for molecular identification of pectinolytic isolate were described elsewhere [15]. Briefly, DNA sample were prepared by adding 25 μL TE, followed by incubated at 95 °C for 2 min. 1 μL of mixture was used as DNA template. PCR reaction was prepared using PCR kit MyTaqTM HS Red Mix (Bioline, London, United Kingdom), PCR was carried out for 30 cycles [15]. The PCR product was sent to commercial sequencing service. Sequences data was analyzed using MEGA 7 software and BLAST algorithm [20] and database from EzBioCloud [21].

3 Results

3.1 Screening Pectinolytic Activity

Microorganism used in this study was previously isolated from cacao bean fermentation. Out of 138 isolates, only 4 isolates have pectinolytic activity, namely IDI-001, IDI-002, IDI-004 and IDI-008 as shown in Table 1.

Based on qualitative and quantitative analysis isolate IDI-001 has higher pectinolytic activity with activity of 1.8 U mL^{-1}, follow by IDI-002, IDI-004 and IDI-008 with activity 0.95, 0.62 and 0.43 U mL^{-1}, respectively, as shown in Table 2.

3.2 Characterization of Pectinase Activity

Isolate IDI-001 were used in enzyme characterization because it has highest pectinase activity than other isolates. Phosphate buffer with pH range of 5–7 with interval of 1 was used to determined pH optimum of pectinase. In parallel the optimum time for pectinase enzyme production were records Fig. 1. Based on Fig. 1, pectinase enzyme

Table 1 Screening pectinolytic activity

No	ID	Clear zone	No	ID	Clear zone	No	ID	Clear zone
1	IDI-Y001	–	47	IDI-Y047	–	93	IDI-033	–
2	IDI-Y002	–	48	IDI-Y048	–	94	IDI-045	–
3	IDI-Y003	–	49	IDI-Y049	–	95	IDI-049	–
4	IDI-Y004	–	50	IDI-Y050	–	96	IDI-056	–
5	IDI-Y005	–	51	IDI-Y051	–	97	IDI-057	–
6	IDI-Y006	–	52	IDI-Y052	–	98	IDI-058	–
7	IDI-Y007	–	53	IDI-Y053	–	99	IDI-060	–
8	IDI-Y008	–	54	IDI-Y054	–	100	IDI-072	–
9	IDI-Y009	–	55	IDI-Y055	–	101	IDI-077	–
10	IDI-Y010	–	56	IDI-Y056	–	102	IDI-078	–
11	IDI-Y011	–	57	IDI-Y057	–	103	IDI-079	–
12	IDI-Y012	–	58	IDI-Y058	–	104	IDI-081	–
13	IDI-Y013	–	59	IDI-Y059	–	105	IDI-082	–
14	IDI-Y014	–	60	IDI-Y060	–	106	IDI-083	–
15	IDI-Y015	–	61	IDI-Y061	–	107	IDI-084	–
16	IDI-Y016	–	62	IDE-Y025	–	108	IDI-080	–
17	IDI-Y017	–	63	IDE-Y062	–	109	5	–
18	IDI-Y018	–	64	IDE-Y001	–	110	6	–
19	IDI-Y019	–	65	IDE-Y015	–	111	8	–
20	IDI-Y020	–	66	IDE-007	–	112	40	–
21	IDI-Y021	–	67	IDE-Y048	–	113	12	–
22	IDI-Y022	–	68	IDE-Y056	–	114	14	–
23	IDI-Y023	–	69	IDE-Y271	–	115	16	–
24	IDI-Y024	–	70	IDE-Y065	–	116	17	–
25	IDI-Y025	–	71	IDE-Y087	–	117	18	–
26	IDI-Y026	–	72	IDE-Y091	–	118	22	–
27	IDI-Y027	–	73	IDT-Y004	–	119	26	–
28	IDI-Y028	–	74	IDT-Y013	–	120	30	–
29	IDI-Y029	–	75	**IDI-001**	+ +	121	K01	–
30	IDI-Y030	–	76	**IDI-002**	+ +	122	K03	–
31	IDI-Y031	–	77	IDI-003	+	123	K05	–
32	IDI-Y032	–	78	**IDI-004**	–	124	K07	–
33	IDI-Y033	–	79	IDI-007	–	125	K09	–
34	IDI-Y034	–	80	**IDI-008**	+	126	K011	–
35	IDI-Y035	–	81	IDI-010	–	127	K013	–
36	IDI-Y036	–	82	IDI-013	–	128	K016	–
37	IDI-Y037	–	83	IDI-015	–	129	K017	–

(continued)

Table 1 (continued)

No	ID	Clear zone	No	ID	Clear zone	No	ID	Clear zone
38	IDI-Y038	–	84	IDI-019	–	130	K019	–
39	IDI-Y039	–	85	IDI-020	–	131	K021	–
40	IDI-Y040	–	86	IDI-021	–	132	K023	–
41	IDI-Y041	–	87	IDI-023	–	133	K025	–
42	IDI-Y042	–	88	IDI-024	–	134	K029	–
43	IDI-Y043	–	89	IDI-025	–	135	K031	–
44	IDI-Y044	–	90	IDI-026	–	136	K035	–
45	IDI-Y045	–	91	IDI-027	–	137	K037	–
46	IDI-Y046	–	92	IDI-028	–	138	K049	–

Table 2 Qualitative and quantitative activity of pectinolytic isolates

ID	Substrate pectin (%)	Incubation time (h)	Total diameter (mm)	Colony diameter (mm)	Clear zone diameter (mm)	Pectinase activity (U mL^{-1})*
IDI-001	1	24	1.57	0.76	1.06	1.80
IDI-002	1	24	1.35	0.67	1.00	0.95
IDI-004	1	24	1.38	0.75	0.84	0.62
IDI-008	1	24	1.35	0.74	0.81	0.43

* Determination using protocol in method 2.4

from IDI-001 was optimum at pH 7.0 with production time at 48 h with activity of 1.81 U mL^{-1}. Meanwhile for pH 5 and 6 was optimum at 72 h with activity 1.01 and 0.86 U mL^{-1}, respectively.

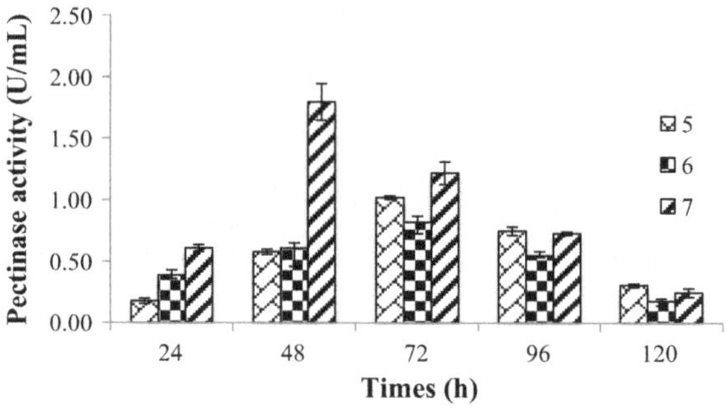

Fig. 1 Optimum pH and time for enzyme production

Table 3 Table captions should be placed above the tables

ID	BLAST result	Homology
IDI-001	*Bacillus subtilis* strain TLO3, complete genome	100%

Fig. 2 PCR result **a**, sequencing result **b** using 27F and 1492 R primers

3.3 Identification of Pectinolytic Isolate

Molecular identification of pectinolytic isolates was performed using 16S rRNA using primer 27F and 1492R primer [15]. PCR product was analyzed to commercial sequencing services 1st Base, Malaysia. Sequencing result were subjected to BLAST in NCBI Gen Bank (Table 3). Based on BLAST result IDI-001 is 100% similarity to *Bacillus subtilis* strain TLO3. The highest score of BLAST for IDI-001 was *B. subtilis* strain NCIB 3610 with score of 2558 and the lowest score was *B. pumilus* strain NBRC- 12,092 with score of 2283 (Fig. 2).

MEGA 7 software was used for constructed of phylogenetic tree, using Kimura-2 parameter and neighbor-joining algorithm, with 1000 bootstrap. As shown in Fig. 3. IDI 001 has close relationship to *Bacillus subtilis* strain SBMP4 with bootstrap value of 100%.

4 Discussion

Fermentation is the essential process in chocolate manufacturing. The activities of microbes and enzymes during cacao beans fermentation was affected fermented cacao beans. Pectinase enzyme is one of crucial enzyme active during cacao bean

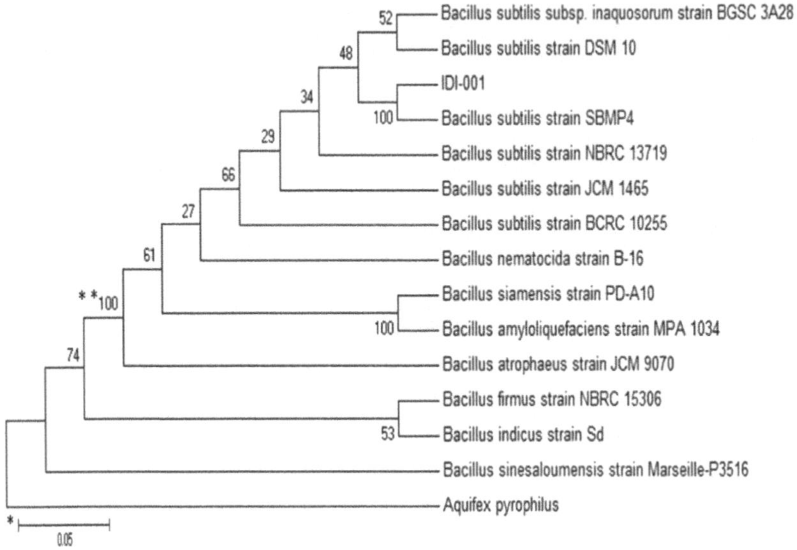

Fig. 3 Phylogenetic tree of *Bacillus subtilis* IDI-001

fermentation. This enzyme marked to maceration of pectin content in cacao bean pulp during cacao bean mass fermentation, leading cacao bean pulp drainage of sweating and degraded [2]. By knowing the microbial diversity and enzymes active during cocoa bean fermentation the quality of cocoa bean could be estimated, thus consistent product could be applied and monitoring [2, 15, 22].

Microbial producing pectinase enzyme has been reported elsewhere [11, 14, 23–25]. More of previous study showed that pectinase enzymes were produced by yeast strain e.g. *Candida, Pichia, Saccharomyces, Zygosaccharomyces, Torulopsis candida, T. holmii,* and *Kluyveromyces marxianus* [11, 23, 24]. Another researcher declaired that a pectinase enzyme was produced from bacteria strain e.q. *Bacillus cereus, Bacillus thuringensis, Bacillus subtilis, Bacillus fusiformis, Bacillus sphaericus, Bacillus pumilus* [13, 14, 25]. Our research reported the production of pectinase enzymes produced by *Bacillus* sp. from cacao bean mass fermentation.

In previous result showed that wide range of yeast strain during cacao bean fermentation process for instant *Hanseniaspora guilliermondii, H. opuntiae, H. thailandica, Saccharomyces cerevisiae, Candida tropicalis, Torulaspora delbrueckii, Pichia kudriavzevii, Issatchenkia orientalis,* and *Sporobolomyces roseus* [26]. In addition, bacterial strains *Lactobacillus plantarum, L. pentosus, L. fermentum, Acetobacter pasteurianus, A. ghanensis, Enterobacter* and *Bacillus* sp. [15]. In this study we used this yeast and bacteria strains to screen pectinolytic producing strains. Out of 138 yeast and bacteria strains only 4 isolates can hydrolysis citrus pectin and produced pectinase enzyme namely IDI-001, IDI-002, IDI-004 and IDI-008. Out of 4 pectinolytic strains only IDI-001 was used as pectinase producing strain, because of the highest pectinase activity. This strain IDI-001 can produce pectinase with activity of

1.80 U mL^{-1}. Another strain IDI-002, IDI-004 and IDI-008 was only produced 0.95, 0.62 and 0.43 U mL^{-1}, respectively (Table 2). Another study was also reported that pectinase enzyme from yeast *Kluyveromyces marxianus* CCT 3172, *K. thermotolerans* CCT 1698, *Saccharymyces cerevisiae* var. chevalieri CCT 1698, and *Candida rugopeliculata* CCT 1707 with activity 21.69, 12.03, 7.16 and 4.05 μmol galacturonic acid μg^{-1} protein min $^{-1}$, respectively [24]. Furthermore, from *Bacillus* sp. BS-66 and *Bacillus* sp. BS-06 with activity 5.01 and 3.27 U mg^{-1} of protein, respectively [25] and *B. subtilis* BS 66, *B. fusiformis* BS 90, and *B. pumilus* BS 22 with activity 8.61, 12.81 and 5.99 U mg^{-1} of bacterial dry weight, respectively [13].

Characterization of pectinase enzyme showed that pectinase enzyme from *Bacillus subtilis* IDI-001 was optimum produced at pH 7.0 and 48 h incubation. The optimum pectinase was produce at 72 h incubation for pH 6.0 and 5.0 (Fig. 1). The optimum condition for pectinase enzyme produced by *K. marxianus* was pH 5.0 and 40 °C, whereas the optimum condition for *S. cerevisiae* was 4.5 and 30 °C [24]. *Bacillus* sp. BS-6 was optimum at pH 4 and 50 °C with activity 3.27 and 5.01 U mg^{-1} of protein mL^{-1}, respectively [25]. Pectinase from *Bacillus* sp. was produced at optimum condition of pH 5.0 and 40 °C, with activity of 0.82 U mL^{-1} and 0.99 U mL^{-1}, respectively [14].

Identification of IDI-001 isolate showed this strain belongs to *Bacillus subtilis* with 100% similarity. More researchers were reported that *Bacillus* sp. was founded during cocoa beans fermentation [1, 2]. But the role of *Bacillus* sp. during cocoa bean fermentation is not yet discussed in detailed. Another study by Outtarra was reported the important of *Bacillus* sp. during cocoa bean fermentation as pectinolytic producing strains [13, 25] as well as reported by Berutu et al. [14]. Outtarra reported that some species from *Bacillus* sp. from cocoa bean fermentation can produce pectinase enzymes such as *B. cereus, B. subtilis, B. pumilus B. fusiformis, B. thuringensis, B. sphaericus,* [25].

Although this study is preliminary study of *Bacillus* in produce pectinolytic strain from cacao bean fermentation. However, this study described the important of *Bacillus* sp. during cacao bean fermentation. Forture research is needed for more screening and characterization of *Bacillus* strain as pectinolytic producer as well as characterization of their enzymes for application of cacao bean fermentation process.

5 Conclusion

Bacillus subtilis IDI-001 is strain pectinolytic from cacao bean fermentation. This study described the important role of *Bacillus* sp. during cacao bean fermentation process as pectinolytic producer strain. Furthermore, pectinase is one key process during cacao bean fermentation. Further research is needed to understand the characteristic of pectinase enzymes produce by *Bacillus* sp. and their application for cacao bean fermentation.

References

1. Thompson SS, Miller KB, Lopez AS, Camu N (2012) Cocoa and coffee. Food Microbiol Fundam Front 881–899
2. Schwan RF, Wheals AE (2004) The microbiology of cocoa fermentation and its role in chocolate quality. Crit Rev Food Sci Nutr 44(4):205–221
3. De Vuyst L, Weckx S (2016) The cocoa bean fermentation process: from ecosystem analysis to starter culture development. J Appl Microbiol 121(1):5–17
4. Schwan RF (1998) Cocoa fermentations conducted with a defined microbial cocktail inoculum. Appl Environ Microbiol 64(4):1477–148
5. Puerari C, Magalhães KT, Schwan RF (2012) New cocoa pulp-based kefir beverages: Microbiological, chemical composition and sensory analysis. Food Res Int 48(2):634–640
6. Fogarty WM, Ward OP (1974) Pectinases and pectic polysaccharides. Prog Ind Microbiol 13:61–119
7. Rexová-Benková Ľ, Markovič O (1976) Pectic enzymes. Adv Carbohydr Chem Biochem 33:323–385
8. Blanco P, Sieiro C, Villa TG (1999) Production of pectic enzymes in yeasts. FEMS Microbiol Lett 175(1):1–9
9. Roelofsen PA (1953) Polygalacturonase activity in yeast, Neurospora and tomato extract. Biochimica et Biophys Act 10:410–413
10. Gauthier B, Guiraud J, Vincent JC, Porvais JP, Galzy P (1977) Comments on yeast flora from traditional fermentation of cocoa in the ivory coast. Rev Ferment Ind Aliment 32:160–163
11. Sanchez J, Guiraud JP, Galzy P (1984) A study of the polygalacturonase activity of several yeast strains isolated from cocoa. Appl Microbiol Biotechnol 20(4):262–267
12. Sanchez J, Daguenet G, Guiraud JP, Vincent JC, Galzy P (1985) A study of the yeast flora and the effect of pure culture seeding during the fermentation process of cocoa beans. Lebensmittel-Wissenschaft+ Technol 18(2):69–75
13. Ouattara HG, Reverchon S, Niamke SL, Nasser W (2011) Molecular identification and pectate lyase production by Bacillus strains involved in cocoa fermentation. Food Microbiol 28(1):1–8
14. Berutu CAM, Fahrurrozi F, Meryandini A (2017) Pectinase production and clarification treatments of apple (Malus domestica) juice. Ann Bogorienses 21(2):63–68
15. Fahrurrozi F, Stepanie S, Purwanto AP, Sofia DY, Ratnakomala S (2020) Identification of lactic acid bacteria and acetic acid bacteria of lampung cocoa beans fermentation. In: IOP Conference series: earth and environmental science 439 (1), pp 012061. IOP Publishing, Bogor
16. Abdollahzadeh R, Pazhang M, Najavand S, Fallahzadeh-Mamaghani V, Amani-Ghadim AR (2020) Screening of pectinase-producing bacteria from farmlands and optimization of enzyme production from selected strain by RSM. Folia Microbiol 65(4):705–719
17. Oumer OJ, Abate D (2017) Characterization of pectinase from Bacillus subtilis strain Btk 27 and its potential application in removal of mucilage from coffee beans. Enzym Res
18. Miller GL (1959) Use of dinitrosalicylic acid reagent for determination of reducing sugar. Anal Chem 31(3):426–428
19. Qureshi AS, Bhutto MA, Chisti Y, Khushk I, Dahot MU, Bano S (2012) Production of pectinase by bacillus subtilis EFRL 01 in a date syrup medium. Afr J Biotech 11(62):12563–12570
20. Altschul SF, Madden TL, Schäffer AA, Zhang J, Zhang Z, Miller W, Lipman DJ (1997) Gapped BLAST and PSI-BLAST: a new generation of protein database search programs. Nucl Acids Res 25(17):3389–3402
21. Yoon SH, Ha SM, Kwon S, Lim J, Kim Y, Seo H, Chun J (2017) Introducing EzBioCloud: a taxonomically united database of 16S rRNA gene sequences and whole-genome assemblies. Int J Syst Evol Microbiol 67(5):1613
22. Thompson SS, Miller KB, Lopez AS (2007) Cocoa and Coffee. In Doyle MP, Beuchat LR (3rd) 2007, Food microbiology: fundamentals and frontiers, ASM Press, Washington D. C
23. Schwan RF, Cooper RM, Wheals AE (1996) Endo-polygalacturonase of the yeast Kluyveromyces marxianus is constitutive, highly active on native pectin and is the main extracellular protein. Prog Biotechnol 14:861–868

24. Schwan RF, Cooper RM, Wheals AE (1997) Endopolygalacturonase secretion by Kluyveromyces marxianus and other cocoa pulp-degrading yeasts. Enzyme Microb Technol 21(4):234–244
25. Ouattara HG, Reverchon S, Niamke SL, Nasser W (2010) Biochemical properties of pectate lyases produced by three different Bacillus strains isolated from fermenting cocoa beans and characterization of their cloned genes. Appl Environ Microbiol 76(15):5214–20
26. Fahrurrozi (2015) Microbiological and biochemical investigations of cocoa bean fermentation (Doctoral dissertation, Staats-und Universitätsbibliothek Hamburg)

Chemical Bond Analysis of Activated Carbon from Oil Palm Waste Using FTIR and Raman Spectroscopy

Resti Marlina, Ria Yolanda Arundina, Ester Ria Togatorop, Sefrico Agung Saifulloh, Ismadi, Deni Purnomo, Ismail Budiman, Yudi Darma, and Subyakto

Abstract Biomass-derived carbon has become a fascinating topic along with renewable, low-cost, and environmentally friendly carbon source issues. In this study, chemical analysis was carried out on activated carbon from oil palm waste, such as oil palm empty fruit bunches (OPEFB) and oil palm shells (OPS), using hydrothermal and pyrolysis techniques. Hydro-char as a precursor is produced from hydrothermal carbonization by using oil palm empty fruit bunches and oil palm shells. The sample from raw material to activated carbon was analyzed using (Fourier Transform Infrared) FTIR and Raman spectroscopies to observe the chemical mechanism for converting biomass to activated carbon. Proximate analysis was conducted to see the quality of our activated carbon samples. Moreover, the FTIR spectra have successfully observed functional groups in lignocellulosic materials like cellulose, hemicellulose, and lignin. In addition, D band and G band as fingerprint of carbon materials has been clearly observed from the Raman spectra. The quality of the carbon material has been calculated by determining the value of I_D/I_G ratio. The calculation of I_D/I_G gives a value of 0.93, indicating that our carbon material has a good degree of carbonization. In the end, the results of the proximate test showed that the activated carbon produced appropriately with the SNI 63–3730 standard regarding technical activated carbon.

Keywords Activated carbon · FTIR spectroscopy · Oil palm empty fruit bunch · Oil palm shell · Raman spectroscopy

R. Marlina (✉) · Ismadi · D. Purnomo · I. Budiman · Subyakto
Research Center for Biomass and Bioproducts, National Research and Innovation Agency, Cibinong, West Java, Indonesia
e-mail: resti.marlina@brin.go.id

R. Y. Arundina · E. R. Togatorop
Department of Bioprocess Technology, Faculty of Agricultural Technology, Brawijaya University, Malang, Indonesia

S. A. Saifulloh · Y. Darma
Department of Physics, Faculty of Mathematics and Natural Sciences, Bandung Institute of Technology, Bandung, Indonesia

M. A. Rahandi Lubis (ed.), *Proceedings of the 1st International Conference of Lignocellulose*, Green Energy and Technology,
https://doi.org/10.1007/978-981-19-1596-3_27

1 Introduction

Oil palm (*Elaesis oleifera*) is the largest export commodity in Indonesia. According to data from the Directorate of Plantations, the Indonesian Ministry of Agriculture, it was stated that the area of oil palm plantations in Indonesia reached 15 million hectares in 2020 with total annual production attains 49 million tons [1]. However, in order to produce crude palm oil (CPO), the palm oil industries also generate many industrial by-products in the form of solids, liquids, and gases during their process. The mainly by-product of palm oil production are solid products such as empty oil palm fruit bunches (23%) and oil palm shells (6.6%) [2]. In fact, the decomposition process of these by-products will become an environmental problem if it is not handled properly. Therefore, many studies have been conducted to solve this problem by converting oil palm waste (by-products) into high economic value products like activated carbon [2].

Activated carbon is a black porous solid material which contains 85–95% carbon. Considering the characteristics of a large pore surface area, activated carbon is widely used in various applications such as water purification and pollutant absorption for environmental issue [3], catalysts for industrial issue [4] and energy storage devices as electrode in supercapacitor or batteries for renewable energy issue [5]. Utilization of biomass waste, in this case is oil palm waste, becomes a precursor in the manufacture of activated charcoal is based on the components that arrange the biomass, such as cellulose, hemicellulose, and lignin. According to Gaol et al. [6], oil palm empty fruit bunches (OPEFB) components are consisting of cellulose (28%), hemicellulose (25.9%), holocellulose (20.27%), and lignin (25.83%) [6] while oil palm shell (OPS) has constituent components such as cellulose (23.96%), hemicellulose (19.81%), holocellulose (43.77%), and lignin (45.82%) [3]. Based on that information, OPEFB and OPS have a great potential to be used as activated carbon products.

There are two general methods of carbonization that are usually used to produce the activated carbon, such as hydrothermal and pyrolysis carbonization. Hydrothermal converts lignocellulosic biomass into several carbonaceous materials and by-products using a thermochemical conversion process [7]. This process is environmentally friendly, non-toxic, and low-cost process, because using water as a medium, as well as the use of low temperature (150 °C–350 °C) and self-generated pressure (until 10 MPa) [8]. Furthermore, pyrolysis is a type of decomposition reaction in which organic materials break down in the absence of oxygen. This process usually occurs above 450 °C in the near absence of oxygen by flowing nitrogen gas [9]. In this study, activated carbon synthesis has been conducted using a hydrothermal process followed by a pyrolysis process. The mechanism of degradation or conversion of biomass from raw material into activated carbon will be explained through FTIR and Raman spectroscopic characterization to provide better knowledge related to chemical bonding and the chemical decomposition of oil palm empty fruit bunches and palm shells into activated carbon.

2 Materials and Methods

2.1 *Materials*

Commercial products of Oil Palm Empty Fruit Bunches (OPEFB) and Oil Palm Shell (OPS) from local market based on oil palm plantation (Riau, Sumatera) were used as the raw material in this study. Potassium Hydroxide (KOH) analytical grade (Merck, CAS No 1310-58-3) was used as activation agent for our material and Hydrochloric Acid 37% analytical grade (Merck, CAS No 7647-01-0) was used for neutralization process. Furthermore, deionized water (DI) was utilized in hydrothermal and neutralization process.

2.2 *Synthesis of Activated Carbon*

The activated carbons (ACs) were prepared through three stage process involved pre-treatment, hydrothermal, and chemical activation. Hydrothermal and chemical activation processes were modified from previous work [10] that using Lignite as carbon precursor [10]. In the pre-treatment stage, OPEFB and OPS were grinded using ring flaker (Fritsch GmBH, Germany) to reduce the particle size and sieved using 10–20 mesh. Then, the materials were washed in fresh water and dried in the oven at 105 °C for 24 h to reduce the moisture content. The grinded samples were packed in seal bag and store in room temperature. This sample was named as raw material (RAW).

The second stage of hydrothermal process were carried out using autoclave mini reactor for 1:8 proportion of sample and DI water at 220 °C for 8 h. Then, hydro-char product was dried in oven at 105 °C for 24 h. The sample was grinded and sieved using 100 mesh-sieve. This sample was named as hydro-char (HTC).

The third stage of chemical activation was executed by immersing carbon powder in KOH 20% (w/w) with the proportion of carbon/KOH solution was 1:4. Activation process was performed throughout 2 h stirring in chemical solution. Then, the sample was pyrolyzed in standard muffle furnace at 800 °C for 1 h. The activated sample was washed using 1 M HCl and subsequently washed several times with DI water until pH neutral. Then the sample was dried at 105 °C for 24 h which is defined as activated carbon (AC).

2.3 *Characterizations*

Proximate analysis was carried out according to the SNI 63–3730 standard including yield, moisture content, and ash content. Yield parameter was obtained by calculating the ratio of final and initial weight of sample. Moisture content was determined by drying the samples in oven at 115 °C for 3 h and calculating the difference between the initial and final weight. The last, ash content was determined by heating one gram of our sample in crucible using muffle furnace at 850 °C for 2 h and calculating comparison between final and initial weight as well. All of the procedures are corresponded to the SNI 63–3730 standard procedures.

For the characterizations, FTIR spectra were obtained using UATR-FTIR instrument (Spectrum Two, Perkin Elmer, USA) to detect the chemical bond from functional group of lignocellulosic components in our samples. The FTIR spectra was scanned in the range of 4000–400 cm^{-1} with the following parameters such number of scanning 16, resolution 4 cm^{-1}, and interval data 1 cm^{-1}. Furthermore, Raman analysis were executed on Micro Confocal Raman Spectroscopy (LabRAM HR Evolution, Horiba, Japan) with 532 nm laser excitation, 200–3000 cm^{-1} Raman shift range, grating 1800, hole 600, objective lens $\times$ 100, acquisition time 15 s, and accumulation 3 as parameter measurement.

3 Results and Discussions

3.1 Proximate Analysis

The picture of our samples from raw material to the activated carbon are shown in Fig. 1. The difference appearance from the raw material after pre-treatment process (a), after hydrothermal carbonization (HTC) (b), and after pyrolysis carbonization (AC) was clearly shown. HTC products (hydro-char) show brighter color than pyrolysis products (activated carbon). It is obviously due to the partial degradation of lignocellulosic compound in hydro-char. Not all of the lignocellulosic components are degraded at same temperatures. There are different temperatures to degrade the lignocellulosic compound into carbon-rich charcoal [11]. Hemicellulose and cellulose are decomposed at 220–315 °C and 300–400 °C, respectively, while lignin that has three-dimensional amorphous macromolecules is decomposed at higher temperature around 500–600 °C [12]. The fully degradation through carbonization was achieved after pyrolysis process which is shown as perfect black powder for AC sample.

Proximate analysis was conducted to define yield, moisture content, and ash content according to SNI 63-3730. Briefly procedure was mentioned before (in Sect. 2.3) and the results are shown in Table 1. From these data, OPS's yield has almost 3 times higher than OPEFB with a value of 24.29%. This phenomenon is due to the density of OPS that has much greater and higher hardness level than OPEFB. This characteristic is related to the amount of lignin content in OPS compared to OPEFB which is lignin has an important role to provide the hardness properties of wood. The moisture content shows that the OPS material has a high moisture content with the value of 20.67% compared to the OPEFB material which is only 13.00%. This might be occurred because the formation of pores during the activation process

Fig. 1 The samples of OPEFB and OPS: **a** raw material, **b** hydro-char, and **c** activated carbon

Table 1 Proximate Analyses of OPEFB and OPS

Materials	Yield (%)	Moisture (%)	Ash Content (%)
OPEFB	8.72	13.00 ± 0.01	6.50 ± 0.01
OPS	24.29	20.67 ± 0.02	3.50 ± 0.01

in OPS material is more difficult compared to the formation in OPEFB's pores. Eventually, the water molecules are more difficult to evaporate in fewer pores, so it still remains in OPS's material.

The good results of ash content were shown with a value of 6.50% and 3.50% for OPEFB and OPS, respectively. These values are appropriate to the SNI 63-3730 standard for activated carbon with 10% maximum value. The OPEFB has a higher ash content because the components of OPEFB material tend to be more easily decomposed than OPS material. OPEFB material is dominated by cellulose and hemicellulose components which are degraded at 450 °C, while OPS material has more lignin components which are decomposed at higher temperatures (> 500 °C). Therefore, the ash content is more produced in OPEFB material than OPS material.

3.2 *FTIR Spectroscopy*

The FTIR spectra of our samples was plotted using normalization data measurement, as shown in Fig. 2. From FTIR spectra of OPEFB (Fig. 2a), a medium-strong and broad absorption was observed at 3360 cm^{-1} wavenumber which is related to the stretching of O–H bonded in hydroxyl group of cellulose [13]. Another typical peak occurs around wavenumber of 2900 cm^{-1} due to the C–H stretching from the CH_2-OH group in cellulose [14]. The 1700–500 cm^{-1} wavenumber range of FTIR spectra is called the fingerprint area. In fingerprint area, it was observed many sharp and well-defined absorption bands as consequence of the existence of several functional group in each component of OPEFB.

From the fingerprint area, it is shown that there is an absorption peak at wavenumber 1705 cm^{-1} which comes from the C = O stretching bond that has

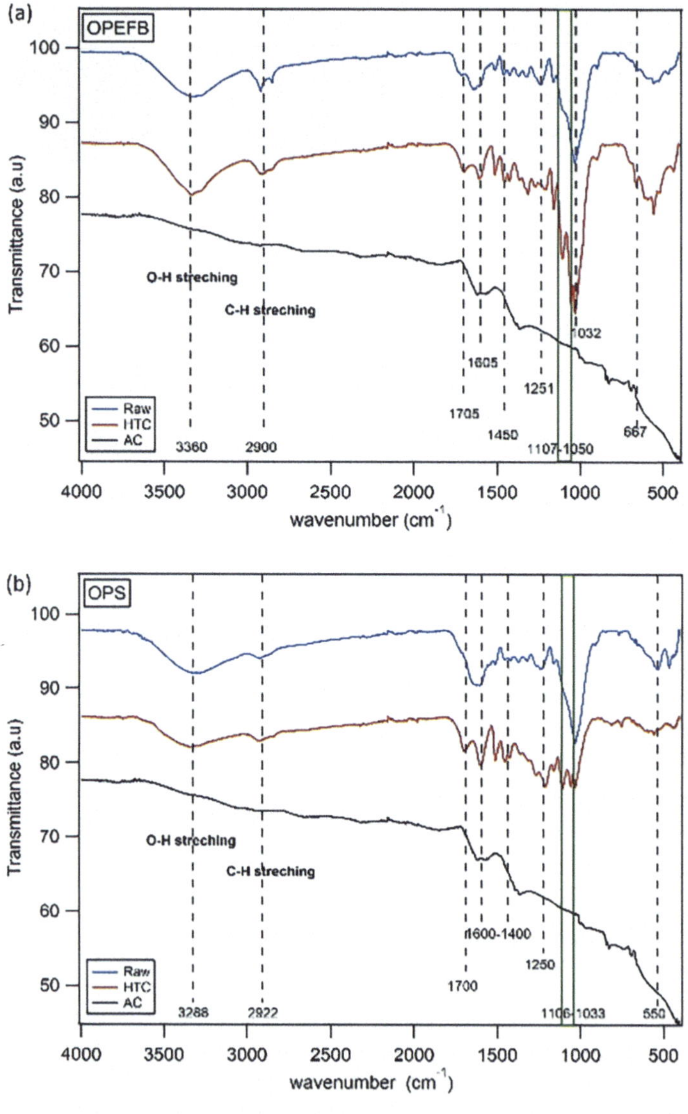

Fig. 2 FTIR spectra of OPEFB **a** and OPS **b**

been reported as contribution of the acetyl and uronic ester groups of hemicellulose ester compounds [15]. Furthermore, the C = C bond of the cellulose group has been shown in absorption peak at 1605 cm^{-1} wavenumber. The absorption peak also appears at 1450 cm^{-1} wavenumber which is contribution from the C–H bending in cellulose and hemicellulose compounds. The C–O stretching bonds were detected at 1251 cm^{-1} wavenumber which were identified as aromatic ester bonds in lignin compounds.

The sharp prominent peaks have been observed between 1107 and 1050 cm^{-1} wavenumber. The absorption peak at 1107 cm^{-1} wavenumber indicates C–O bond in the tertiary alcohol functional group and the wavenumber 1050 cm^{-1} peak indicates C–O bond in the primary alcohol functional group. The transformation of the tertiary alcohol functional group into a primary alcohol is characterized by the breaking the C–O bond attached to the phenol –OH functional group during the hydrolysis process. It has been reported that this absorption is due to the contribution of C–H aromatic compounds in the field of O deformation (deformation in primary alcohols) [13]. The sharp absorption peak at 1032 cm^{-1} wavenumber was identified as a stretching CO–O-CO bond of anhydride compound while the absorption around 667 cm^{-1} wavenumber is a contribution from the C–O bond out of plane bending mode. The absorption peaks at 1450, 1032 and 650 cm^{-1} wavenumbers are mainly due to carbohydrates and have no significant contribution from lignin [13].

For the FTIR spectra of OPS (Fig. 2b), the absorption peaks of –OH stretching and C–H stretching at 3288 $cm^{-1}$and 2922 cm^{-1} wavenumber are still visible as in the OPEFB material. The hydroxyl group is the contribution of cellulose, hemicellulose, and lignin components, while the alkane group is mostly contributed by lignin compounds [3, 16]. Comparing to the OPEFB spectra, the hydroxyl groups in the OPS material tend to be less than the OPEFB material which is depicted by a smoother curve. This phenomenon is appropriated with the chemical component of OPEFB that has more cellulose and hemicellulose content compared to OPS material.

Distinguishable peaks were also seen in pure OPS material at 1610 cm^{-1} wavenumber which is identified as C = C bond from unsaturated ketones. C = C alkene bond at 1600 cm^{-1} wavenumber and C = O aldehyde bond at 1700 cm^{-1} wavenumber was also observed due to the contribution of hemicellulose compounds during hydrothermal process. In addition, the absorption peaks in the range 1600–1400 cm^{-1} wavenumber have been contributed from aromatic lignin compounds in the C = C double bond. The contribution of the C–O stretching was occurred in 1250–1000 cm^{-1} wavenumber. Similar with OPEFB material, the absorption peak in the range 1106–1033 cm^{-1} wavenumber is a process of transformation of the tertiary to the primary alcohol functional group due to the hydrolysis process was determined as well. Moreover, tabulation of the FTIR absorption spectra can be seen in Table 2.

In general, the FTIR spectra of HTC samples shows that the component of lignin, in lignocellulosic materials, was less decomposed than hemicellulose and cellulose components which are almost completely decomposed [3]. On the activated carbon material after pyrolysis process, the FTIR spectra shows no significant absorption peak for both OPEFB and OPS materials or almost all the absorption peaks of the

Table 2 FTIR wavenumber characterization of OPEFB and OPS [10–12, 16, 17]

Wavenumber (cm-1)	Group/compound class
3360–3350	O–H stretching cellulose
2900	C–H stretching cellulose
1705	C = O stretching hemicellulose
1610	C = C stretching cellulose/lignin
1450	C–H bending cellulose/hemicellulose
1250	C–O stretching aromatic lignin
1106–1050	C–O stretching tertiary-primary alcohol
1032	CO–O–CO bond anhydride
667–550	C–O out of plane bending mode

functional groups disappeared. This happens because the functional groups of the raw material have been evaporated as volatile material during the pyrolysis stage [15]. This also indicates that the sample has been successfully converted into carbon material which corresponds to the theory that carbon materials are not detected in the IR mode but actively detected in Raman mode.

3.3 Raman Spectroscopy

Raman spectroscopy is a technique that widely used for analyzing carbon-based materials due to its sensitivity to different carbon structures. It can produce any distinctive Raman peaks for various forms of materials, especially carbon, such as diamond, graphite, and amorphous forms [13]. The fingerprint of carbon materials in Raman spectra are observed at Raman shift 1000–2000 cm^{-1}. It consists of two intense broad bands which is indicated as D and G band. D band is attributed to the disordered graphite and G band is considered to be phonon mode with E_{2g} symmetry of graphite [14]. Moreover, G peak was identified due to the C–C bond stretching of all pairs of sp^2 atoms in both rings and chains while D peak correspond to breathing mode of sp^3 in rings [13]. The presence of sp^2—sp^3 bonds determines the (amorphous) carbon structure in the material [15]. The D band is activated by defects and its intensity can be used to quantify disorders.

It has been accepted that the relationship between D and D band intensities is proportional to the degree of organization of carbonaceous materials. For biomass-derived carbon, which has a more disordered structure, the D and G band are more likely to relate to the aromatic ring systems and do not relate to its graphitic structure [15]. D band is attributed to condense structure of large aromatic ring systems, which has more than six fused rings, while G band indicates amorphous structure with small aromatic systems which has 3–5 fused rings [13].

Raman analysis was successfully done using OriginPro 2021 software from OriginLab Corporation (USA). Savitzky-Golay method for second order polynomial was performed to smoothing our Raman spectra before the fitting method. Fitting method was conducted by using multiple peak fit with gaussian line shape. The fitted peaks are shown in Fig. 3 and the parameter were given in Table 3. It can be seen that the spectra for the OPEFB and OPS samples were almost familiar with slightly different value. From OPEFB spectra, D and G band peaks appear around 1351 cm^{-1} and 1585 cm^{-1}Raman shift, respectively, with the ratio of I_D/I_G is 0.9318. Almost same value was also found for OPS spectra that has 0.9311 for I_D/I_G ratio while D band and G band peaks were observed at 1333 cm^{-1} and 1590 cm^{-1} Raman shift. These I_D/I_G ratios show that our samples have good degree of carbonization where the disordered structures quite balance with the graphitic structure. Moreover, in the amorphous carbon materials, our results expressed that the vibration molecule from large aromatic ring systems are quite balance with the small aromatic ring systems. In the other hand, we have compared our Raman spectra with Raman Database from Bio-Rad Laboratories informatics Division. From that database, it can be concluded that our samples are perfectly match with Ivory Black (CAS 9021-99-6) substance type amorphous carbon (C) with the value is 99.10%.

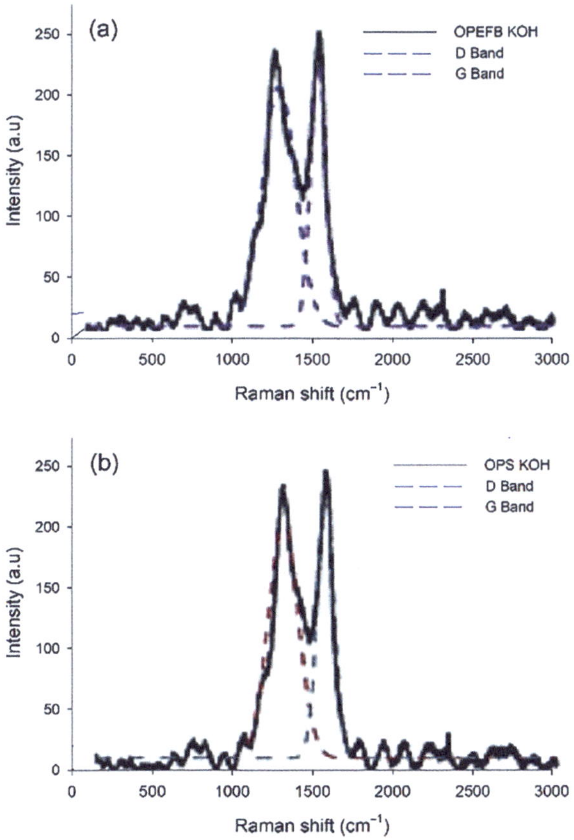

Fig. 3 Raman Spectra of OPEFB **a** and OPS **b**

Table 3 Fitting parameters of Raman spectra

Sample	Line Shape	$\omega_D(\text{cm}^{-1})$	$\omega_G(\text{cm}^{-1})$	$FWHM_D$	$FWHM_G$	I_D/I_G
OPEFB-KOH	Gaussian	1351	1585	232	107	0.9318
OPS-KOH	Gaussian	1333	1590	212	122	0.9311

4 Conclusions

Activated carbon has been successfully made using OPEFB and OPS materials with hydrothermal and pyrolysis techniques. Proximate analysis shows that carbon yield from OPS have a higher than OPEFB. Moreover, our samples had met the requirement of SNI 63-3730 standard regarding technical activated carbon for the moisture and ash content. The results of the FTIR spectra show that the bond represents a lignocellulosic material with the mechanism of degradation or conversion of biomass from raw material to hydro-char and to activated carbon. Furthermore, Raman analysis shows D band and G band as the fingerprint of carbon materials. The I_D/I_G ratio gives the number of ~ 0.93 that indicates our samples have quite balance disordered and graphitic structure for our amorphous carbon. It can also express that the vibration molecules from large aromatic ring systems are quite balance with the small aromatic ring systems.

Acknowledgements Author would like to thank DIPA Indonesian Institute of Sciences (LIPI) and the Research Centre of Biomaterials-LIPI for financial and the facilities' support and Japan Society for the Promotion of Science (JSPS) for sharing publication fee.

References

1. Ministry of Agriculture (2019) Tree crop estate statistics of Indonesia 2018–2020. Secretariate of directorate general of estates, Jakarta
2. Haryanti A, Norsamsi N, Sholiha PSF, Putri NP (2014) Studi Pemanfaatan Limbah Padat Kelapa Sawit. Konversi 3(2):20
3. Budiman I, Hermawan D, Febrianto F, Pari G, Subyakto (2019) Char properties and pollutant adsorption capability of oil palm shell using hydrothermal process. Biomass Convers Biorefinery 9(4), 681–8
4. Lin F, Liu X, Ma M, Qi F, Pan Y, Wang L, Ma P, Zhang Y (2021) Real-time monitoring the carbonization and activation process of activated carbon prepared from Chinese parasol via zinc chloride activation. J Anal Appl Pyrol 155:105089
5. Jiang L, Sheng L, Fan Z (2018) Biomass-derived carbon materials with structural diversities and their applications in energy storage. Sci China Mater 61(2):133–58
6. Gaol MRLL, Sitorus R, Yanthi S, Surya I, Manurung R (2013) Pembuatan Selulosa Asetat Dari A—Selulosa Tandan Kosong Kelapa Sawit. J Teknik Kimia USU 2(3):33–9
7. Xiao LP, Shi ZJ, Xu F, Sun RC (2012) Hydrothermal carbonization of lignocellulosic biomass. Biores Technol 118:619–623
8. Parshetti GK, Hoekman SK, Balasubramanian R (2013) Chemical, structural and combustion characteristics of carbonaceous products obtained by hydrothermal carbonization of palm empty fruit bunches. Biores Technol 135:683–689

9. Bakhtiar MHABM, Sari NBA, Yaacob A, Yunus MFBM, Ismail K (2019) Characterization of oil palm empty fruit bunch (EFB) biochar activated with potassium hydroxide under different pyrolysis temperature. J Eng Sci Technol 14(5):2792–2807
10. Wu Y, Cao JP, Zhao XY, Hao ZQ, Zhuang QQ, Zhu JS, Wang XY, Wei XY (2017) Preparation of porous carbons by hydrothermal carbonization and KOH activation of lignite and their performance for electric double layer capacitor. Electrochim Acta 252:397–407
11. Balat M (2008) Mechanisms of thermochemical biomass conversion processes. Part 1: reactions of pyrolysis. Energy Sour Part A Recov Util Environ Effects 30:620–635
12. Waters CL, Janupala RR, Mallinson RG, Lobban LL (2017) Staged thermal fractionation for segregation of lignin and cellulose pyrolysis products: An experimental study of residence time and temperature effects. J Anal Appl Pyrol 126:380–9
13. Asadullah M, Zhang S, Li CZ (2010) Evaluation of structural features of chars from pyrolysis of biomass of different particle sizes. Fuel Process Technol 91(8):877–881
14. Sadezky A, Muckenhuber H, Grothe H, Niessner R, Pöschl U (2005) Raman microspectroscopy of soot and related carbonaceous materials: Spectralanalysis and structural information. Carbon 43:1731–1742
15. Amdani RZ, Ossler F, Brackmann C (2021) Raman spectroscopy for characterizing porous carbon. Lund University, Sweden
16. Xiao X, Bian J, Li MF, Xu H, Xiao B, Sun RC (2014) Enhanced enzymatic hydrolysis of bamboo (Dendrocalamus gigantus munro) culm by hydrothermal pre-treatment. Biores Technol 159:41–47
17. Khalil HS, Ismail H, Rozman HD, Ahmad MN (2001) The effect of acetylation on interfacial shear strength between plant fibres and various matrices. Eur Polymer J 37(5):1037–1045
18. Nazir MS, Wahjoedi BA, Yussof AW, Abdullah MA (2013) Eco-friendly extraction and characterization from oil palm empty fruit bunch. BioResources 8(2):2161–2172
19. Ching YC, Ng TS (2014) Effect of preparation conditions on cellulose from oil palm empty fruit bunch fiber. Biores Technol 9(4):6373–6385
20. Rosli NS, Harun S, Jahim JM, Othaman R (2017) Chemical and physical characterization of oil palm empty fruit bunch. Malays J Anal Sci 21(1):188–196
21. Yacob A (2008) Comparison of various sources of high surface area carbon prepared by different types of activation. Malays J Anal Sci 12:264
22. do Amaral Junior MA, Matsushima JT, Rezende MC, Gonçalves ES, Marcuzzo JS, Baldan MR, (2017) Produc-tion and characterization of activated carbon fiber from textile PAN Fiber. J Aerosp Technol Manag 9(4), 423–30

Microscopic Analysis of Activated Carbon Addition to the Pore Structure of Bamboo Paper

Novitri Hastuti, Dian Anggraini Indrawan, Lisna Efiyanti, Santiyo Wibowo, and Saptadi Darmawan

Abstract Bamboo fibres are currently an option to reduce the use of wood fibres as paper material. This study aims to integrate bamboo fibres and activated carbon as a paper matrix and analyse its microscopic characteristic. Activated carbon were mixed in a paper matrix made of bamboo fibres. Microscopic analysis was carried out to determine the effect of activated carbon addition to the paper pores. Pore size is known as a determinant of the physical properties of paper in its use for various purposes. Microscopic analysis using a scanning electron microscope and Brunauer–Emmett–Teller (BET) showed that the addition of activated carbon had an effect on the pore size, microspore volume and surface area. The ratio of activated carbon and fibres of 1:1 revealed the largest surface area with a microspore volume of 0.04 g/cm^3 and the smallest average pore size of about 6 nm. The larger ratio of activated carbon to fibres decreased the surface area significantly with an increase in microspore volume and pore size.

Keywords Bamboo · Pore · Activated carbon · Microscopic · Paper

1 Introduction

Bamboo fibers are known to be able to be used as a composite material such as paper and fibersboard [1, 2]. Bamboo fibers are known to have rapid and easy growth; having superior physical properties such as its low-density, high stiffness and strength [3]. As the raw material of paper, bamboo fibers orientation is a concern because fibers were randomly deposited without any preferred orientation with large pores

N. Hastuti (✉) · D. A. Indrawan · L. Efiyanti · S. Darmawan
Reserch Center for Biomass and Bioproducts, National Research and Innovation Agency (BRIN), 16911 Bogor, Indonesia
e-mail: novi043@brin.go.id

S. Wibowo
Research Center for Applied Botany, National Research and Innovation Agency (BRIN), 16911 Bogor, Indonesia

M. A. Rahandi Lubis (ed.), *Proceedings of the 1st International Conference of Lignocellulose*, Green Energy and Technology,
https://doi.org/10.1007/978-981-19-1596-3_28

between the fibers; resulting in more cavities and the appearance of the large holes [4]. However, the orientation of the fibers will affect the physical and mechanical properties of the paper. As it is known that the elastic modulus of fibers, the degree of bonding, and the presence of curl, kinks, crimps, and micro-compressions in the fibers have a dominant role in determining paper strength [5].

Paper as a packaging material requires certain characters to maintain food quality. Previous research has shown that bamboo paper added with activated carbon was able to maintain the quality of food ingredients seen from the analysis of its nutritional content compared to pure bamboo paper [2, 6]. A large surface area of activated carbon is an effective adsorbent because it is a highly porous material and allows contaminant particles to be adsorbed [7], which caused differences in the quality of packaged food. However, an in-depth analysis of the effect of activated carbon on the microstructure of paper has not been analyzed. Previous study on bamboo paper mentioned that the structure of pure bamboo paper has a large pore/cavities due to the weak and irregular bonding of bamboo fibers caused the fibers randomly deposited without any preferred orientation [4]. The weakness of bamboo paper with its large cavities can be enhanced by mixing the carbon matrix in form of activated carbon into paper matrices. Activated carbon addition and its microstructure in the fibers interface could determine the mechanical and thermal properties of composites which is attributed to the high preferred orientation of carbon layers [8]. Carbon fibers are uniformly dispersed in the reinforced matrix and bridged each other to form penetrating pores with different sizes, whereas the pore structure has an critical impact on physical properties of paper-based friction materials [9]. Therefore, the aim of this study was to observe the microscopic changes of paper pore structure after activated carbon addition into bamboo fibers-based paper.

2 Materials and Method

2.1 Activated Carbon and Bamboo Paper Preparation

Activated carbon was produced from sawdust waste carbonized in 400–500 °C for 4–5 h and activated by water vapor in 800 °C for 70 min. The resultant activated carbon was pulverized using grinder and filtered by mesh filter. Activated carbon was added into pulp as additive in ratio pulp to activated carbon were 1:0 (control); 1:1 and 1:2 and designated to get paper density of 60 g/m^2. Characterization of activated carbon included moisture content, ash content, fixed carbon and iodine number referred to Indonesian Standard of Activated Carbon (SNI 06-3730-1995) [2].

Pulp for paper which used in this study were made from *ampel* bamboo (*Bambusa vulgaris*). Paper was made from *ampel* bamboo fibers produced by semi-chemical method using hot soda process. First, *ampel* bamboo were cleaned, chopped and dried. After reaching the air dry moisture content, bamboo chips were pulped. Pulping

Table 1 Ratio of fibers and activated carbon in bamboo paper

Sample	Remarks
F1C0	Ratio of fibers and activated carbon 1:0
F1C1	Ratio of fibers and activated carbon 1:1
F1C2	Ratio of fibers and activated carbon 1:2
F2C1	Ratio of fibers and activated carbon 2:1

was carried out using a cooking boiler with a capacity (per batch) of 1000 g of oven-dried flakes, with the following conditions: the ratio of raw fibers to the cooking solution is 1:8 (w/v), the maximum cooking temperature (100 °C) for 2 h, and NaOH 10.5%. Paper with activated carbon was prepared through series of process as described in our previous study. Prior to the manufacture of paper sheets, activated carbon were added to the bamboo pulp in a proportion (percentage) of 20% with a target grammage (base weight) of 60 g/m^2 [2]. The representative paper had been denoted as F1C0/control, F1C1, F1C2 and F2C1 which were attributed to the ratios of bamboo fibers and activated carbon in the paper matrices (Table 1).

2.2 Microscopic and Surface Area Analysis

Microscopic analysis was performed using Scanning Electron Microscope (SEM) of (Zeiss Type EVO 50, Germany) with acceleration voltage of 10 kV. Surface area analysis was carried out using Surface Area and Pore Analyzer Meso 112 (JWGB Science and Technology, China) operated in 77.35 K under nitrogen atmosphere. Surface area analysis included Brunauer–Emmett–Teller (BET) surface area and average pore sizes of each sample.

3 Results and Discussion

3.1 Activated Carbon Characteristics and Bamboo Paper Morphology

Characteristics of activated carbon as described in our previous study [2]. The yield of resultant activated carbon was 48% with the moisture content, ash content and fixed carbon were 1.5%, 24% and 71%, respectively. The value of adsorption of iodine was 900.35 mg/g. In general, characteristics of activated carbon meet the requirements according to technical activated carbon quality standards (SNI 06-7370-1995), except for the ash content which requires maximum value of 10% [10]. Microscopic appearance of fully bamboo paper and activated carbon-bamboo fibers as paper matrices as described in Fig. 1.

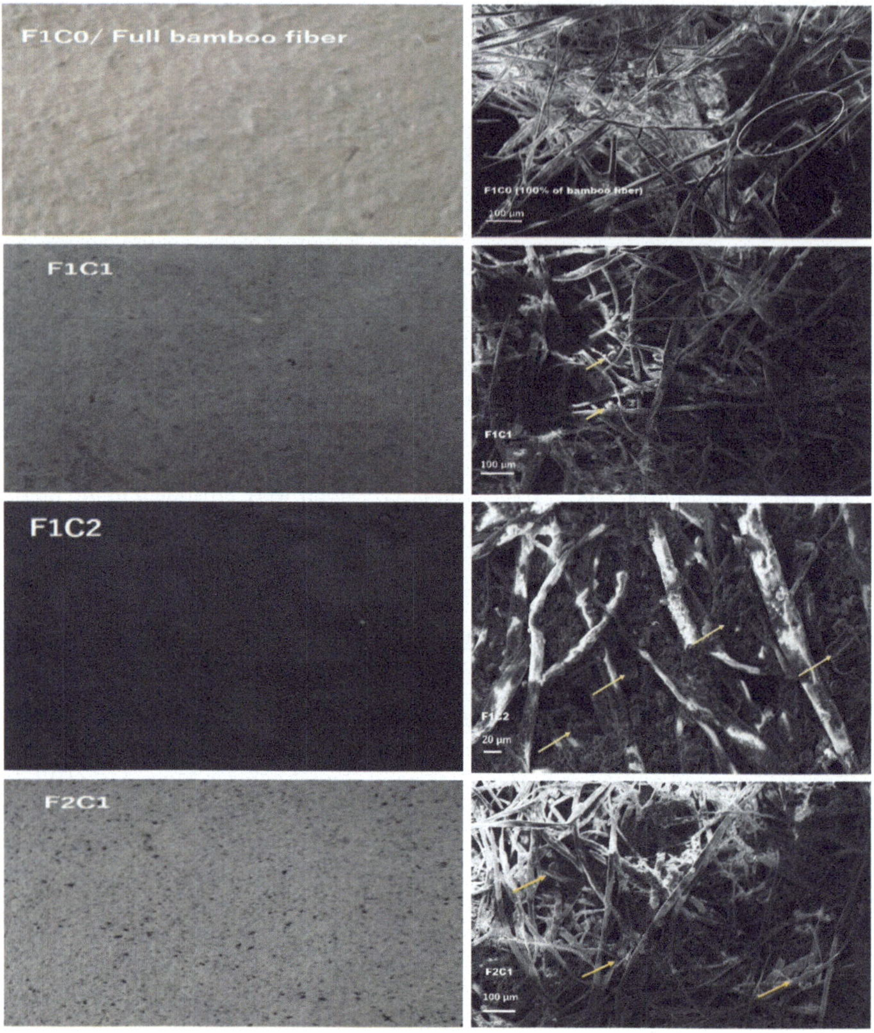

Fig. 1 Macro and microscopic images by scanning electron microscope/SEM in 100 × magnification (upper to bottom) of: F1C0; F1C1; F1C2; F2C1 (yellow narrows appointed to activated carbon)

The addition of activated carbon increased the surface area of the paper and reduce the holes between the fibers. Microscopic images by SEM of paper with 100% bamboo fibers revealed a 100 µm holes (pores) between the fibers (upper image of Fig. 1). Paper with activated carbon showed smaller pores or holes between the fibers (F1C1 and F1C2). These results shows that the pores structure of the paper is influenced by the composition of the paper, the combination ratio of its constituent materials and compression rate during the bonding process [11].

Paper with a higher composition of activated carbon (F1C2) resulting larger micropore compared to paper with a higher composition of fibers (F2C1). Nevertheless, higher composition of fibers (F2C1) seemed had more pores between the fibers in smaller size compared to the paper with lower fibers ratio/F1C1. Paper with lower fibers ratio (F1C1) showed larger size of pores between the fibers, but those pores had been filled by activated carbon completely. It could be supposed that less amount of fibers gives disadvantages with bigger pores and a non-uniform pores size distribution [12]. The larger pore size of native bamboo paper (F1C0) was in good accordance with the previous study which found that fibers of paper made by 100% of bamboo were deposited randomly without any preferred orientation with large pores between the fibers [4]. The properties of bamboo fibers bundles are expectedly affected by the variation of the self-construction properties of fibers cells, such as cellulose content, microfibril angle and crystallinity, on the radial direction of bamboo stem [13].

Activated carbon addition increased the surface area of bamboo paper by its porosity. The porosity is defined as the ratio of the volume of measurable pores and voids in a certain amount of solids to the total volume occupied by the solids [9]. Higher porosity by the presence of activated carbon gives benefit for paper because the ability of paper to bind water can be improved [14]. In addition, pores of activated carbon are also expectedly serve as an adsorbent of particular molecules and unwanted items, which gives benefit for paper for wrapping purposes [6].

3.2 Surface Area Analysis

Surface area analysis included the Brunauer–Emmett–Teller (BET) surface area, average pore diameter and micropore volume. The results as described in Table 2.

Table 2 revealed that the largest BET surface area was belong to the bamboo paper of F1C1 for 961.43 m^2/g. The largest to the smallest surface area in the order of BET surface area were F1C1 > F1C2 > F2C1 > F1C0. Fully bamboo paper (F1C0) which made by 100% of bamboo fibers has the smallest surface area, average pore diameter and micropore volume compared to the paper with activated carbon. The results indicated that the presence of activated carbon in the same ratio (F1C1) increased the paper's surface area and micropore volume with the smallest pore

Table 2 Surface area analysis of bamboo-activated carbon paper

No	Sample	BET surface area (m^2/g)	Average pore diameter (nm)	Micropore volume (cm^3/g)
1	F1C0 (100% bamboo fibers)	82.63	22.46	0.02
2	F1C1	961.43	5.88	0.04
3	F1C2	432.20	17.56	0.08
4	F2C1	171.49	16.97	0.07

diameter. When the ratio of activated carbon became higher (F1C2) surface area decreased, in contrast, micropore volume and pore diameter increased. This fact might be caused by higher porous structure of activated carbon in the paper matrices. Activated carbon which added to the fibers had high value of adsorption of iodine which attributed to its porous structure. The value of the iodine adsorption capacity of activated carbon attributed to the micropore structure formed during the activation process which associated to the formation more hexagonal crystallites and affect the pore structure [14, 15]. This fact was enhanced by paper with a higher activated carbon ratio (F1C2) had higher micropore volume. In this study, activated carbon was obtained by water vapour in 800 °C for 70 min after carbonizing the sawdust in 400–500 °C. Carbonizing the cellulosic material in high temperature is a general method for obtaining three dimensional porous biomass-activated carbon [16].

In spite of paper with higher fibers ratio than activated carbon (F2C1) showed higher surface area than paper made by 100% of bamboo fibers (F1C0), its surface area much lower than paper with lower fibers ratio containing activated carbon (F1C1 and F1C2). It can be concluded that the presence of activated carbon improved the paper's surface area significantly. Based on the IUPAC pore size classification, all paper samples are classified as having mesoporous which are between 2 and 50 nm [17, 18].

4 Conclusions

The addition of activated carbon gives a larger surface area to the bamboo paper matrices. The ratio of activated carbon and bamboo fibers has an important role in the pore size and volume which affects the surface area. The ratio of activated carbon and bamboo fibers which gives the largest surface area and smallest micropore volume was 1:1. When a higher ratio of activated carbon had been applied, surface area became smaller but pore diameter and micropore volume increased. In addition, when ratio of fibers higher than activated carbon, surface area drastically decreased and both of micropore volume and pore diameter slightly decreased. These findings provide a new insight on the application of activated carbon to paper-based materials for a variety of purposes.

Acknowledgements We thank the Japan Society for the Promotion of Science (JSPS) for partial funding of the publication fee.

References

1. Indrawan D, Roliadi H, Margareth Tampubolon R, Pari G, Santoso A, Iqbal M (2015) Penyempurnaan Sifat Papan Serat Berkerapatan Tinggi Dari Campuran Rumput Gelagah, Tandan Kosong Kelapa Sawit Dan Bambu. J Penelit Has Hutan 33(3):193–214

2. Indrawan DA, Hastuti N, Efiyanti L, Pari G (2018) Pemanfaatan Teknologi Kertas Nano Karbon sebagai Pembungkus Wortel (utilization of nano carbon paper technology as carrot wrapping). J Penelit Has Hutan 36(2):139–58
3. Chin SC (2020) Thermal and mechanical properties of bamboo fiber reinforced composites. Mater Today Commun Internet 23:100876. https://doi.org/10.1016/j.mtcomm.2019.100876
4. Tanpichai S, Witayakran S, Srimarut Y (2019) Porosity, density and mechanical properties of the paper of steam exploded bamboo microfibers controlled by nanofibrillated cellulose. Integr Med Res Internet 8(4):3612–22. https://doi.org/10.1016/j.jmrt.2019.05.024
5. Page D, Seth R, De Grace J (1979) Elastic modulus of paper-1. Control Mech Tappi 62(9):99–102
6. Efiyanti L, Hastuti N, Indrawan DA (2018) Potential utilization of nano carbon wrapping paper from bamboo for packaging of brownies cake. In: Yetti E, Palar R, Anugerah MD, editors. Proceeding 5th international symposium on innovative bio-production Indonesia. Bogor: Research Center for Biotechnology LIPI
7. Johnson C (2014) Advances in pretreatment and clarification technologies. In: Ahuja S, editor. Comprehensive Water Quality and Purification. Elsevier B.V. p 60–74
8. Liu X, Deng H, Zheng J, Sun M, Cui H, Zhang X (2020) Mechanical and thermal conduction properties of carbon / carbon composites with different carbon matrix microstructures. New Carbon Mater Internet 35(5):576–84. https://doi.org/10.1016/S1872-5805(20)60511-X
9. Li C, Fu Y, Wang B, Zhang W, Bai Y, Zhang L et al (2020) Effect of pore structure on mechanical and tribological properties of paper-based friction materials. Tribol Int Internet 148:106307. https://doi.org/10.1016/j.triboint.2020.106307
10. Badan Standardisasi Nasional (1995) Mutu dan Cara Uji Arang Aktif Teknis. (SNI 06–7370–1995)
11. Kitahara S, Matsumoto T (1996) The relationship between porosity and mechanical strength in paper-based friction materials. SAE Trans [Internet]. 105:1353–9. Available from: http://www.jstor.org/stable/44720854
12. Maheshwari PH, Mathur RB, Dhami TL (2008) The influence of the pore size and its distribution in a carbon paper electrode on the performance of a PEM Fuel cell. 54:655–9
13. Wang F, Shao Z (2020) Study on the variation law of bamboo fi bers tensile properties and the organization structure on the radial direction of bamboo stem. Ind Crop Prod Internet 152(130):112521. https://doi.org/10.1016/j.indcrop.2020.112521
14. Darmawan S, Pari G, Santoso A (2010) Sifat Papan Serat MDF dengan Penambahan Arang. J Penelit Has Hutan. 28(4):395–405
15. Pari G, Hendra D, Pasaribu RA (2006) Pengaruh Lama Waktu Aktivasi dan Konsentrasi Asam Fosfat terhadap Mutu Arang Aktif Kulit Kayu Acacia mangium. J Penelit Has Hutan 24(1):33–46
16. Wan H, Hu X (2019) From biomass-derived wastes (bagasse, wheat straw and shavings) to activated carbon with three-dimensional connected architecture and porous structure for Li-ion batteries. Chem Phys Internet 521(2018):108–14. https://doi.org/10.1016/j.chemphys.2019.01.012
17. Zdravkov B, Cermak J, Sefara M, Janku J (2007) Pore classification in the characterization of porous materials: A perspectiveˇ. Cent Eur J Chem 5(2):385–95
18. Purbaningtias TE, Kurniawati P, Wiyantoko B, Prasetyoko D (2017) Pengaruh Waktu Aging pada Modifikasi Pori Zeolit Alam dengan Cetyltrimethylammonium Bromide (CTABr). J Sains dan Technol 6(2):321–30

Ethanol Production from Sweet Sorghum Juice by *Saccharomyces Cerevisiae* InaCC Y586 and *Saccharomyces Cerevisiae* InaCC Y653 Immobilized on Alginate Matrix

Sita Heris Anita, Ismu Purnaningsih, Atit Kanti, and Euis Hermiati

Abstract Sweet sorghum has the prospect of being used as a source of food, feed, and fuel (ethanol). The high sugar content in sweet sorghum stalk can be fermented to ethanol through yeast fermentation. In this study the performance of two yeast strains (*Saccharomyces cerevisiae* InaCC Y653 and *Saccharomyces cerevisiae* InaCC Y586) in enhancing the ethanol production either using a free cell or an immobilized cell system was evaluated. Calcium alginate was used as the matrix for cell immobilization. Fermentation was carried out in the batch process for 72 h, the concentration of ethanol and sugars were determined by Gas Chromatography (GC) and High-Performance Liquid Chromatography (HPLC), respectively. The highest bioethanol production was obtained from sweet sorghum juice fermented using *S. cerevisiae* Y586 in a free cell system, which was 35.01 ± 0.01 g/L or 0.48 ± 0.06 g/g in ethanol yield. The free cell system gives the best result than the immobilized cell system. The immobilization system by using Ca-alginate matrix did not improve the ethanol production. Optimization of some immobilization parameters should be done for further study.

Keywords Fermentation · Immobilization · Sweet sorghum juice

S. H. Anita (✉) · I. Purnaningsih
Research Center for Applied Microbiology, National Research and Innovation Agency, Cibinong, Indonesia
e-mail: sita001@brin.go.id

A. Kanti
Research Center for Biosystematics and Evolution, National Research and Innovation Agency, Cibinong, Indonesia

E. Hermiati
Research Center for Biomass and Bioproducts, National Research and Innovation Agency, Cibinong, Indonesia

M. A. Rahandi Lubis (ed.), *Proceedings of the 1st International Conference of Lignocellulose*, Green Energy and Technology,
https://doi.org/10.1007/978-981-19-1596-3_29

1 Introduction

The world's crude oil production is estimated to decrease in line with the rapidly increasing of world energy consumption, which is around 5 billion barrels in 2050. Therefore, it is necessary to discover a new alternative source of non-fossil fuel. Bioethanol produced from plants is indeed interesting because of its good combustion quality. Bioethanol is a promising biofuel because it has several advantages, including high octane number, good gas emissions production, and greenhouse effect reduction [1, 2].

Bioethanol is produced by conversion of starchy, sugar rich, or lignocellulosic biomass via fermentation using microorganisms [3, 4]. Currently, sweet sorghum is widely used as a raw material for ethanol production. Sweet sorghum has advantages such as a fast life cycle (3–5 months), rapid growth, high sugar accumulation in stems (12–21%), as well as drought and salinity resistance [1, 3, 5].

Sweet sorghum juice has approximately 12–18% fermentable sugars, consist of sucrose, glucose, and fructose. However, sugar content in sweet sorghum juice should not be more than 20% [6, 7]. The high sugar content will inhibit yeast performance, resulting in imperfect fructose fermentation. Bioethanol production from raw materials with fermentable sugars such as sweet sorghum juice is more promising than using starch or lignocellulosic biomass due to the simple process steps without any pretreatments or hydrolysis to obtain fermentable sugars [8].

Several studies have shown that the ethanol concentration and sugar conversion efficiency from fermented sweet sorghum juice are not optimal from the maximum that can be achieved [2, 3]. Therefore, it is necessary to increase the ethanol yield by improving fermentation method, such as using the immobilization method.

Some of the benefits obtained with immobilized cell systems include reducing contaminant levels, increasing substrate absorption, increasing fermentation activity, being more tolerant to high substrate concentrations, reducing end-product inhibition, protecting cells from inhibitors, and being reusable [9, 10]. Anita et al. [11] reported that immobilized cells resulted in a higher ethanol yield than free cells system. Entrapment in calcium alginate beads is commonly used for immobilization method due to its ease of preparation, relatively low cost, and non-toxic.

In this study the performance of two yeast strains (*Saccharomyces cerevisiae* InaCC Y586 and *Saccharomyces cerevisiae* InaCC Y653) in enhancing the ethanol production from sweet sorghum juice was evaluated. The free cell and immobilized cell system were used for fermentation process.

2 Materials and Methods

2.1 Materials

Sweet sorghum juice (SSJ) was extracted from Super 2 cultivar grown in a demonstration plantation of sorghum at Cibinong Science Center, Bogor, Indonesia. Culture medium such as yeast extract and peptone were purchased from Hi-Media. Dextrose, KH_2PO_4, $(NH_4)_2HPO_4$, $MgSO_4.7H_2O$, $CaCl_2$, and NaCl were purchased from Merck, Germany. Sodium alginate was purchased from Wako Pure Chemical Industries, Ltd., Japan. *Saccharomyces cerevisiae* InaCC Y586 and *Saccharomyces cerevisiae* InaCC Y653 were isolate collection from LIPI.

2.2 Methods

Microorganism. *S. cerevisiae* InaCC Y586 and *S. cerevisiae* InaCC Y653 were cultured on yeast extract-peptone-dextrose (YPD) agar medium composed of 10 g/L yeast extract, 20 g/L peptone, 20 g/L dextrose, and 25 g/L agar. The culture was preserved at temperature of 4 °C.

Yeast Biomass Production. As much as five loops of 24 h *S. cerevisiae* was inoculated into medium production (100 mL) that consists of 10 g/L yeast extract, 20 g/L peptone, 20 g/L glucose, 5 g/L KH_2PO_4, 5 g/L $(NH_4)_2HPO_4$, and 0.5 g/L $MgSO_4.7H_2O$ in 300 mL Erlenmeyer flask. The incubation process was carried out in a shaker incubator with 120 rpm in a speed, temperature of 27 ± 3 °C for 24 h. After incubation was finished, the yeast cells were separated by centrifugation the production medium at 8000 rpm, 4 °C for 10 min. Initial yeast biomass for the immobilization process was prepared by dissolving 30 mg of yeast cells (wet weight) into 30 mL sterilized NaCl 0.9% (w/v).

Immobilization. *S. cerevisiae* InaCC Y586 and *S. cerevisiae* InaCC Y653 were immobilized on an alginate matrix. Thirty milliliters of yeast biomass with 10^7–10^8 cells/mL of initial cell count were added into 120 mL sterilized sodium alginate 2% (w/v) solution. The alginate-cell solution was then dripped down into 0.1 M $CaCl_2$ solution, and was stirred slowly at 100 rpm, 4 °C for 1 h. The alginate-cell matrix was then rinsed with sterile distilled water up to 4 times to release excess of Ca^{2+} ions and unentrapped cells. The Ca-alginate-cell matrix was decanted and ready to use for the fermentation process.

Fermentation. The SSJ was added with the micronutrients: 0.2 g/L $(NH_4)_2SO_4$, 0.125 g/L K_2HPO_4, and 0.05 g/L $MgSO_4.7H_2O$, and the solution was adjusted to pH 4.5. As much as 100 mL of supplemented SSJ was then added into Erlenmeyer flask and sterilized at 110 °C for 15 min. Fermentation was performed by inoculating 10% (w/v) of the immobilized yeast into sterilized SSJ. The flask was then sealed by a gas trap. The incubation was done at room temperature (27 ± 3 °C) for 72 h. The sampling process was carried out every 24 h to analyze the sugar and ethanol content. The free cell was used as a control.

Analytical Methods. Samples that have been taken were measured for their sugar and ethanol contents. The sugar contents, including of glucose, sucrose, and fructose, were determined using High-Pressure Liquid Chromatography (HPLC) (Shimadzu, Japan) with a RID 20A detector. Samples were filtered with 0.4 μm filters. The column used was a Shim-pack NH_2 150 mm, acetonitrile: water eluent in a ratio 74:26 and with a flow rate of 1 mL min $^{-1}$. The column temperature was 30 °C. The number of samples injected was 10 μL. The sugar concentrations were expressed in g/L for glucose, sucrose, and fructose. The ethanol content was measured by using Gas Chromatography (GC) (Shimadzu, Japan) with a Rxi-1 ms (dimethyl polysiloxane) column and a flame ionization detector. The temperatures of the injector, detector and column oven were 150 °C, 200 °C, and 115 °C, respectively. The number of samples injected was 1 μL.

Evaluation of Fermentation Parameters. Sugar consumption was calculated as in Eq. (1) [11]. The ethanol concentration was calculated based on Eq. (2):

$$\text{Sugar consumption } (\%) = \frac{\text{amount of sugar used } \left(\frac{\text{g}}{\text{L}}\right)}{\text{initial amount of sugar } \left(\frac{\text{g}}{\text{L}}\right)} \times 100 \quad (1)$$

$$\text{Ethanol concentration (g/L): ethanol content (v/v) (\%)} \times 10 \times 0.789 \quad (2)$$

The ethanol yield was calculated as ethanol produced (g) per sugar consumed (g). The efficiency of sugar conversion to ethanol (SCE) was calculated by the ratio of the ethanol yield sample to the theoretical ethanol yield (0.51 g/g) multiplied by 100 [11].

3 Results and Discussion

In this study, the total sugar of SSJ was about 73.58 g/L that consists of 36.21 g/L glucose, 28.19 g/L fructose, and 9.18 g/L sucrose (Fig. 1). The sugar profile of SSJ in this study contained more glucose and fructose than sucrose. Widianto et al. [12] reported that the amount of glucose and fructose in sweet sorghum juice cultivar FS501 was higher compared to sucrose content while other cultivars, KCS105 and FS902, was vice versa. Sweet sorghums are mainly composed of fermentable sugar such as sucrose, glucose, and fructose. Higher sucrose concentrations were observed in sweet sorghum juice at three months for the USA 1 and the Sugar Graze cultivar and Hunni Green at six months of harvesting time [13]. Pabendon et al. [14] also reported that sugar concentration in sweet sorghum juice cultivar Super-1 and Super-2 was higher in the dry season than that in rainy season. The sugar composition

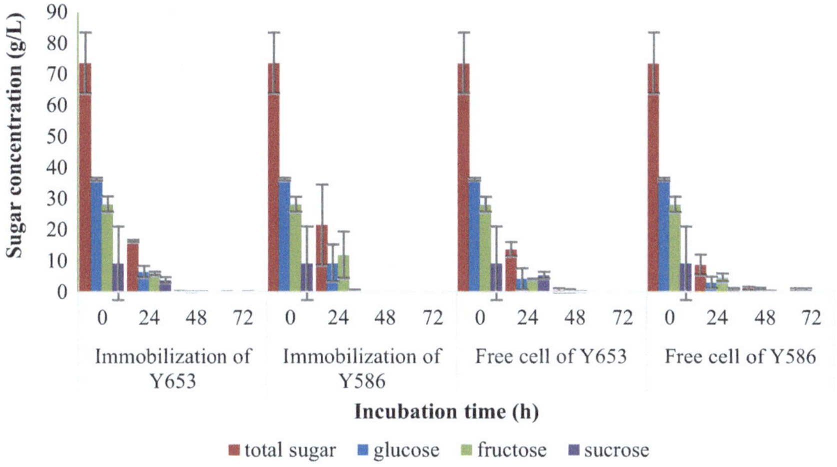

Fig. 1 Sugar concentration of two *S. cerevisiae* strains during fermentation

and concentration in SSJ varied based on cultivar type, planting time, maturity or harvesting time, temperature, planting space, and fertilizer [2, 12, 13].

The sugar concentration decreased significantly at the first 24 h of fermentation time, either in free cells or immobilized cells system, for both yeast strains (Fig. 1). However, the strain Y586 consumed sucrose more rapidly than glucose and fructose, while Y653 consumed all types of sugars at the same rate. Previous research reported that *S. cerevisiae* strains usually first utilized or consumed sucrose and glucose and converted it to ethanol before they consumed fructose. The differences in sugar utilization preferences can affect the growth of yeast cells and their capacity to produce ethanol [12].

The sugar uptake or sugar consumption in the fermentation medium was significantly high at the first 24 h. The highest sugar consumption was about 91.22% obtained by a free cell of Y586 followed by a free cell of Y653, immobilized cells of Y653, and immobilized cells of Y586, which were 81.42%, 77.567%, and 71.65%, respectively (Fig. 2). The sugars were completely consumed after 48 h of fermentation in all the fermentation systems. The results indicated that the sugar consumption rate of the free cells was faster than that of the immobilized cells. The carrier system slightly limits the sugar utilization of immobilized cells. As much as 2% (w/v) of sodium alginate slows down the diffusion of the substrate, thereby limiting the mass transfer process. In contrast, the concentration of sugars remaining in the fermentation medium of immobilized cells on alginate-loofah matrix was lower than those in the free cell cultures [10]. This shows that alginate-loofah matrix provided good properties in the mass transfer process, including the diffusion of substrate into the cells.

In Fig. 3 we can see the fermentation performance using *S. cerevisiae* InaCC Y586 and InaCC *S. cerevisiae* Y653 in free and immobilized cells systems. The results show that fermentation using free cells of Y586 for 72 h resulted in the

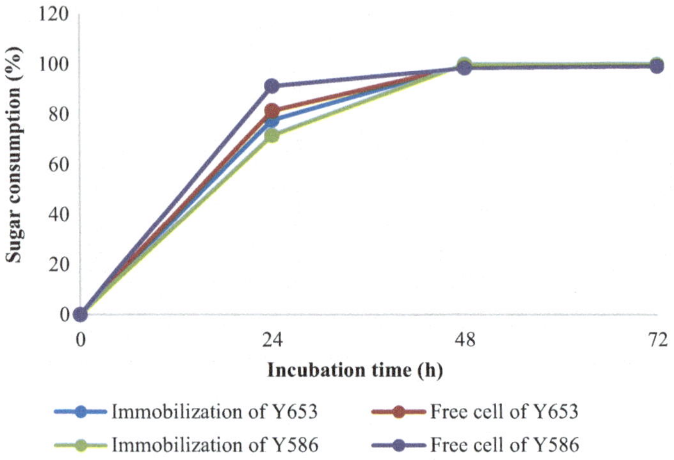

Fig. 2 Sugar consumption of two *S. cerevisiae* strains during fermentation

highest ethanol concentration, ethanol yield, and SCE, which were of 35.01 g/L, 0.48 g/g, and 94.98%, respectively. Of the two strains, the Y586 strain performed better in ethanol production. It might be related to the sugar utilization preferences, since Y586 consumed sucrose more rapidly than other sugars. Widianto et al. [12] reported that *S. cerevisiae* commonly first used sucrose and glucose and converted it into ethanol before it used fructose. Yeast with high SCE can be applied to enhance the ethanol yield from SSJ. Y586 strain has an SCE value greater than 90% (Fig. 3c). These fermentation performances indicated that the Y586 strain is a promising for fermentation of ethanol using sweet sorghum juice as a substrate.

In this study, an immobilized cell system did not give improvement in ethanol production. Sodium alginate in a concentration of 2% (w/v) exhibited low substrate inhibition, which was proved by the complete sugar consumption after 48-h fermentation. However, this matrix seems could not facilitate product diffusion. Calcium alginate concentration affects ethanol production, so that it should be optimized. The sodium alginate concentration of 3% and 4% (w/v) gives the highest ethanol production [15]. Calcium alginate matrix has been widely used in the immobilization system by some researchers and showed good results [10, 15–17].

In general, *S. cerevisiae* InaCC Y586 showed better performance than *S. cerevisiae* InaCC Y653 on ethanol production, either in free cells system or in immobilized cells system. The free cell system gives a better result than the immobilized cell system. The highest ethanol yield obtained in this study (0.48 g/g) is relatively close to previous reports (0.36 g/g-0.49 g/g) [2, 12]. In this research, immobilized cell system has good prospects for application. However, there are still many parameters that need to be optimized. Zain et al. [16] reported that Ca-alginate bead with 0.5 cm diameter obtained higher ethanol production than bead with 0.3 cm diameter. The concentration of $CaCl_2$ in Ca-alginate matrix preparation also affects ethanol production [15]. The immobilized yeast bead's weight to fermentation solution weight gives an

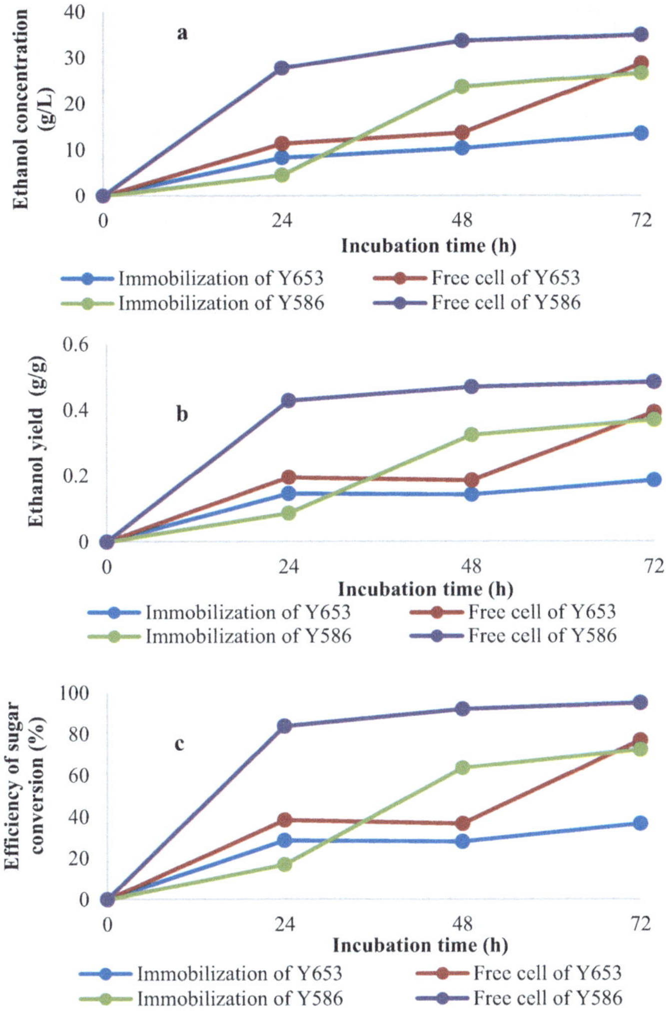

Fig. 3 Performance of ethanol concentration **a**, ethanol yield **b**, and efficiency of sugar conversion **c** of two *S. cerevisiae* strains during fermentation

effect on ethanol production as well [9, 16]. Elbashiti et al. [15] explained that many factors such as an immobilization type, matrix type, and calcium alginate concentration must be optimized to improve the ethanol yield. The appropriate condition of immobilized cell system is very favorable for ethanol production applications in the matter of increasing ethanol yield, shortening ethanol fermentation intervals, and reducing production costs [9].

4 Conclusions

Fermentation of sweet sorghum juice in free cells and immobilized cells systems has been successfully performed using two strains of *S. cerevisiae.* Of the two, *S. cerevisiae* InaCC Y586 is the more potential strain for fermentation of sweet sorghum juice to produce ethanol. The free cell system gives the better result than the immobilized cell system. The highest ethanol concentration, ethanol yield, and sugar conversion efficiency in this study are 35.01 g/L, 0.48 g/g, and 94.98%, respectively, and are obtained from free cells system using *S. cerevisiae* InaCC Y586 at 72 h of fermentation. Further study to optimize the immobilization parameters such as the size of beads, sodium alginate concentration, and ratio volume of beads to the substrate needs to be conducted.

Acknowledgements The authors gratefully acknowledge the Program Science and Technology Research Partnership for Sustainable Development "Revegetation of Alang-Alang (*Imperata cylindrica*) Field Combined with Sustainable Production and Utilization of Biomass for Food and Energy Solution for funding this research. The authors also acknowledge facilities, scientific and technical support from Advanced Characterization Laboratories Cibinong—Integrated Laboratory of Bioproducts, Indonesian Institute of Sciences.

References

1. Khalil SRA, Abdelhafez AA, Amer EAM (2015) Evaluation of bioethanol production from juice and bagasse of some sweet sorghum varieties. Ann Agric Sci 60(2):317–324
2. Jasman J, Prijambada ID, Widianto D, Hidayat C (2017) Increasing bioethanol yield from fermentation of sweet sorghum (Sorghum bicolor L. Moench) sap by mixed culture composed of two yeast strains. Indian J Sci Technol 10(21):1–8
3. Buruiana CT, Vizireanu C, Furdui B (2018) Bioethanol production from sweet sorghum stalk juice by ethanol-tolerant *Saccharomyces cerevisiae* strains: an overview. Ann Univ Dunarea Jos Galati Fascicle VI Food Technol 42(2):153–167
4. Batog J, Frankowski J, Wawro A, Lacka A (2020) Bioethanol production from biomass of selected sorghum varieties cultivated as main and second crop. Energies 13:6291
5. Almodares A, Hadi MR (2009) Production of bioethanol from sweet sorghum: a review. Afr J Agric Res 4(9):772–780
6. Nasidi M, Akunna J, Deeni Y, Blackwood D, Walker G (2010) Bioethanol in Nigeria: comparative analysis of sugarcane and sweet sorghum as feedstock sources. Energy Environ Sci 3:1447–1457
7. Jia F, Chawhuaymak J, Riley MR, Zimmt W, Ogden KL (2013) Efficient extraction method to collect sugar from sweet sorghum. J Biol Eng 7(1):1–8
8. Zabed H, Faruq G, Sahu JN, Azirun MS, Hashim R, Boyce N (2014) Bioethanol production from fermentable sugar juice. Sci World J 957102:1–11
9. Shen F, Zeng Y, Deng S, Liu R (2011) Bioethanol production from sweet sorghum stalk juice with immobilized yeast. Proced Environ Sci 11:782–789
10. Nuanpeng S, Thanonkeo S, Klanrit P, Thanonkeo P (2018) Ethanol production from sweet sorghum by *Saccharomyces cerevisiae* DBKKUY-53 immobilized on alginate-loofah matrices. Braz J Microbiol 49:140–150

11. Anita SH, Mangunwardoyo W, Yopi Y (2016) Sugarcane bagasse as a carrier for the immobilization of *Saccharomyces cerevisiae* in bioethanol production. Makara J Technol 20(2):73–81
12. Widianto D, Arofatullah A, Yuwono T, Prijambada ID (2010) Ethanol production by fermentation of various sweet-stalk sorghum juices using various yeast strains. Indonesian J Biotechnol 15(2):86–93
13. Mutepe RD (2012) Ethanol production from sweet sorghum. Northwest University, South Africa
14. Pabendon MB, Efendi R, Santoso SB, Prastowo B (2017) Varieties of sweet sorghum super-1 and super-2 and its equipment for bioethanol in Indonesia. IOP Conf Series Earth Environ Sci 65:012054
15. Elbashiti T, Alkafarna A, Elkahlout K (2017) Bioethanol production by immobilized *Saccharomyces cerevisiae* using different lignocellulosic materials. IUG J Nat Stud 25(1):30–39
16. Zain MM, Kofli NT, Yahya SRS (2011) Bioethanol production by calcium alginate immobilised St1 Yeast system: effect of size beads, ratio and concentration. IIUM Eng J 12(4):11–19
17. Damayanti A, Kamoro AC, Bahlawan ZAS (2021) Review calcium alginate beads as immobilizing matrix of functional cells: extrusion dripping method, characteristics, and application. IOP Conf Series Mater Sci Eng 1053(1):012017

Antifeedant Activity of the Mixture of Liquid Smoke, Neem Oil and Biosurfactant Against *Spodoptera Frugiperda*

A. H. Prianto, Budiawan, Y. Yulizar, D. Tarmadi, I. G. Rivo, S. Vani, P. Simanjuntak, and V. Septiani

Abstract Biopesticides provide an alternative advantage over synthetic pesticides because they are environmentally friendly, have low toxicity, and have specific targets. Neem seed extract (*Azadirachta indica.* A. Juss) can be pest control. The main components of neem oil are azadirachtin, salanin, and nimbin. These components synergistically act as insect growth inhibition, antifeedant, and repellant. On the other hand, wood vinegar is a by-product of the pyrolysis of biomass which has insecticidal properties. Therefore, this study aimed to evaluate the effect of biosurfactant, neem oil, and wood vinegar composition and their efficacy on *Spodoptera frugiperda.* Neem oil was added at 10–30%, biosurfactant 10–15%, and wood vinegar was 55–80%. The highest antifeedant effectivity (91.62%) was obtained in a formulation containing.

Keywords Biosurfactant · Neem oil · Wood vinegar · *Spodoptera frugiperda*

A. H. Prianto (✉) · Budiawan · Y. Yulizar
Department of Chemistry, Faculty of Mathematic and Natural Sciences, Universitas Indonesia, Kota Depok, Indonesia
e-mail: ariefprianto@gmail.com

A. H. Prianto
Research Center for Biomass and Bioproducts, BRIN, Bogor, Indonesia

D. Tarmadi · I. G. Rivo · S. Vani
Research Center for Applied Zoology, BRIN, Bogor, Indonesia

P. Simanjuntak
Research Center for Chemistry, BRIN, Tangerang, Indonesia

V. Septiani
Faculty of Chemistry, Universitas STMIPA, Bogor, Indonesia

M. A. Rahandi Lubis (ed.), *Proceedings of the 1st International Conference of Lignocellulose*, Green Energy and Technology,
https://doi.org/10.1007/978-981-19-1596-3_30

1 Introduction

Fall Armyworm is a pest that attacks more than 80 plant species, including corn, rice, sorghum, sugarcane, vegetables, and cotton. The species of armyworm, *Spodoptera frugiperda* can result in a very significant loss of corn yields. This pest is first discovered in Central and West Africa in early 2016. In 2018, the armyworm (S. *frugiperda*) has spread to almost all countries of Sub-Saharan Africa. The pest is predicted to spread more widely throughout the world. This pest is a transboundary destructive pest that will continue to spread because it has unique biological characteristics. In addition, it is also supported by the high volume of trade goods exchange between countries [1].

The fall army worm first discovered in Indonesia has damaged maize crops with a heavy attack rate in West Pasaman district, West Sumatra. Caterpillar larvae can damage almost all parts of the corn plant as reported in Lampung [1]. The life cycle of this pest varies depending on the season, which is 30 days in summer, 60 days in spring and 80–90 days in autumn. In fact, the control of fall armyworm, *S. frugiperda* has not been maximized and only relies on chemical pesticides, in which these pesticides are feared that long-term use can damage the environment and endanger humans. Therefore, alternative materials of pesticide that are more environmentally safe and effective are still needed.

It has been well known that biopesticides or pesticides derived from plant extracts and microorganisms can be an alternative control. Prianto et al. [2] reported that active compounds from neem and wood vinegar exhibited high potential to control pest of the genus Spodoptera. Neem (*Azadirachta Indica* A. Juss) contains several active compounds including azadirachtin, nimbin and salanine, while the main components in wood vinegar are organic acids and phenolic derivatives. Even though several earlier studies on the use of biopesticides as control of agricultural pests have been carried out, but few have reported their effectiveness against *S. frugiperda*. Neem and wood vinegar have potential as biopesticides due to the activity of active compounds as antifeedant, repellent and disrupt insect growth inhibitor. Suryani [3] reported that a mixture of wood vinegar and neem oil was able to kill leafhoppers (*Nilaparvata lugens*) found on cocoa stems. Therefore, this study aimed to evaluate the effectiveness of the biopesticide formulation from a mixture of neem seed oil, biosurfactant, and wood vinegar against fall armyworm (*S. frugiperda*) that attacks maize.

2 Material and Methods

2.1 Materials

Neem seeds, rice husks, and the commercial biosurfactants of Castor oil polyoxyethylene ether were purchased from the local market in Bogor, Indonesia, while distilled water was acquired from PT. Ikapharmindo Putramas (Jakarta, Indonesia).

In addition, S. frugiperda was obtained from the Research Center for Biology, BRIN, Indonesia.

2.2 *Methods*

2.2.1 Preparation Neem Oil and Wood Vinegar

Dwarf Late Napier Grass stems cut into small pieces of 1-2 cm size, dried in the sun to dry and then ground with cruiser until smooth and screened with 60 mesh (0.2 mm). The preparation of neem oil used the cold press method. Neem seeds are used in air-dry conditions with moisture content below 15%. The pressing process uses a screw press machine at a rotation speed of 120 rpm. The produced oil was then collected. Wood vinegar was made from rice husks. It was burned in the carbonize reactor and heated at 400 °C. The smoke generated was then condensed and collected. Furthermore, wood vinegar was separated to get tar as a by-product.

2.2.2 Preparation of Emulsions

Neem oil (1–3 mL) was added to the surfactant and homogenized for 10 min. Furthermore, the wood vinegar was prepared in various volumes according to the proportions in the emulsion. The emulsions were composed of neem oil, wood vinegar, and surfactant in the ratio of 1:8:1; 2:7:1; and 3:6:1. Wood vinegar and a mixture of neem oil and biosurfactant were homogenized with a homogenizer at a speed of 6000 rpm. In formulations with a tar composition, it was added at the end of the process while being homogenized.

2.2.3 Characterization of Emulsions

The emulsion formed was analyzed based on droplet size, polidiversity index (PDI) and zetapotential, which was measured using the Zetasizer ZSP (Malvern Instruments, UK). Therefore, the smallest was determined to be the best, while a zetasizer was used to evaluate the diversity of droplets (polidiversity index) and the zeta potential value.

2.2.4 Antifeedant Activity

Insect colonies of *S. fungiperda* were reared at room temperature (25 + 1 °C), relative humidity > 70%, and fed with corn leaves. Antifeedant activity using the choice feeding test method with a feed size of 1.5 × 2.0 cm. Corn leaves were dipped in the emulsion for 3 s and dried in the air. Then, each leaf sample was placed in a plastic container (5.5 × 6 cm), which was then added to the third instar *S. frugiperda* larvae. Observations were made after 24 h with 7 replications. The area consumed was calculated and the antifeedant activity was calculated using the following formula:

$$\text{Antifeedant index }\% = \frac{a}{b} \times 100\%$$

A: leaf area consumed (control-treatment)
B: leaf area consumed (control+treatment)

3 Results and Discussion

3.1 Determination of Lignin, Cellulose, and Hemicellulose

Antifeedant is a condition that inhibits the pest from consuming the treated feed (corn leaves). These conditions are caused by the composition of chemical compounds in the biopesticide formulation derived from a mixture of neem seed oil and wood vinegar, in which this biopesticide can disrupt the sensory system (primary antifeedant effect), or affect the insect's central nervous system, which can affect the feeding process (secondary antifeedant effect) [4]. Table 2 shows the percentage effectiveness of biopesticide formulation from a mixture of neem seed oil and wood vinegar.

In this study, there were four groups of different biopesticide formulations, namely groups A, B, C and D with concentrations of neem seed oil were 10%, 20%, and 30% for each group, while the concentration of biosurfactants used in each treatment group

Table 1 The composition of emulsion formulations

Formulation	Biosurfactant (%)	Liquid smoke (%)	Neem oil(%)
A_1	10	80.0	10
A_2	10	70.0	20
A_3	10	60.0	30
B_1	12.5	78.5	10
B_2	12.5	68.5	20
B_3	12.5	58.5	30
C_1	15	75.0	10
C_2	15	65.0	20
C_3	15	55.0	30
D_1	10	80.0	10
D_2	10	70.0	20
D_3	10	60.0	30

Table 2 Percentage effectiveness of biopesticide formulation from a mixture of neem seed oil and wood vinegar

Formulation	Composition (%)				
	NO	S	LS	T	Antifeedant activity (%)
A_1	10	10	80		72.62
A_2	20	10	70		79.53
A_3	30	10	60		83.86
B_1	10	12.5	77.5		61.14
B_2	20	12.5	67.5		69.12
B_3	30	12.5	57.5		71.63
C_1	10	15	75		70.96
C_2	20	15	65		74.09
C_3	30	15	55		81.51
D_1	10	10	79	1	79.85
D_3	30	10	59	1	91.62
D_2	20	10	69	1	86.46

NO: Neem Oil, S: Biosurfactan, LS: Liquid smoke, T: Tar

was 10% for formulation A and D, 12.5% for formulation B and 15% for formulation C. Formulation A was a biopesticide formulation that used a biosurfactant with a concentration of 10%, while formulations A1, A2, A3 used 10%, 20%, and 30% of neem seed oil, respectively. Based on the results, the highest effectiveness, second effectiveness, and third effectiveness were obtained from formulations A3, A2, and A1, respectively. In which, the percentages of effectiveness from formulations of A3, A2, and A1 were 83.86%, 79.53%, 74.62%, respectively (Table 2). These results suggested that the increasing the concentration of neem seed oil affects the antifeedant effectiveness for formulation. Previous reported that azadirachtin as the main active component in neem seed oil can act as an inhibitor for reproductive, mating, sexual communication, chitin, feeding (antifeedant) of insect [5].

Formulation B used 12,5% of biosurfactant. Similar to formulation A, formulation B also used 10%, 20%, and 30% of neem seed oil for formulations of B1, B2, and B3, respectively. The results showed that formulation B3 exhibited the highest antifeedant effectiveness followed by B2, and B1 formulations with values were 71.63%, 69.12%, and 61.14% for B3, B2, and B1, respectively. Based on observations, the highest antifeedant effectiveness in group C was found in formulation C_3 with a value of 81.51%, in which this formulation used lowest wood vinegar concentration compared to that in the C_1 and C_2 formulations. These data suggested that the antifeedant effectiveness was affected by the concentration of neem seed oil. In formulation D, tar of wood vinegar was added. The results showed that formulation D3 showed the highest antifeedant effectiveness followed by formulations D2 and D1, with 91.62%, 86.46%, and 79.85% for D3, D2, and D1, respectively. Interestingly, formulation D3 showed higher antifeedant effectiveness than other formulations. This phenomenon may be addressed to the pungent smell of phenol from wood vinegar tar that has an important role as a repellent, causing the fall armyworm to dislike consuming the treated corn leaf by biopesticide formulations. As an insecticide, phenol acts as an antimicrobial and antifeedant [6]. A previous studies revealed that antifeedants act as an inhibitor of the consumption-ability of pests even though the insects themselves have not died [7].

Table 3 shows the results of antifeedant activity analysis Statistical analysis showed that the treatment of biopesticide formulations was significantly different at the 95% confidence level ($p < 0.05$). Duncan's test showed that the same letter indicated not significantly different. The results clearly showed that formulations A3, C3, D1, D2, and D3 exhibited high antifeedant activity. The treatment in group A with 10% biosurfactant, 30% neem oil, and 60% wood vinegar showed comparable effectiveness with 12.3% biosurfactant concentration and reduced wood vinegar (57.5%) in formulation B, and various neem concentrations (10–30%) with the addition of 1% tar in treatment D. These results indicted that the constituent components in the formulation can be substitute each other. However, formulation A3 was ideal because it generated high effectiveness, had low biosurfactant, and did not use use tar. The addition of tar becomes less profitable because the production process is more complicated than wood vinegar. Therefore, the formulations in group A were analyzed for their characteristics by using a zetasizer.

The addition of wood vinegar tar composition in the biopesticide formulation affects the antifeedant activity, this can be seen in column group four where the percentage effectiveness of the antifeedant formulation D1 (10% neem seed oil), in which this formulation showed higher antifeedant activity that that for formulation A_2 (20% neem seed oil). Biopesticide formulation with 10% neem seed oil and addition of tar showed higher antifeedant activity compared to biopesticide formulation made from 20% neem seed oil without the addition of tar. Smoked tar is fraction that is insoluble in water, highly acidic and has higher specific gravity than liquid smoke [8, 9].

Table 3 Analysis of antifeedant activity

Type	Formulation			
	A	B	C	D
1	72.62^{abc}	61.14^{a}	70.97^{ab}	79.86^{bcd}
2	79.53^{bcd}	69.12^{ab}	74.10^{bc}	86.47^{d}
3	83.86^{cd}	71.64^{abc}	81.52^{bcd}	91.62^{d}

Table 4 Zeta potential analysis

Formulation	Zeta potencial (mV)	Size (nm)	Stability
A_1	− 36.90 + 6.33	183.0 + 16.2	V
A_2	− 31.70 + 5.83	188.0 + 43.8	V
A_3	− 47.50 + 11.70	296.3 + 39.5	V

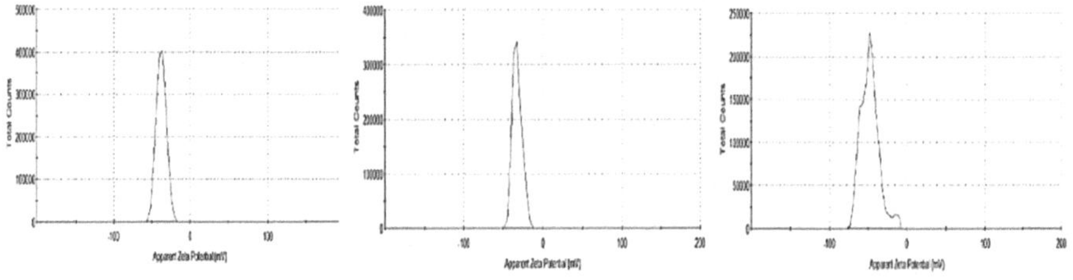

(a) (b) (c)

Fig. 1 Zeta potential distribution **a** formulatio A_1, **b**, formulation A_2, and, **c** formulation A_3

The zeta potential measurement of the emulsion formulation had a value between 31.7 and 47.5 mV, which indicated food emulsion stability after 7 days. Formulation A_3 had the best zeta potentioal value of 47.5 mV, which showed the best stability in the formulation within 30% neem oil composition, 60% liquid smode, and 10% biosurfactant.

The zeta potential distribution of the emulsion of formulations A_1, A_2, and A_3 is shown in Fig. 1. Based on their distribution, the A_1 and A_2 had the least variation, while the A_3 had the highest variabilitu, but the A_3 showed the best emulsion stability. However, the droplet size in emulsion of the A_3 showed the largest size compared to that of the A_1 and A_2.

4 Conclusion

The composition of neem seed oil contained in a formulation showed a strong influence on antifeedant activity. The composition of the formulation that had high and optimal antifeedant activity was found in the formulation containing 30% neem seed oil, 10% biosurfactant, and 60% liquid smode and antifeedant activity by 83.86%.

References

1. Nonci N, Kalqutny SH, Mirsam H, Muis A, Azrai M, Aqil M (2019) Pengenalan fall army warm (*Spodoptera frugiperda*) Hama Baru Pada Tanaman Jagung di Indonesia. Balai Penelitian Tanaman Serealia
2. Prianto AH, Fahlawati FA, Sukandar D (2018) J Lignocellulose Tech. 2, 24–28
3. Suryani R (2019) Pemanfaatan asap cair dan minyak biji mimba sebagai biopestisida pengendalian hama wereng batang coklat (Nilaparvata lugens). Skripsi. Sekolah Tinggi MIPA Bogor: Bogor
4. Ramadhan RAM, Puspasari LT, Meliansyah R, Maharani M, Hidayat Y, Dono D (2016) Bioaktivitas formulasi minyak biji azadirachta indica (A. Juss) Terhadap Spodoptera litura F. J Agrikultura. 1–8

5. Prianto AH, Dewi AS, Basri H dan Yusuf, S.: Efektifitas Formulasi Minyak Mimba Hasil Screw Press Terhadap Rayap Tanah (Coptotermes gestroi). Prosiding Seminar Lignoselulosa 2016. 44–46 (2016)
6. Utami S, Syaufina L, Haneda NF (2010) Daya racun ekstrak kasar daun bintaro (*Cerbera odollam gaertn.*). Terhadap larva spodoptera litura Fabricius. J Ilmu Pertanian Indonesia 15(2):96–100
7. Alfan DR (2018) Efektivitas minyak biji mimba (Azadirachta indica A. Juss) dan asap cair dari sekam padi pada pengendalian hama dan pertumbuhan tanaman padi (Oryza sativa L.). Skripsi. Sekolah Tinggi MIPA Bogor: Bogor
8. Rusydi SM (2019) Prinsip dasar teknologi pirolisa biomassa. Unimal Press, Lhokseumawe
9. Rusdy A (2009) Efektivitas ekstrak nimba dalam pengendalian ulat grayak (Spodoptera litura F.) Pada Tanaman Selada. J Floratek

The manufacturer's authorised representative in the EU is Springer Nature Customer Service Centre GmbH, Europaplatz 3, 69115 Heidelberg, Germany. If you have any concerns regarding our products, please contact ProductSafety@springernature.com

Printed and bound by CPI Group (UK) Ltd, Croydon, CR0 4YY
01/10/2026
02225930-0010